AF356609

Intelligent Point Cloud Processing, Sensing and Understanding

Intelligent Point Cloud Processing, Sensing and Understanding

Editors

Miaohui Wang
Guanghui Yue
Jian Xiong
Sukun Tian

Basel • Beijing • Wuhan • Barcelona • Belgrade • Novi Sad • Cluj • Manchester

Editors

Miaohui Wang
Shenzhen University
Shenzhen
China

Guanghui Yue
Shenzhen University Health
Science Center
Shenzhen
China

Jian Xiong
Nanjing University of Post
and Telecommunications
Nanjing
China

Sukun Tian
Shandong University
Jinan
China

Editorial Office
MDPI
St. Alban-Anlage 66
4052 Basel, Switzerland

This is a reprint of articles from the Special Issue published online in the open access journal *Sensors* (ISSN 1424-8220) (available at: https://www.mdpi.com/journal/sensors/special_issues/IX18KRFUQ1).

For citation purposes, cite each article independently as indicated on the article page online and as indicated below:

Lastname, A.A.; Lastname, B.B. Article Title. *Journal Name* **Year**, *Volume Number*, Page Range.

ISBN 978-3-7258-0241-8 (Hbk)
ISBN 978-3-7258-0242-5 (PDF)
doi.org/10.3390/books978-3-7258-0242-5

Contents

About the Editors

Miaohui Wang

Miaohui Wang (*Senior member of IEEE*) received his Ph.D. from the Department of Electronic Engineering at *The Chinese University of Hong Kong* (CUHK). Between 2014 and 2015, he was a researcher at the Innovation Laboratory at *InterDigital Inc.* in San Diego, CA, focusing on the standardization of video coding. Between 2015 and 2017, he worked as a Senior researcher, specializing in computer vision and machine learning at *The Creative Life* (TCL) Research Institute of Hong Kong. Currently, he is a tenured Associate Professor at the College of Electronics and Information Engineering at *Shenzhen University* (SZU), China. He has authored or co-authored more than one hundred peer-reviewed papers in high-quality international journals and conferences. His research interests cover a wide range of topics in the fields of high-dimension visual data compression and understanding, medical image analysis, computer vision, and machine learning. Dr. Wang is the recipient of the Best Thesis Award from the *Ministry of Education of Shanghai City* and *Fudan University* (FDU). He also received the Best Paper Award from the *International Conference on Advanced Hybrid Information Processing* (2018), and received the Outstanding Reviewer Award from the *IEEE International Conference on Multimedia & Expo* (2021).

Guanghui Yue

Guanghui Yue (*IEEE Member*) is currently an Associate Professor (Master's Supervisor and Post-Doctoral Tutor) at the School of Biomedical Engineering at *Shenzhen University*, China, and also at the National–Regional Key Technology Engineering Laboratory for Medical Ultrasound in China. He received his Ph.D. (2019) in Information and Communication Engineering from *Tianjin University*, China. From September 2017 to January 2019, he spent seventeen months visiting the Multimedia Lab affiliated with the School of Computer Science and Engineering at *Nanyang Technological University* (NTU), Singapore. Dr. Yue researches broadly topics in the fields of 3D image/video processing, image/video quality assessment and enhancement, medical image analysis, computer vision, and machine learning and its applications, among others. He has published over fifty papers in prestigious journals and conferences, including more than twenty in IEEE/ACM Transactions. Dr. Yue serves as an academic member of the artificial intelligence branch of the Guangdong Medical Association. He also serves as a reviewer of many high-quality journals, such as IEEE TMM/TCSVT/TII/TIE/TIM/SPL, Elsevier NEUCOM/SP/DSP/SPIC/JVCI, and IET IP/EL. He is also a recipient of the First Prize for Scientific and Technological Progress Award of the Tianjin Municipality.

Jian Xiong

Jian Xiong (*IEEE Member*) received his Ph.D. from the *University of Electronic Science and Technology of China*, Chengdu, China, in 2015. In 2014, he was a Research Assistant at the Image and Video Processing Laboratory at *The Chinese University of Hong Kong*. He is currently an Associate Professor at the School of Communications and Information Engineering at the *Nanjing University of Posts and Telecommunications* in China. He has authored or co-authored more than fifty peer-reviewed papers in international journals and conferences. He received the Best Paper Award from the *International Conference on Advanced Hybrid Information Processing* (2018). His research interests include image and video processing, point cloud compression, and computer vision.

Sukun Tian

Sukun Tian received his Ph.D. in the Manufacture Engineering of Aeronautics and Astronautics in 2020, from *Nanjing University of Aeronautics and Astronautics* (NUAA) in China, and became a Post-Doctoral Fellow at the School of Mechanical Engineering at *Shandong University* (SDU), Jinan, China, in 2023. He is currently an Assistant Professor, Associate Researcher, and PhD Supervisor at the Center of Digital Dentistry at the *Peking University* School and Hospital of Stomatology. Dr. Tian has authored or co-authored over forty peer-reviewed papers in journals and conferences (e.g., IEEE TMI/ JBHI/ TIM, Mater. Design., Comput. Biol. Med., Clin. Oral Implants Res.) His current research interests cover a wide range of topics in the fields of biomedical engineering, medical image analysis, intelligent manufacturing, and artificial intelligence (AI) techniques in healthcare and medical applications.

Editorial

Intelligent Point Cloud Processing, Sensing, and Understanding

Miaohui Wang [1],*, Guanghui Yue [2], Jian Xiong [3] and Sukun Tian [4],*

[1]   Guangdong Key Laboratory of Intelligent Information Processing, College of Electronics and Information Engineering, Shenzhen University, Shenzhen 518052, China
[2]   School of Biomedical Engineering, Shenzhen University Health Science Center, Shenzhen 518037, China; yueguanghui@szu.edu.cn
[3]   School of Communications and Information Engineering, Nanjing University of Post and Telecommunications, Nanjing 210042, China; jxiong@njupt.edu.cn
[4]   School of Stomatology, Peking University, Beijing 100081, China
*   Correspondence: wang.miaohui@gmail.com (M.W.); sukhum169@bjmu.edu.cn (S.T.)

1. Introduction

Point clouds are considered one of the fundamental pillars for representing the 3D digital landscape [1], despite the irregular topology between discrete data points. Recent advances in sensor technology [2] that acquire point cloud data to enable flexible and scalable geometric representations have paved the way for the development of new ideas, methodologies, and solutions in ubiquitous sensing and understanding applications. Existing sensor technologies, such as *LiDAR, stereo cameras*, and *laser scanners* [3], can be used from a variety of platforms (e.g., satellites, aerial, drones, vehicle-mounted, backpacks, handheld, and static terrestrial) [4,5], viewpoints (e.g., nadir, oblique, and side view) [6], spectra (e.g., multispectral) [7], and granularities (e.g., point density and completeness) [8]. Meanwhile, many promising methods have been developed based on computer vision and deep learning to process the point cloud data [9,10]. However, the expanding applications of point clouds in complex and diverse scenarios, such as autonomous driving [11], robotics [12], augmented reality [13], and urban planning [14], pose new challenges [15] to existing intelligent point cloud approaches.

Recently, artificial intelligence has greatly facilitated the extraction of valuable information from complex point cloud data [16]. Deep learning-based models [16] have shown impressive performance in various point cloud tasks, such as completion [17], compression [18], 3D reconstruction [19], semantic segmentation [19], and object detection [20]. However, as we face increasingly complex and dynamic 3D application scenarios, more accurate, efficient, and effective methods are becoming more and more urgent [21]. Therefore, further investigation on improving intelligent point cloud processing, sensing, and understanding capabilities is of great significance.

This Special Issue collects promising approaches that develop innovative technologies for generating, processing, and analyzing various formats of point cloud data. A total of ten contributions (nine regular articles and one survey) from China, Turkey, Romania, Portugal, the USA, Italy, and the Republic of Korea have been ultimately accepted for publication. These contributions delve into diverse aspects of point clouds, including structural analysis, instance segmentation, registration, texture mapping of 3D meshes, model acceleration and deployment, 3D modeling, up-sampling, plant part segmentation, image-to-point-cloud reconstruction, and LiDAR point cloud (LPC) object detection. The next section provides a concise introduction to each contribution collected in this Special Issue.

2. Overview of Contributions

Contribution 1 explored the application of graph kernels in the structural analysis of point clouds, emphasizing their effectiveness in preserving topological structures and

Citation: Wang, M.; Yue, G.; Xiong, J.; Tian, S. Intelligent Point Cloud Processing, Sensing, and Understanding. *Sensors* **2024**, *24*, 283. https://doi.org/10.3390/s24010283

Received: 25 December 2023
Accepted: 28 December 2023
Published: 3 January 2024

enabling machine learning methods on evolving vector data represented as graphs. Specifically, a unique kernel function was introduced to tailor for similarity determination in the point cloud data. To reflect the underlying discrete geometry, the kernel was further formulated based on the proximity of geodesic route distributions in graphs. By demonstrating the effectiveness of the kernel function in supervised classification using a convolutional neural network (CNN), experimental results validated the efficiency of the proposed kernel function for understanding the geometric and topological aspects of 3D point clouds.

Contribution 2 presented a weakly supervised instance segmentation approach for point clouds, addressing the challenge of inaccurate bounding box annotations. To avoid labor-intensive point-level annotations, they first developed a self-distillation architecture that leveraged the consistency regularization, and then utilized data perturbation and historical predictions to enhance generalization, as well as prevent over-fitting to noisy labels. Later, they selected reliable samples and corrected labels based on historical consistency. Experimental results on the benchmark dataset demonstrated the effectiveness and robustness of their approach, achieving comparable performance to existing supervised methods and outperforming recent weakly supervised methods.

Contribution 3 proposed a robust alignment scheme for point clouds, where the rotation and translation coefficients were calculated using the angle of the normal vector of the building facade and the distance between outer endpoints. Experimental results demonstrated the feasibility and robustness of their alignment method on homologous and cross-source point clouds. In addition, they also pointed out that the future work can further optimize the efficiency of parameter-dependent building facade point extraction and explore applications to point cloud registration with varying sensor qualities.

Contribution 4 developed a novel sequential pairwise color-correction approach to mitigate texture seams generated from multiple images. By selecting a reference image and computing the color correction paths through a weighted graph, this approach could effectively enhance the color similarities among different images, resulting in high-quality textured meshes. Experimental results show that the proposed method outperforms existing schemes in both qualitative and quantitative evaluations on an indoor dataset, especially in scenarios with high triangle transitions.

Contribution 5 designed a light-weight CNN model for moving object segmentation in LPCs, addressing the challenge of real-time processing on embedded platforms. The proposed network achieved a reduction in parameters compared to the state of the art, demonstrating efficient processing on the *RTX 3090 GPU*. In addition, it has been also successfully implemented on an FPGA platform, achieving 32 fps for moving-object segmentation, meeting the real-time requirements in autonomous driving. Despite its comparable error performance with significantly fewer parameters, this light-weight model faced potential challenges, such as simplifying the network structure without compromising performance and addressing the sacrifice of low-level details for computational acceleration.

Contribution 6 addressed the challenge of accurately representing cultural heritage objects for finite element analysis (FEA) to understand their mechanical behavior. Unlike the use of traditional CAD 3D models and non-uniform rational B-spline surfaces (NURBS), they employed an alternative method utilizing the re-topology procedure to create simplified yet accurate 3D models for FEA. This study emphasized the importance of retaining the formal definition compatible with FEA software, demonstrating its effectiveness for morphologically complex objects. Experimental results demonstrate that the proposed method can reduce the mesh size, while maintaining high accuracy compared to high-resolution reality-based models. Future work can be developed to improve interoperability, material segmentation, and detailed parameterization for a more comprehensive understanding of the structural behavior of cultural heritage objects.

Contribution 7 proposed a point cloud up-sampling via multi-scale features attention (PU-MFA) method, leveraging the U-Net structure to combine multi-scale features and cross-attention mechanism. PU-MFA was developed to adaptively and effectively use multi-scale features, demonstrating superior performance in generating high-quality dense points.

Experimental validations on synthetic and real-scanned datasets show the effectiveness of PU-MFA. It is worth noting that PU-MFA currently has limitations in addressing arbitrary up-sampling ratios.

Contribution 8 introduced the MASPC_Transform, a segmentation network for plant point clouds designed to address the challenges posed by the intricate and small-scale nature of plant organs. Leveraging a multi-head attention separation and a spatially grounded attention separation loss, MASPC_Transform established connections for similar point clouds scattered across different areas in the point cloud space. Additionally, a position-coding method was proposed to enhance the feature extraction in the presence of disordering point clouds. Experimental results demonstrated that MASPC_Transform outperformed existing approaches on the plant segmentation. Finally, they also emphasized the need for further testing on new open-source datasets to validate the generalizability of the MASPC_Transform.

Contribution 9 presented a novel 3D-SSRecNet network for efficient 3D point cloud reconstruction from a single image. 3D-SSRecNet was composed of a 2D image feature extraction network based on a backbone network for object detection, and a point cloud prediction network for minimizing the reconstruction loss. The specially chosen activation function was then employed for better shape prediction and lower reconstructed error. Experimental results on two datasets demonstrated the promising performance of 3D-SSRecNet. Although 3D-SSRecNet can be considered as a computationally effective solution for point cloud reconstruction, future work can be investigated to further improve local reconstruction effects while maintaining computational efficiency.

Contribution 10 provided a comprehensive survey on deep learning-based LiDAR 3D object detection for autonomous driving. It summarized the commonly used feature extraction and processing techniques for LPCs, the coordinate systems in LiDAR object detection, and the stages of autonomous driving. Furthermore, a deep learning-based LPC object detection methods were classified into three categories: projection, voxel, and raw point clouds. They have also conducted in-depth analyses, comparisons, and summaries of the advantages and disadvantages of existing LPC object detection methods. Finally, they pointed out that there are still many open issues in improving model speed and accuracy to achieve real-time processing for level-4 to level-5 autonomous driving.

3. Conclusions

This Special Issue serves as a portfolio, bringing together a wide range of contributions that address crucial challenges and advancements in the region of point cloud processing, sensing, and understanding. The selected papers represent a collective endeavor to push the boundaries of point cloud knowledge, offering intelligent solutions to existing challenges, while also unlocking new applications for 3D point clouds. We believe that the above papers will provide valuable insights for researchers and practitioners in this field, stimulating ongoing evolution towards academic and industrial solutions that are not only more accurate, but also more efficient and effective.

Author Contributions: Original draft preparation, M.W.; review and editing, G.Y., J.X. and S.T. All authors have read and agreed to the published version of the manuscript.

Data Availability Statement: The related datasets can be referred to each contribution in this Editorial.

Acknowledgments: The authors express their sincere gratitude to Runnan Huang for his extensive support and assistance in the preparation of this Editorial.

Conflicts of Interest: The authors declare no conflicts of interest.

List of Contributions

1. Balcı, M.A.; Akgüller, Ö.; Batrancea, L.M.; Gaban, L. Discrete Geodesic Distribution-Based Graph Kernel for 3D Point Clouds. *Sensors* **2023**, *23*, 2398.
2. Peng, Y.; Feng, H.; Chen, T.; Hu, B. Point Cloud Instance Segmentation with Inaccurate Bounding-Box Annotations. *Sensors* **2023**, *23*, 2343.
3. Pang, L.; Liu, D.; Li, C.; Zhang, F. Automatic Registration of Homogeneous and Cross-Source TomoSAR Point Clouds in Urban Areas. *Sensors* **2023**, *23*, 852.
4. Dal'Col, L.; Coelho, D.; Madeira, T.; Dias, P.; Oliveira, M. A Sequential Color Correction Approach for Texture Mapping of 3D Meshes. *Sensors* **2023**, *23*, 607.
5. Xie, X.; Wei, H.; Yang, Y. Real-Time LiDAR Point-Cloud Moving Object Segmentation for Autonomous Driving. *Sensors* **2023**, *23*, 547.
6. Gonizzi Barsanti, S.; Guagliano, M.; Rossi, A. 3D Reality-Based Survey and Retopology for Structural Analysis of Cultural Heritage. *Sensors* **2022**, *22*, 9593.
7. Lee, H.; Lim, S. PU-MFA: Point Cloud Up-Sampling via Multi-Scale Features Attention. *Sensors* **2022**, *22*, 9308.
8. Li, B.; Guo, C. MASPC_Transform: A Plant Point Cloud Segmentation Network Based on Multi-Head Attention Separation and Position Code. *Sensors* **2022**, *22*, 9225.
9. Li, B.; Zhu, S.; Lu, Y. A single stage and single view 3D point cloud reconstruction network based on DetNet. *Sensors* **2022**, *22*, 8235.
10. Alaba, S.Y.; Ball, J.E. A survey on deep-learning-based lidar 3d object detection for autonomous driving. *Sensors* **2022**, *22*, 9577.

References

1. Xu, H.; Li, C.; Hu, Y.; Li, S.; Kong, R.; Zhang, Z. Quantifying the effects of 2D/3D urban landscape patterns on land surface temperature: A perspective from cities of different sizes. *Build. Environ.* **2023**, *233*, 110085. [CrossRef]
2. Rogers, C.; Piggott, A.Y.; Thomson, D.J.; Wiser, R.F.; Opris, I.E.; Fortune, S.A.; Compston, A.J.; Gondarenko, A.; Meng, F.; Chen, X.; et al. A universal 3D imaging sensor on a silicon photonics platform. *Nature* **2021**, *590*, 256–261. [CrossRef] [PubMed]
3. Schwarz, S.; Preda, M.; Baroncini, V.; Budagavi, M.; Cesar, P.; Chou, P.A.; Cohen, R.A.; Krivokuća, M.; Lasserre, S.; Li, Z.; et al. Emerging MPEG standards for point cloud compression. *IEEE J. Emerg. Sel. Top. Circuits Syst.* **2018**, *9*, 133–148. [CrossRef]
4. Wang, Q.; Tan, Y.; Mei, Z. Computational methods of acquisition and processing of 3D point cloud data for construction applications. *Arch. Comput. Methods Eng.* **2020**, *27*, 479–499. [CrossRef]
5. Amarasingam, N.; Salgadoe, A.S.A.; Powell, K.; Gonzalez, L.F.; Natarajan, S. A review of UAV platforms, sensors, and applications for monitoring of sugarcane crops. *Remote Sens. Appl. Soc. Environ.* **2022**, *26*, 100712. [CrossRef]
6. Lei, J.; Song, J.; Peng, B.; Li, W.; Pan, Z.; Huang, Q. C2FNet: A coarse-to-fine network for multi-view 3D point cloud generation. *IEEE Trans. Image Process.* **2022**, *31*, 6707–6718. [CrossRef] [PubMed]
7. Wang, C.; Gu, Y.; Li, X. A Robust Multispectral Point Cloud Generation Method Based on 3D Reconstruction from Multispectral Images. *IEEE Trans. Geosci. Remote Sens.* **2023**, *62*, 5407612.
8. Rebolj, D.; Pučko, Z.; Babič, N.Č.; Bizjak, M.; Mongus, D. Point cloud quality requirements for Scan-vs-BIM based automated construction progress monitoring. *Autom. Constr.* **2017**, *84*, 323–334. [CrossRef]
9. Xie, W.; Wang, M.; Lin, D.; Shi, B.; Jiang, J. Surface Geometry Processing: An Efficient Normal-based Detail Representation. *IEEE Trans. Pattern Anal. Mach. Intell.* **2023**, *45*, 13749–13765. [CrossRef] [PubMed]
10. Mirzaei, K.; Arashpour, M.; Asadi, E.; Masoumi, H.; Bai, Y.; Behnood, A. 3D point cloud data processing with machine learning for construction and infrastructure applications: A comprehensive review. *Adv. Eng. Inform.* **2022**, *51*, 101501. [CrossRef]
11. Sun, X.; Wang, M.; Du, J.; Sun, Y.; Cheng, S.S.; Xie, W. A Task-Driven Scene-Aware LiDAR Point Cloud Coding Framework for Autonomous Vehicles. *IEEE Trans. Ind. Inform.* **2022**, *1*, 1–11. [CrossRef]
12. Pomerleau, F.; Colas, F.; Siegwart, R. *A Review of Point Cloud Registration Algorithms for Mobile Robotics*; Now Publishers Foundations and Trends® in Robotics: Hanover, MA, USA, 2015; Volume 4, pp. 1–104.
13. Chen, Y.; Wang, Q.; Chen, H.; Song, X.; Tang, H.; Tian, M. An overview of augmented reality technology. *J. Phys. Conf. Ser.* **2019**, *1237*, 022082. [CrossRef]
14. Alexander, C.; Tansey, K.; Kaduk, J.; Holland, D.; Tate, N.J. An approach to classification of airborne laser scanning point cloud data in an urban environment. *Int. J. Remote Sens.* **2011**, *32*, 9151–9169. [CrossRef]
15. Li, Y.; Ma, L.; Zhong, Z.; Liu, F.; Chapman, M.A.; Cao, D.; Li, J. Deep learning for LiDAR point clouds in autonomous driving: A review. *IEEE Trans. Neural Netw. Learn. Syst.* **2020**, *32*, 3412–3432. [CrossRef] [PubMed]
16. Guo, Y.; Wang, H.; Hu, Q.; Liu, H.; Liu, L.; Bennamoun, M. Deep learning for 3D point clouds: A survey. *IEEE Trans. Pattern Anal. Mach. Intell.* **2020**, *43*, 4338–4364. [CrossRef] [PubMed]
17. Huang, Z.; Yu, Y.; Xu, J.; Ni, F.; Le, X. Pf-net: Point fractal network for 3d point cloud completion. In Proceedings of the IEEE/CVF Conference on Computer Vision and Pattern Recognition (CVPR), Seattle, WA, USA, 13–19 June 2020; pp. 7662–7670.

18. Huang, R.; Wang, M. Patch-Wise LiDAR Point Cloud Geometry Compression Based on Autoencoder. In Proceedings of the International Conference on Image and Graphics (ICIG), Nanjing, China, 22–24 September 2023; pp. 299–310.
19. Ma, B.; Liu, Y.S.; Zwicker, M.; Han, Z. Surface reconstruction from point clouds by learning predictive context priors. In Proceedings of the IEEE/CVF Conference on Computer Vision and Pattern Recognition (CVPR), New Orleans, LA, USA, 18–24 June 2022; pp. 6326–6337.
20. Zhou, Y.; Sun, P.; Zhang, Y.; Anguelov, D.; Gao, J.; Ouyang, T.; Guo, J.; Ngiam, J.; Vasudevan, V. End-to-end multi-view fusion for 3D object detection in lidar point clouds. In Proceedings of the Conference on Robot Learning (CoRL), Virtual, 16–18 November 2020; pp. 923–932.
21. Yang, B.; Haala, N.; Dong, Z. Progress and perspectives of point cloud intelligence. *Geo-Spat. Inf. Sci.* **2023**, *26*, 189–205. [CrossRef]

Article

Discrete Geodesic Distribution-Based Graph Kernel for 3D Point Clouds

Mehmet Ali Balcı [1], Ömer Akgüller [1,*], Larissa M. Batrancea [2,*] and Lucian Gaban [3]

[1] Department of Mathematics, Faculty of Science, Muğla Sıtkı Koçman University, 48000 Muğla, Turkey
[2] Department of Business, Babeş-Bolyai University, 7 Horea Street, 400174 Cluj-Napoca, Romania
[3] Faculty of Economics, "1 Decembrie 1918" University of Alba Iulia, 510009 Alba Iulia, Romania
* Correspondence: oakguller@mu.edu.tr (Ö.A.); larissa.batrancea@ubbcluj.ro (L.M.B.)

Abstract: In the structural analysis of discrete geometric data, graph kernels have a great track record of performance. Using graph kernel functions provides two significant advantages. First, a graph kernel is capable of preserving the graph's topological structures by describing graph properties in a high-dimensional space. Second, graph kernels allow the application of machine learning methods to vector data that are rapidly evolving into graphs. In this paper, the unique kernel function for similarity determination procedures of point cloud data structures, which are crucial for several applications, is formulated. This function is determined by the proximity of the geodesic route distributions in graphs reflecting the discrete geometry underlying the point cloud. This research demonstrates the efficiency of this unique kernel for similarity measures and the categorization of point clouds.

Keywords: simplicial complex; Wasserstein distance; Kullback–Leibler information; point cloud processing

Citation: Balcı, M.A.; Akgüller, Ö.; Batrancea, L.M.; Gaban, L. Discrete Geodesic Distribution-Based Graph Kernel for 3D Point Clouds. *Sensors* **2023**, *23*, 2398. https://doi.org/10.3390/s23052398

Academic Editors: Miaohui Wang, Guanghui Yue, Jian Xiong and Sukun Tian

Received: 12 January 2023
Revised: 16 February 2023
Accepted: 17 February 2023
Published: 21 February 2023

1. Introduction

Point clouds are one of the most direct representations of geometric datasets. One of the sources for obtaining point clouds is 3D shape acquisition devices such as laser range scanners, which also have applications in many disciplines. These scanners provide generally noisy raw data in the form of disorganized point clouds representing surface samples. Given the growing popularity and very wide applications of this data source, it is essential to work directly with this representation without having to go through an intermediate step that can add computational complexity and fitting errors. Another important area where point clouds are frequently used is the representation of high-dimensional manifolds. Such high-dimensional and general isodimensional data are found in nearly all disciplines, from computational biology to image analysis and financial data [1–5]. In this case, due to the high dimensionality, manifold reconstruction is challenging from a technological standpoint, and the relevant calculations have to be performed directly on the raw data, i.e., the point cloud.

Graph structures are important tools for representing discrete geometric data and the discrete manifold underlying the data, as they can reflect the structural and relational arrangements of objects [6–9]. In the classification of discrete graph-based geometric structures, the problem of accurately and effectively calculating the similarity of these data sets arises. Graph kernel functions are widely used to solve this type of problem [10–14]. Graph kernels have proven to be powerful tools for the structural analysis of discrete geometric data. There are two main advantages to using graph kernel functions. First, graph kernels can characterize graph properties in a high-dimensional space and therefore have the capacity to preserve the topological structures of the graph. Second, graph kernels make rapidly evolving machine-learning methods for vector data applicable to graphs.

The goal of this study was to construct a kernel function that generalizes across the topology and geometry of a 3D point cloud. A geodesic is a curve in differential geometry that depicts the shortest route between two points on a surface, or more broadly, the shortest path on a Riemannian manifold. In addition, it is intended to expand the idea of a "straight line" on any differentiable manifold coupled to a more generic medium. Geodesics on piecewise linear manifolds, the foundation of discrete differential geometry, were first defined by [15] and then extended to polyhedral surfaces by [16]. In discrete differential geometric techniques, the underlying geometry of the point cloud is assumed to be known, and triangulation is used to include the locations between points into the geometry. Thus, noise reduction and subdivision techniques may also be implemented while using a raw 3D point collection. In the graph structure to be constructed from the 3D point cloud, the idea of a geodesic becomes the problem of the shortest path [17,18].

In general, point clouds should intensively sample the border of a smooth surface rebuilt using moving least squares [19–21], implicit [22], or Voronoi/Delaunay [23,24] methods. Since point clouds may describe 3D forms using graphs without the requirement for the explicit storing of the manifold connection, they have become a popular alternative surface representation to polygonal meshes [25–28]. Despite the fact that numerous particular graph types, such as the k-nearest neighbor graph [29], the Reeb graph [30], and the Gabriel graph [31], give methods to the geometry and topology of point clouds, higher-order topologies and submanifold topologies disregard topological characteristics. Graph representations of simplicial complex skeletons that preserve the submanifold topologies of point clouds and may potentially integrate higher-order topological information are used in this study. In general, skeleton-based representations provide a compact and expressive form abstraction that aims to imitate human intuition. Using concise, informative, and easily computable skeletal representations as opposed to full models may facilitate the comparison process. In practice, it may be hard to locate a query-like item in a database by comparing point clouds or stacks of hundreds of triangles.

This work presents an approach to the geodesic curves of the manifold by calculating the shortest paths on the graph structures defined by the simplicial complex skeleton of the submanifold from which the point cloud is taken. If the graph structures created by the simplicial complex skeleton are altered by noise, the approaches to the geodesic distributions of the submanifold are not significantly impacted. Consequently, a kernel function to be defined by the Wasserstein similarity of the distributions of discrete geodesics in the skeletons of 3D point clouds has shown to be an excellent assessment tool for point cloud similarity. In a variety of 3D applications, such as 3D object retrieval and inverse procedural modeling, measuring the similarity between 3D geometric objects is crucial. This study aimed to find and demonstrate the effectiveness of a kernel function that calculates the similarity of 3D point clouds while taking into consideration the discrete geometry and topology of the point cloud. The effective kernel function is obtained using the Wasserstein-1 distance by comparing the geodesic distributions in the graphs to those that are isomorphic to the skeleton of the simplicial complexes on the point cloud.

Direct comparison and classification are utilized to compare this newly constructed kernel function to graph models. Using the values of the kernel function, the intra-class and inter-class similarity matrices of the point clouds were constructed during the direct comparison procedure. Considering the geometric and topological aspects of the studied models, it has been noted that the Alpha complex skeletons are the networks on which the kernel function operates most effectively. Using the kernel function, the similarity of a point cloud to itself was stored in the matrices used in the classification procedures. At this, the graph communities acquired for each point are used. Using a convolutional neural network, supervised classification is performed according to the airplane, car, person, plant, and vase classes.

Related Works

As a collection of large points in three dimensions, point clouds may stand in for an object's spatial distribution and surface properties. Model reconstruction [32,33], terrain monitoring [34,35], and resource monitoring and exploitation [36,37] are just a few examples of the various research and application sectors that rely on 3D point clouds obtained by stationary laser scanning. Accurate and efficient registration is crucial for obtaining a full scene or object from a collection of point clouds obtained via stationary laser scanning [34,38], and this has a knock-on effect on subsequent processing and applications such as segmenting and classifying the cloud as well as detecting and tracking objects within it. Moreover, semantic segmentation using point cloud data has made significant strides in recent years [39–43], thanks in large part to innovations such as point-cloud compression techniques [44,45]. When separating moving objects from LiDAR point clouds, semantic segmentation is essential. Existing semantic segmentation convolutional neural networks are good at predicting the semantic labels of point clouds such as automobiles, buildings, and people. Hence, new computational techniques on point clouds are needed because of their widespread use in many disciplines.

The kernel function of a pair of graphs may be described by decomposing the graphs and comparing the specified pairings of isomorphic substructures. Commonly used substructures are walks, paths, and spanning trees [46–48]. In [49], the authors evaluated the complex motions as decomposed spatio-temporal parts for each video using kernel functions defined with subtrees and created the corresponding binary trees. The resulting kernel function is defined by calculating the number of isomorphic subtree patterns. A kernel family to compare point clouds is proposed by [50]. These kernels are based on a newly developed local tree-walk kernel between subtrees, which is defined by factoring in the properly defined graph models of subtrees. The authors in [51] defined a graph kernel for motion recognition in videos. First, they describe actions in videos using directed acyclic graphs (DAGs). The resulting kernel is defined as an expanding random walking kernel by counting the number of isomorphic walks of the DAGs. [52] proposed a segmentation graph kernel for image classification. In this method, each image is represented by a segmentation graph; each vertex corresponds to a segmented region, and each edge connects a pair of neighboring regions. The resulting kernel function is calculated by counting the imprecise isomorphic subtree patterns between the segmentation graphs. Furthermore, some kernel functions are also effectively used for computer vision applications, such as the shortest path graph kernel [53], non-backward walking kernel [54], Lovas kernel [55], Weisfeiler–Lehman subtree kernel [56].

Although each of the kernel functions mentioned above gives effective results for different problems, they are very sensitive to noise and outliers in empirically obtained 3D point clouds. Furthermore, they do not generate reliable comparison information between isomorphic substructures. In other words, for graphs abstracted from 3D shapes, most of the available kernels cannot determine whether isomorphic substructures are located in the same regions based on the visual background. To overcome these shortcomings, [57] proposed an aligned subtree kernel. The proposed kernel function is calculated by counting the number of isomorphic subtrees rooted in aligned vertices, thus overcoming the shortcoming of neglecting positional or structural correspondences between isomorphic substructures that arise in most graph kernels. Although an aligned subtree kernel is effective on 3D shape classification problems, it cannot guarantee transitivity between aligned vertices. More specifically, given the vertices u, v, and w, if v and u and u and w align, the kernel function cannot guarantee that v and w are also aligned. On the other hand, [58] shows that the cascading alignment step is necessary to guarantee the positive precision of the vertex alignment kernel. Therefore, the aligned subtree kernel cannot be guaranteed as a positive-definite kernel. Furthermore, all the specified kernels reflect only graph properties for each graph pair under comparison and therefore ignore information from other graphs. These disadvantages limit the precision of kernel-based similarity measures.

This paper is organized as follows: In Section 2, we first present the basics of simplicial complexes and graph data obtained from their 1-skeleton. The method of obtaining the graph defined in this way is very important for the kernel function presented, since it will most effectively use the topological and geometric properties of point clouds. Then, the distributions of discrete geodesic curves on these graph structures are determined using Kullback–Leibler information and then a kernel function using the Wassertein-1 distance is introduced. In Section 3, we give basic computational results for this novel kernel function on the Princeton ModelNet-40 benchmark. The computational results measure the similarity of point clouds with each other and show the classification of point clouds. Finally, in Section 4, we present detailed conclusions.

2. Methodology

2.1. Simplicial Complexes

The convex body of the points with $v_0, v_1, \ldots, v_d \in \mathbb{R}^n$ being $d + 1$-affine independent points is called a d-symplex. The details about simplicial topology can be found in [59]. In this study, a d-simplex is denoted by $\sigma = conv\{v_0, \ldots, v_n\} = [v_0, \ldots, v_n]$. The convex body is simply a polyhedron with $d + 1$-affine independent points as vertices. The face of a σ is defined by $conv\{S\}$ as $S \subset [v_0, \ldots, v_n]$. The finite family of simplexes that provide the following properties is called a K complex:

- For $\sigma \in K$ and τ is face of σ, $\tau \in K$;
- When $\sigma, \sigma' \in K$, $\sigma \cap \sigma'$ is empty or is simultaneously a face of σ and σ'.

A collection $\{\sigma \in K \mid dim(\sigma) \leq j\}$ of a simplicial complex K is called the j-skeleton of K and is denoted by $K^{(j)}$. In this definition, $dim(\sigma)$ denotes the dimension of σ. The 1-skeleton of any K complex is $K^{(1)} = \{\sigma \in K \mid dim(\sigma) \leq 1\}$, that is, the set of vertices and edges of the simplex that form the complex. Hence, for $V = \{v_0, \ldots, v_n\}$ and $E = \{[v_i, v_k] \mid [v_i, v_k] \in K\} \subseteq V \times V$, there exists a simple graph $G = (V, E)$ with $K^{(1)} \equiv G$. An example of a 2-dimensional K complex and the corresponding $K^{(1)}$ skeleton is given in Figure 1. It is straightforward to see that $K^{(1)} \equiv G = (V, E)$ with $V = \{v_1, \ldots, v_{10}\}$.

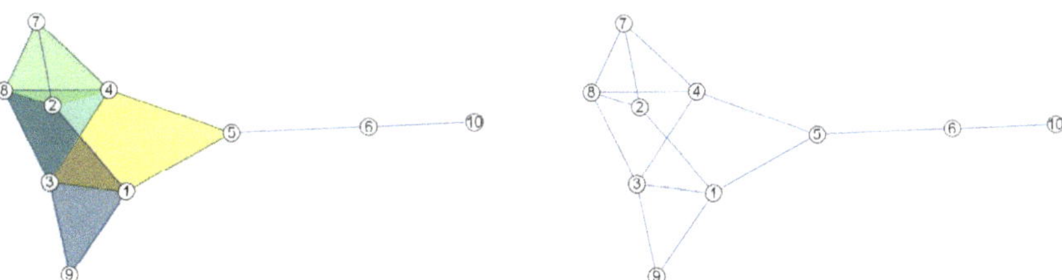

Figure 1. A 2-dimensional complex (**left**) and its 1-skeleton (**right**), where the numbers denote the vertex index.

With a K complex to be formed on a 3D point cloud, it is possible to capture the topological features of the underlying manifold of the point cloud, such as connectivity and holes. Moreover, it is possible to apply various graph algorithms to the simple graph which the skeleton $K^{(1)}$ is isomorphic. Let us consider the shortest graph paths $\ell_1 = v_7, v_8, v_3, v_9$ and $\ell_2 = v_7, v_2, v_1, v_9$ between the vertices v_7 and v_9 on the $K^{(1)}$ skeleton in Figure 1. It is obvious that the ℓ_1 and ℓ_2 are homotopic on $K^{(1)} \equiv G$. Thus, when comparing two skeletons $K_1^{(1)} \equiv G_1$ and $K_2^{(1)} \equiv G_2$ in terms of topological similarity, examining the approximation distributions of discrete geodesics corresponding to the shortest paths yields effective results.

Let us now consider the methods of obtaining various geometric or abstract complexes, which are widely used in practice. In particular, the input is often a series of points from which some hidden field is sampled or approximated. This set of points is called a point cloud. Point clouds have no topology other than a discrete topology, and some connections and some topologies are applied to them. Let a point cloud be $P = \{p_1, \ldots, p_n\}$ and the Euclidean ball be $B(p, r)$ with $p_i \in P$ at the center and r is the radius. We introduce the simplicial complex forming algorithms that we discussed in this study as follows:

Definition 1. *For a point cloud $P \subset \mathbb{R}^3$, a simplex $\sigma = [p_{i_0}, \ldots, p_{i_d}]$ is in the Delaunay complex $Del(P)$ if and only if there is a ball B whose boundary contains the vertices of σ and does not include the other points of the point cloud.*

As an indicative for the simplicial complexes employed in the research, Figure 2 provides a point cloud and an example of the Delaunay complex derived from it.

Figure 2. A point cloud example (**left**) and its Delaunay complex (**right**).

Delaunay complexes have very rich geometric features in 2D and 3D [60,61]. Only computing first-dimensional Delaunay complexes does not seem to be asymptotically faster than the computation of the full Delaunay complex. This makes the complex less attractive for high-dimensional data analysis. In this situation, Delaunay complexes' subcomplex Alpha complexes become a key tool for topological data processing. The Delaunay triangulation of a point cloud includes a subset of the faces, which together make up the Alpha complex $A_r(P)$, a d-dimensional simplicial complex [62].

Definition 2. *For a point cloud $P \subset \mathbb{R}^3$ and the given real number $r > 0$, a simplex $\sigma = [p_{i_0}, \ldots, p_{i_d}]$ is in the Alpha complex $A_r(P)$ if and only if $\bigcap\limits_{p \in Q \subset P} Vor_r(p, P) \neq \varnothing$, where $Vor_r(p, P)$ is the Voronoi ball of p defined by the intersection of the open ball centered at p with radius r and the Voronoi cell of $p \in P$.*

Definition 3. *For a point cloud $P \subset \mathbb{R}^3$ and the given real number $r > 0$, a simplex $\sigma = [p_{i_0}, \ldots, p_{i_d}]$ is in the Čech complex $C_r(P)$ if and only if $\bigcap\limits_{0 \leq j \leq d} B(p_{i_j}, r) \neq \varnothing$.*

It should be noted that the definition of the Čech complex includes a parameter r, which may be useful in practice. Specifically, we can think of creating a sphere at each p_i in the point cloud and looking at the convergence of the spheres on the r scale.

Definition 4. *For a point cloud $P \subset \mathbb{R}^3$ and the given real number $r > 0$, a simplex $\sigma = [p_{i_0}, \ldots, p_{i_d}]$ is in the Vietoris Rips complex $VR_r(P)$ if and only if $\forall j, j' \in [0, d], B(p_{i_j}, r) \cap B(p_{i'_j}, r) \neq \varnothing$.*

In other words, the points $p_{i_0}, \ldots, p_{i_k}$ span a d-simplex if and only if the Euclidean balls with radius r centered at these points have a pairwise intersection.

Definition 5. *Let $\forall i \in [0, d]$ and $q \in Q = P \setminus \{q_0, \ldots, q_d\}$. If $d(q_i, x) \leq d(q, x)$, then it is said that the simplex $\sigma = [q_0, \ldots, q_d]$ is weakly witnessed by point x and if $d(q_i, x) = d(q, x)$, then it is said that the simplex $\sigma = [q_0, \ldots, q_d]$ is strongly witnessed by the point x. For a point cloud $P \subset \mathbb{R}^3$ and $Q \subset P$, the witness complex $W(Q, P)$ is the simplicial complex whose vertices are from the set Q and all faces are weakly witnessed by a point in P.*

Detailed information for these complexes, which are frequently used in the literature, can be found in [63–65]. Moreover, the following features can be given from the same studies:

$$W(Q, P) \subseteq Del(P) \tag{1}$$

and

$$C_r(P) \subseteq VR_r(P) \subseteq C_{2c}(P). \tag{2}$$

A comparison for the complexes used in this study is given in Table 1.

Table 1. Comparison for the complex-forming methods.

K	Dimension	Time Complexity	Guarantee
$Del(P)$	$2^{\mathcal{O}(\lvert P \rvert)}$	$\mathcal{O}(\lvert P \rvert + \lvert P \rvert \log \lvert P \rvert)$	Approx. Geometry
$C_r(P)$	$2^{\mathcal{O}(\lvert P \rvert)}$	$\mathcal{O}(\lvert P \rvert^{d+1})$	Nerve Theorem
$VR_r(P)$	$2^{\mathcal{O}(\lvert P \rvert)}$	$\mathcal{O}(M(\lvert P \rvert))$	Approx. $C_r(P)$
$W(Q, P)$	$2^{\mathcal{O}(\lvert Q \rvert)}$	$\mathcal{O}\left(\dfrac{\lvert P \rvert}{\mu^{d^2}}\right)$	Approx. Geometry

In this study, temporal modified methods introduced by [66,67] are used for Vietoris Rips and Witness complexes, respectively. The time complexities of these methods are given in Table 1. $M(\lvert P \rvert)$ represents the complexity of the product of a matrix of type $\lvert P \rvert \times \lvert P \rvert$ and μ represents the time complexity of the sparsity function of the marker of the subset Q.

2.2. Kernel Function

Modeling and computing object similarity is one of the most difficult tasks in machine learning. When it comes to graphs, graph kernels have gotten a lot of press in recent years and have emerged as the most popular method for learning from graph-structured data. A graph kernel is a symmetric, positive semi-definite function defined on the space of graphs. This function can be expressed as an inner product in some Hilbert spaces. In particular, given a kernel κ, there exists a transformation $\varphi : \mathcal{G} \to \mathcal{H}$ mapping a graph space $\mathcal{G}$ to a Hilbert space $\mathcal{H}$ such that for each $G_1, G_2 \in \mathcal{G}$, $\kappa(G_1, G_2) = < \varphi(G_1), \varphi(G_2) >$.

Graph kernels handle the challenge of graph comparison by attempting to efficiently capture as much of the graph's topology as possible. One of the primary reasons for the widespread adoption of graph kernel methods is that they enable a vast array of techniques to operate directly on graphs. Thus, graph kernels enable the application of machine learning methods to real-world situations using graph-structured data.

First, information on the distributions will be presented before the definition of a kernel function κ defined by the distribution of the geodesics of the graphs $K_1^{(1)} \equiv G_1$ and $K_2^{(1)} \equiv G_2$.

Kullback–Leibler information is a measure of how far apart two pieces of information are in terms of probability [68]. Let $\mathbb{P}$ and $\mathbb{Q}$ be two probability measures with densities of p and q, respectively. Then, the Kullback–Leibler information is defined by

$$\mathcal{L}_X(\mathbb{P}, \mathbb{Q}) = \int \log\left(\frac{p(x)}{q(x)}\right) p(x)\, d\mu(x) = \mathbb{E}_p\left[\log\left(\frac{p(x)}{q(x)}\right)\right]. \tag{3}$$

Since Kullback–Leibler information is not symmetric, it is not a metric. However $\mathcal{L}_X(\mathbb{P}, \mathbb{Q}) + \mathcal{L}_X(\mathbb{Q}, \mathbb{P})$ is symmetric and is called the Kullback–Leibler divergence [69].

Let us consider a random variable X with a mean μ_X and a metric function defined by $d(x; \mu_X) = f(x - \mu_X)$. For $\mu_X, \mu'_X \in \mathbb{R}$, a function defined by $g(x; \mu_X) = d(x; \mu'_X) - d(x; \mu_X)$ is Jensen equal; that is, $\mathbb{E}[g(x)] = g(\mathbb{E}[x])$. Let $c \in \mathbb{R}$ and $(v, \omega) \in \mathbb{R} \times \mathbb{R}^-$. For a function $f(x; \mu_X)$ with $\mathbb{E}[X] = c$, with the density function

$$f(x; \mu_X) = v \exp(\omega d(x; \mu_X)) = g(d(x; \mu_X)) \tag{4}$$

$\mathcal{L}_X((\mathbb{P},\mathbb{Q})$ becomes a metric [70].

Let us look at how to use Kullback–Leibler information to explain geodesic distributions on graphs. The geodesic distance between two vertices u and v in a graph G is defined as the number of edges of the shortest path linking u and v and denoted as $d_G(u,v)$. For example, in Figure 1, $d_{K^{(1)}}(2,5) = 2$. The greatest geodesic distance between u and any other vertex is called the eccentricity of u and denoted by ε. It can be thought of as a measure of how far a vertex is from the furthest vertex in the graph. The eccentricity property allows the radius and diameter of a graph to be defined. The radius of a graph is the minimum eccentricity between the vertices of the graph. The diameter of a graph is the maximum eccentricity of any vertex of the graph. Hence, the diameter is the greatest distance between any pair of vertices. In order to find the diameter of a graph, we first find the shortest path between each pair of vertices using the landmark-based method presented in [71]. The greatest distance of any path is the diameter of the graph. Using the eccentricity of a graph, it is possible to define its two subgraphs. A central subgraph of $K^{(1)}$ is the graph with n vertices of degree α and the smallest eccentricity, and is denoted by $K_{n,\alpha}^{(1),C}$. An orbital subgraph of $K^{(1)}$ is the graph with n vertices of degree β and with n vertices and is denoted by $K_{n,\beta}^{(1),O}$.

Considering the Kullback–Leibler divergence, we can define two types of geodetic distributions for Jensen equal functions:

Definition 6. *Central geodesic density function of $K^{(1)}$ is*

$$f_C(u,\alpha,v) = v\exp\left(-d_G\left(u,K_{n,\alpha}^{(1),C}\right)\right), \tag{5}$$

and the orbital geodesic density function of $K^{(1)}$ is

$$f_O(u,\beta,v) = v\exp\left(-d_G\left(u,K_{n,\beta}^{(1),O}\right)\right), \tag{6}$$

where $v \geq 1$ is a normalization factor.

On a space of probability measures, the Wasserstein distances give a natural metric. They intuitively assess the least amount of effort necessary to change one distribution into another. The Wasserstein distances, in general, do not permit closed-form formulations; however, for $\mathbb{R}$, we have the explicit form as

$$W_1(\rho_1,\rho_2) = \int_{\mathbb{R}} |F_{\rho_1}(t) - F_{\rho_2}(t)|\,dt, \tag{7}$$

where $F_{\rho_1}(t)$ and $F_{\rho_2}(t)$ are the cumulative distribution functions of ρ_1 and ρ_2 [72]. It is feasible to compare the geodesic distributions with the kernel function below using the Wasserstein-1 distance function:

Definition 7. *Let $K_1^{(1)} = (V_1, E_1)$ and $K_2^{(1)} = (V_2, E_2)$. Then, we have a kernel function*

$$\kappa_J\left(K_1^{(1)}, K_2^{(1)}\right) = \frac{W_1\left(\cup_{i=1}^{|V_1|}, f_C(u_i,\alpha_1,v_1), \cup_{j=1}^{|V_2|}, f_C(u_j,\alpha_2,v_2)\right)}{W_1\left(\cup_{i=1}^{|V_1|}, f_O(u_i,\beta_1,v_1), \cup_{j=1}^{|V_2|}, f_O(u_j,\beta_2,v_2)\right)}, \tag{8}$$

where W_1 is Wassterstein-1 distance.

We shall note that the kernel function defined in Equation (8) is symmetric and positive definite.

3. Results

3.1. Data Set

Princeton ModelNet (Internet Access: https://modelnet.cs.princeton.edu/, accessed on 10 January 2023) seeks to give researchers in computer vision, computer graphics, robotics, and cognitive science a thorough and organized library of 3D CAD models. The researchers used information from the SUN database to create a list of the most widespread object types on the globe, which served as the foundation of the dataset. After developing an object vocabulary, the 3D CAD models of each item category were gathered by searching web search engines for each category phrase. Then, using custom-made tools with quality control, human employees were hired at Amazon Mechanical Turk to manually determine whether each CAD model fits into the designated categories [73].

Two methods are used to assess the effectiveness of the graph kernel function given in this paper. The first strategy is to assess the extent to which the topologies of the underlying point cloud geometries have an impact on the kernel function. In this study, the kernel function is used to determine the extent to which point clouds are similar or dissimilar. Point cloud samples containing 10,000 points are taken from the ModelNet-40 dataset for the first method. These point cloud examples belong to the airplane, car, person, plant, and vase classes as they have different inter- and intra-class topologies. Ten sample point clouds are taken from each class, and the kernel function presented in this study is run between these samples.

The second strategy involves using the provided kernel function to create a classification that is based on machine learning. On the ModelNet-40 dataset, this classification methodology is used, and it is compared to other approaches.

3.2. Point Cloud Comparison

To produce graphs from point clouds, each of the simplicial complex derivation techniques described in Section 2.1 are applied to the sample set, and the 1-skeletons of the complexes were selected as the representation graph. Equation (8) was used in this part to generate the graphs for each data group as well as the kernel matrices of the data subgroups. The point clouds used in this approach are presented in Figures A1–A5.

The relevant parameters are selected in the formation of simplicial complexes as the minimum values that connect the 1-skeletons derived from the point cloud. Using a step size of $h = 0.001$ and beginning at 0, the least parameter is found. It is anticipated that different parameters will be found for each sample, both inter- and intra-classes. In addition to this, no discernible variation in the computational time complexity of the simplicial complexes' generation procedures was found. Moreover, better computational times could result from differentiating the approach taken when determining the lowest parameter.

Figures 3–7 contain similarity matrices obtained by using the kernel function to compare point clouds with each other. The samples from each class are complexed using the previously discussed simplex techniques, and the kernel function is applied to the graph structures that were built using the 1-skeletons of the samples.

Similarity matrices generated using the kernel function encode the measurements of the similarity between the point clouds. In the context of the distribution of discrete geodesics, the near-zero values of the kernel function provided by Equation (8), whose inputs are two graphs produced from point clouds, indicate that the two graphs are highly similar. Likewise, the graphs diverge in terms of similarity within the same context for large numbers. This study used four of the most fundamental simplicial complex approaches to determine the similarity metrics within each class. The analysis of the similarity matrices reveals that the common points of the point clouds in the discrete geometry environment within each class are most evident in the graphs produced with Alpha complexes. The geometries underlying the point clouds of the Airplane models are most comparable in the case of the Airplane 5 model. The matrix given in Figure 3 clearly demonstrates that the similarity value derived from the graph of Alpha complexes for this model is close to zero. When a comparable course is followed, the Airplane 4 and 8 models entirely diverge in terms of similarity, as shown by the matrix's high values. In the similarity matrices provided in Figures 4–7, the models most geometrically comparable to other models for the circumstance produced with the Alpha complex have very low values.

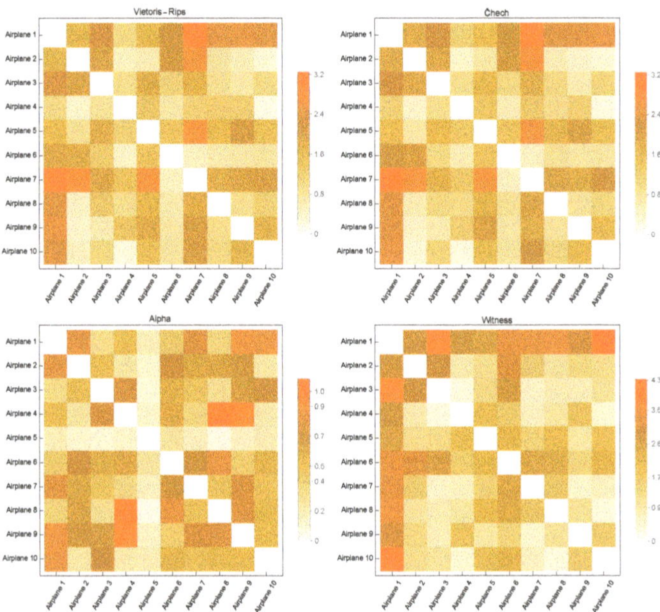

Figure 3. Similarity Matrices for Airplane Models.

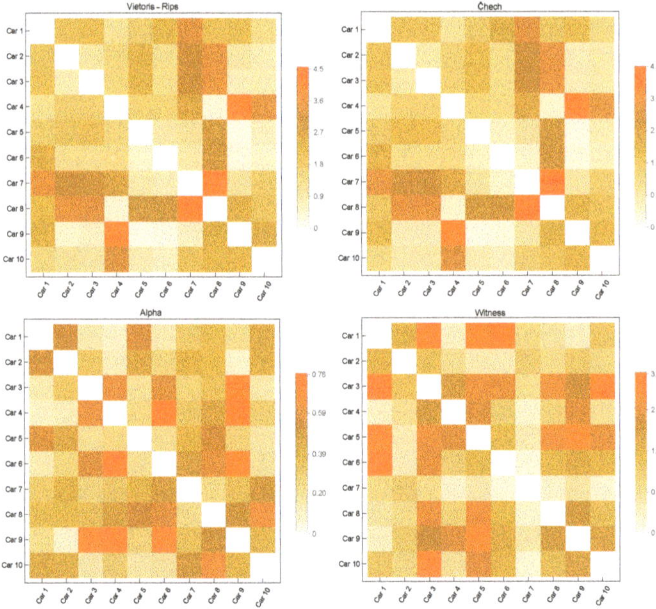

Figure 4. Similarity Matrices for Car Models.

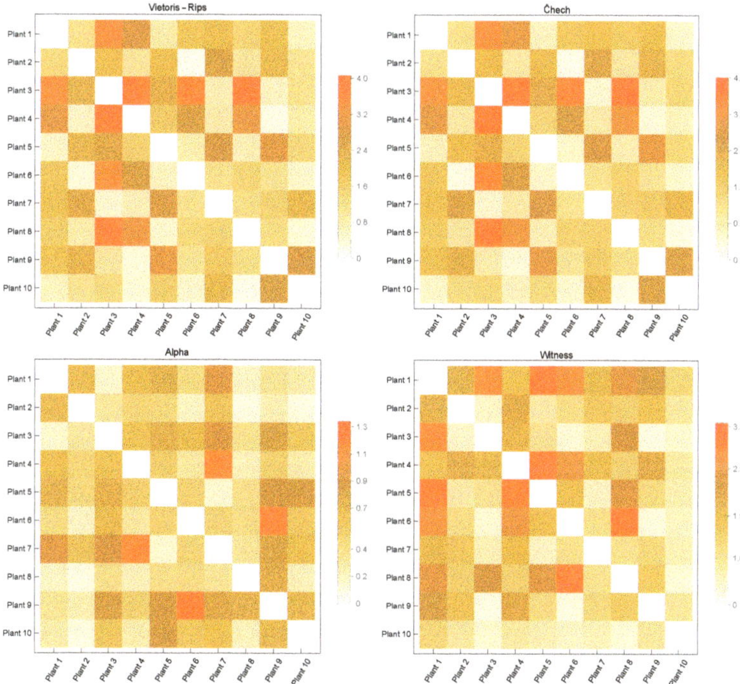

Figure 5. Similarity Matrices for Plant Models.

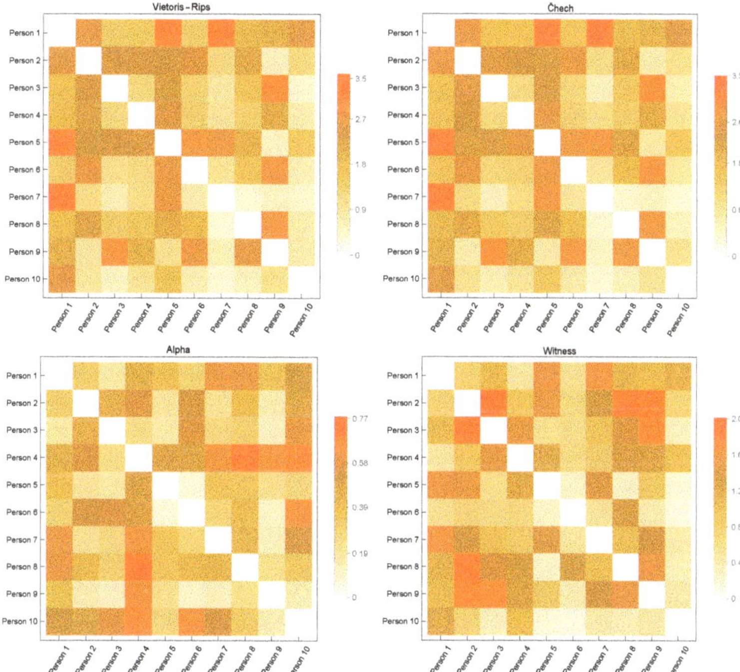

Figure 6. Similarity Matrices for Person Models.

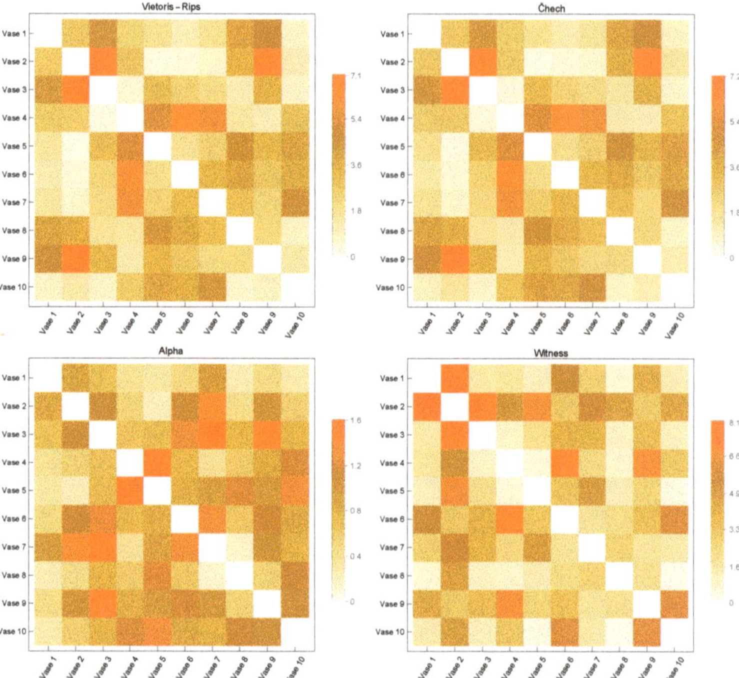

Figure 7. Similarity Matrices for Vase Models.

The graphs generated using the Vietoris Rips and Čech complexes reveal that the kernel function values are quite similar. The graphs formed by the Witness complex have the most inconsistent similarity values among these four techniques. When examining the underlying geometries of each point cloud, the kernel function assigns large values to very comparable geometries. Even if the Vase 2 model is physically comparable to models 1 and 3, the inputs of the Witness complex's similarity matrix are relatively high. The observation that the kernel function computed on the graphs produced with the Vietoris Rips and Čech complexes has a large value when it is low is another result of the calculations performed on the graphs acquired with the Alpha complexes.

Considering all of these circumstances, it is feasible to conclude that the 1-skeleton of Alpha complexes are the graphs in which the kernel function developed in this research performs the best. Additionally, it indicates that these values vary when the underlying geometries have sharp edges. Since geodesics become geometrically singular at sharp endpoints, it is said that the presented kernel function provides better geometrically correct results when the Alpha complex is utilized to generate graphs.

In order to apply this kernel function to assess the similarities between distinct model classes, the similarities between the classes are analyzed in our research using the graphs derived from the Alpha complexes. Within the scope of this precision, Figure 8 depicts the inter-class similarity matrix constructed by applying the kernel function to the graphs generated using the Alpha complexes of five distinct models. This matrix also displays graphs with comparable discrete geodesy distributions whose values are close to zero. There are two remarkable commonalities between the two courses. The Car and Vase models, as well as the Airplane and Person models, are among the most different models. In other words, the kernel function in these models has high values.

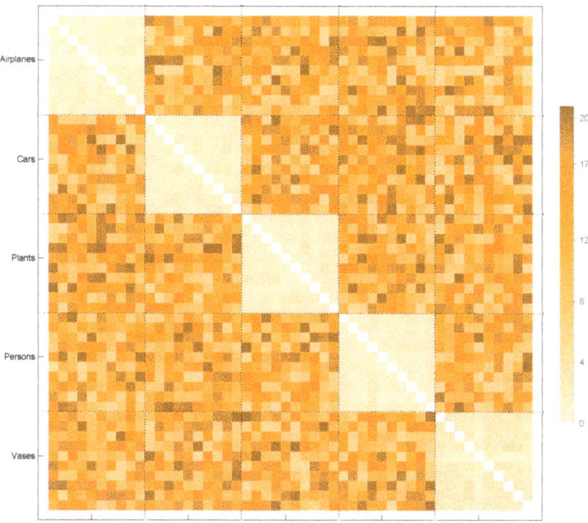

Figure 8. Interclass similarity matrix for graphs obtained by the Alpha complex.

3.3. Classification

Figure 8 illustrates how useful similarity matrices between classes are when they are produced using the graph kernel function discussed in this study. This part describes how to use this kernel function to infer the class to which a given point cloud belongs. The goal is to create a model that learns from the input graphs $K_1^{(1)}, K_2^{(1)}, \ldots, K_N^{(1)}$ and predicts the label of new, unobserved graphs using this dataset of input graphs.

With the usage of graph kernel functions in convolution processes, particularly with the advancements in GCN networks, several deep learning architectures have been built in recent years [74–78]. However, these systems often utilize vectors between graph vertices or edges that correlate with different physical qualities. Additionally, kernel-based classification techniques use kernel functions created across different subgraphs to quantify the similarities between two vertices. It is important to build a matrix between the vertex clusters, namely graph communities, and not between vertex pairs, in the classification process since the kernel function taken into account in this study takes into account the discrete geometries underlying the 3D point clouds.

It is known that the matrices of kernel functions created between two graph classes include the structures utilized to embed graphs in Hilbert space. Therefore, it is inappropriate to apply the in-class kernel function directly to solve the graph classification issue. In this study, communities of graphs that are specific to each point cloud were identified, and matrices with the kernel function were produced among the subgraphs generated by the communities. The Fluid Communities [79] approach was used in this study to find the communities of each graph. The propagation mechanism on which Fluid Communities is based is state of the art in terms of computing cost and scalability. Despite being very effective, Fluid Communities can locate communities in artificial graphs with a degree of accuracy that is competitive with the best options. The first propagation-based method that can recognize a changeable number of communities in the network is Fluid Communities. Figure 9 depicts the 10 graph communities found in the 1-skeleton of the Alpha complex using the Fluid Communities approach.

Figure 9. The 1-skeleton of an Alpha complex emerging from plant point cloud and 10 communities obtained using the Fluid Communities approach.

Each initial point cloud sample contains the communities of graphs derived from Alpha complexes for the classification process. The number of communities is determined to be 10, as the community discovery technique utilized in this study needs that number to be predetermined. The kernel function created in the research among the subgraphs discovered by the communities is then used to realize the commonalities between these communities. In order to represent the inherent geometry of each point cloud, a 10×10 matrix is produced. Then, a CNN network with three convolution layers is used. After the first convolution layer, Max Pooling was performed, followed by Average Pooling for the second and third layers. While ReLU activation functions were used for the first two Fully Connected layers, softmax was utilized for classification after the second layer. The ModelNet-40 dataset's whole sample pool was used in the classification procedure. The ModelNet-40 dataset includes 2468 test models and 9843 training models divided into 40 classes. We uniformly choose 1024 points from each model and normalize them to a unit sphere in order to compare our results with those of other approaches in the literature. We give accurate data from the classification procedure and comparisons to alternative approaches in Table 2.

Table 2. Classification results on ModelNet-40 benchmark.

Method	Input	No. Points	Accuracy
[80]	xyz	1024	% 86.1
[81]	xyz	1024	% 87.1
[82]	xyz	1024	% 87.4
[83]	xyz	1024	% 89.2
[84]	xyz	1024	% 90.0
[85]	xyz	1024	% 90.2
[86]	xyz	1024	% 90.6
[87]	xyz	1024	% 90.7
[88]	xyz	1024	% 91.0
[89]	xyz	1024	% 92.2
[90]	xyz	1024	% 92.3
[91]	xyz	1024	% 93.6
Ours	xyz	1024	% 93.7

4. Discussion and Conclusions

Graphs are maybe one of the most comprehensive data structures used in machine learning. Complex objects may be represented using graphs as a collection of items and their connections, each of which can be annotated with information such as category or vectorial vertex and edge properties. As graph states, both unstructured vector data and structured data types such as time series, images, volumetric data, point clouds, and asset bags may be understood. Most importantly, the extra flexibility that graph-based representations provide is helpful for a variety of applications.

Kernel approaches are being increasingly used as a method for determining how comparably organized items are. In general, kernel approaches compare two objects using a kernel function that corresponds to an inner product in a kernel Hilbert space. Finding an appropriate kernel function that can be traced computationally while capturing the structure's semantics is challenging for kernel approaches. In this paper, a kernel function based on the distribution of discrete geodesics on graphs is used to solve the point cloud comparison problem, one of the most important topics in computer vision. Creating this kernel function requires the description of a technique sensitive to isometries and topological transformations.

The dataset was chosen based on the Princeton ModelNet-40 benchmark. Similarity matrices were generated by applying the kernel function to 10 point clouds across five distinct categories. When the similarity matrices are assessed, it is discovered that the graphs of the 1-skeletons of the Alpha complexes provide the highest internal similarity of the categories. Among the point clouds within each category, the similarity value (lower kernel value) between models that are more topologically connected, i.e., structural holes and bottleneck structures, is rather high. The kernel value distributions of the Airplane, Car, Person, Plant, and Vase models attain lower values (more similarity), notably in the Car and Vase models. When the point clouds of both models are analyzed, it becomes evident that these two groups have several topological similarities. Therefore, we may conclude that the kernel function discussed in this research performs well when applied to topological similarities, particularly when the Alpha complex is considered.

The classification of graphs is an issue that regularly arises in machine learning research. Kernel functions that are defined by structural measurements between the related subgraphs of a graph are commonly used for graph classification problems. In this study, the kernel function suggested for graph classification is studied. Within the graphs of each point cloud, the input matrices for a simple convolution neural network model are generated. Utilizing the aggregates of graphs with vertex clusters, the category to which a graph in vector form belongs was determined. For each network model, the Fluid Community approach was utilized to integrate a fixed cluster size while identifying the communities. The kernel function was used to generate inherent similarity matrices between the subgraphs generated by graph communities. This method yielded classification results with an accuracy rating of 91.1%. Even though this percentage is not the greatest reported in the literature, the results were quite effective in comparison to other approaches.

This kernel function, which was designed to compare point clouds with their geometries and topologies, is independent of the dimension of the point cloud, and therefore may be utilized in future research with 2D or higher dimensional point clouds. In addition, it is anticipated that the deep learning architecture used in the classification method provided in this paper will be refined, resulting in improved classification outcomes.

Author Contributions: Conceptualization, Ö.A. and M.A.B.; Formal analysis, Ö.A., M.A.B. and L.M.B.; Funding acquisition, L.G.; Investigation, Ö.A. and M.A.B.; Methodology, Ö.A., M.A.B. and L.M.B.; Project administration, Ö.A.; Resources, Ö.A., M.A.B., L.M.B. and L.G.; Supervision, Ö.A. and M.A.B.; Writing—original draft, Ö.A., M.A.B., L.M.B. and L.G.; Writing—review and editing, Ö.A., M.A.B., L.M.B. and L.G. All authors have read and agreed to the published version of the manuscript.

Funding: This study was conducted with financial support from the scientific research funds of "1 Decembrie 1918" University of Alba Iulia, Romania. Moreover, analyses and results were funded by TUBITAK grant number 121E031.

Institutional Review Board Statement: Not applicable.

Informed Consent Statement: Not applicable.

Data Availability Statement: Not applicable.

Conflicts of Interest: The funders had no role in the design of the study; in the collection, analyses, or interpretation of data; in the writing of the manuscript; or in the decision to publish the results.

Appendix A

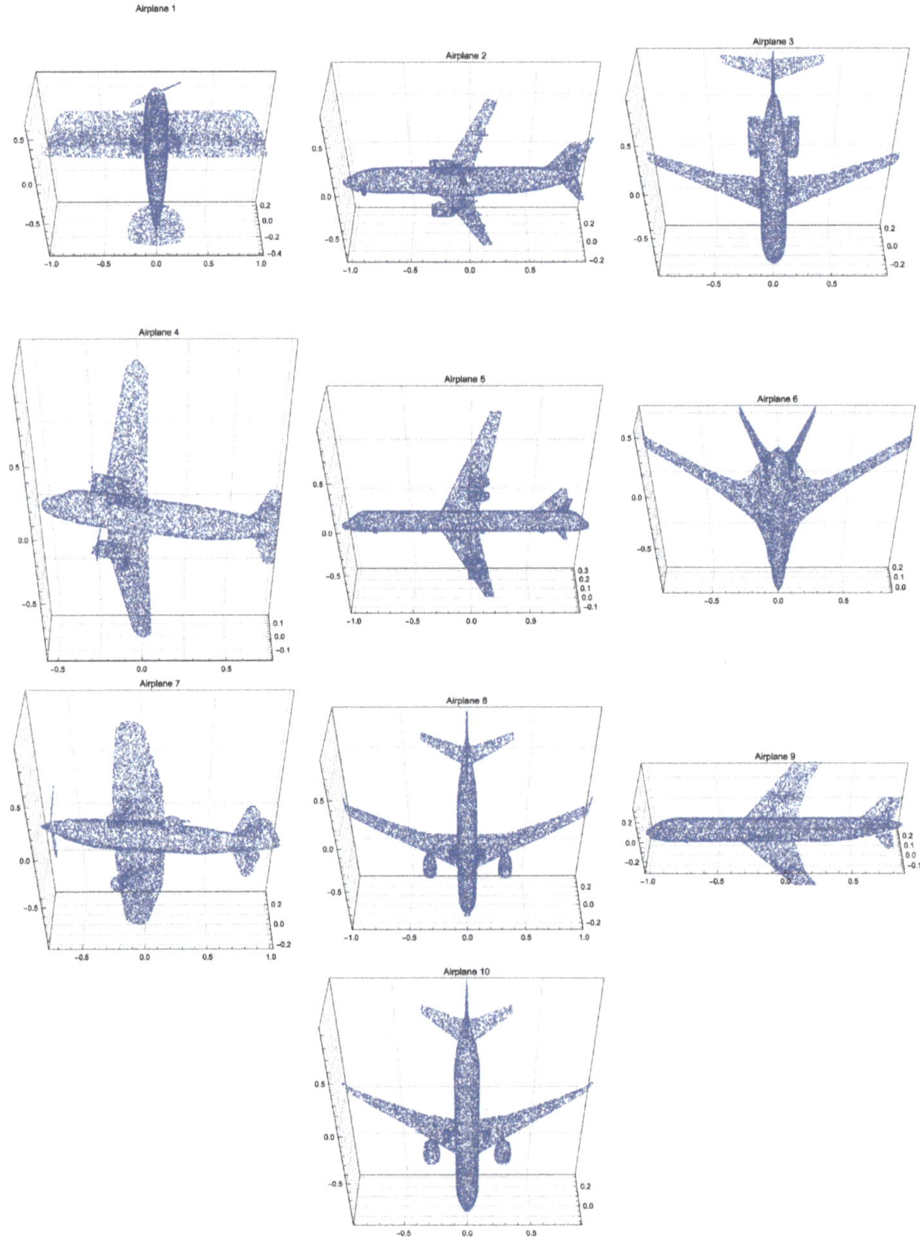

Figure A1. Airplane Models.

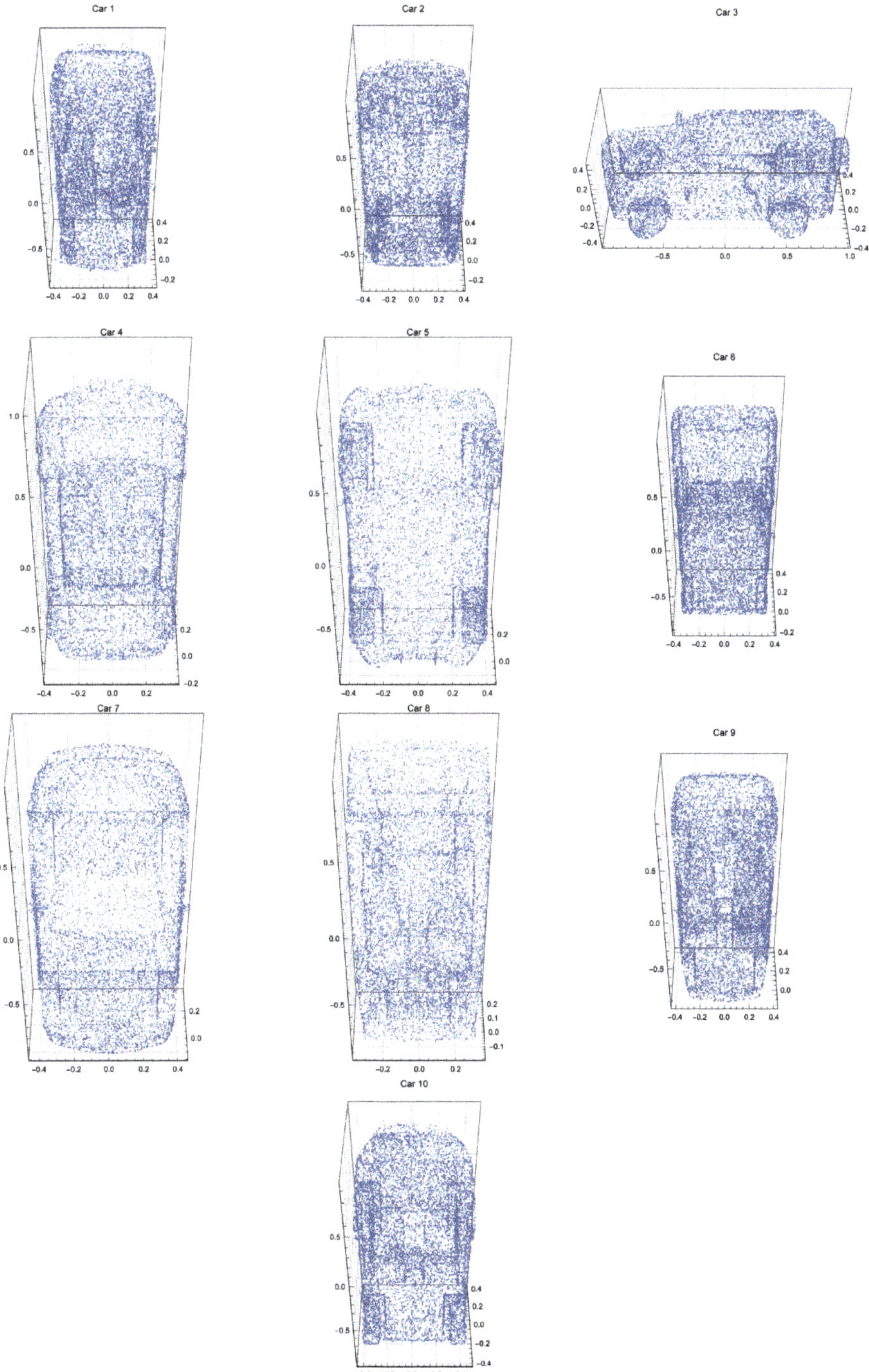

Figure A2. Car Models.

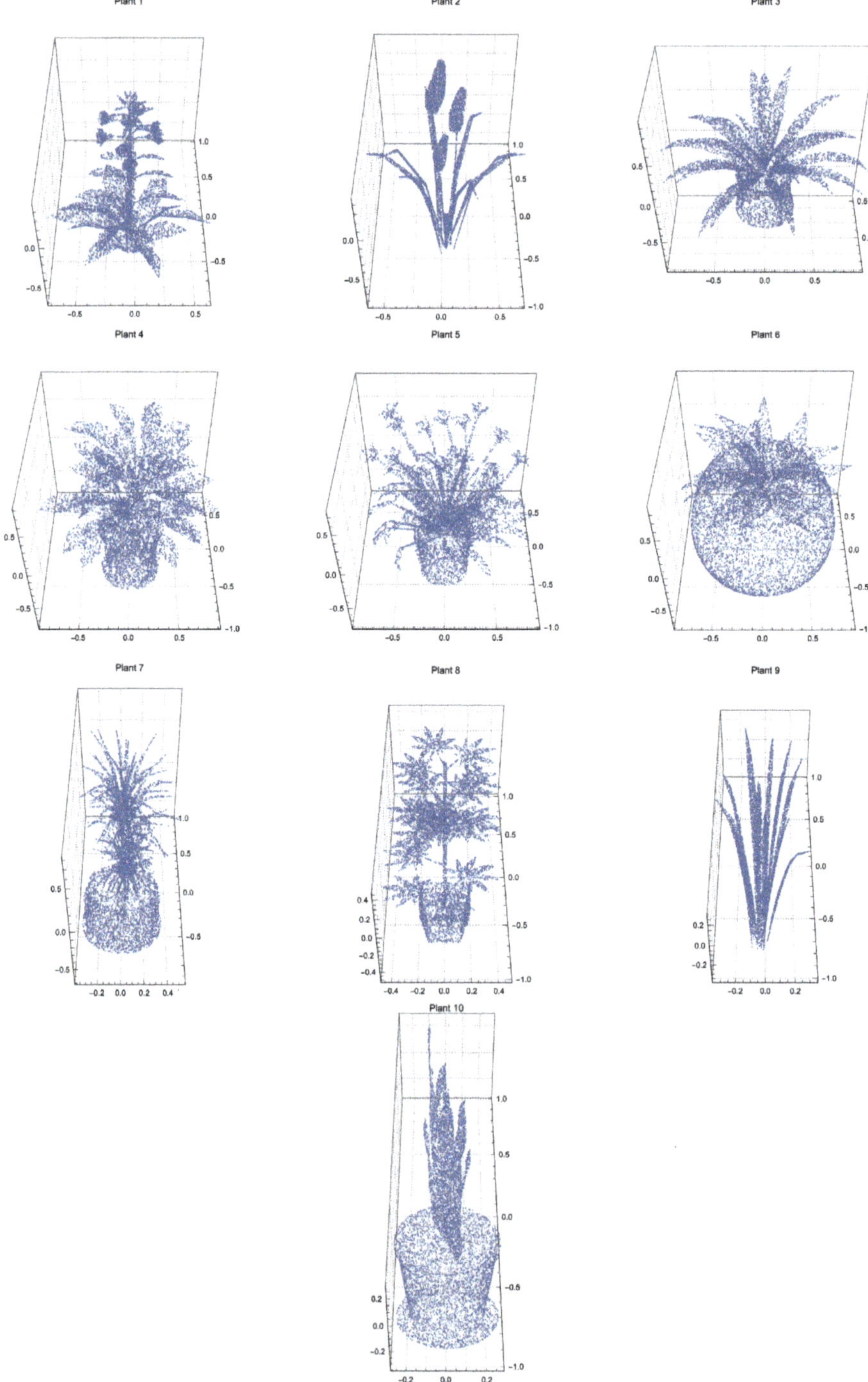

Figure A3. Plant Models.

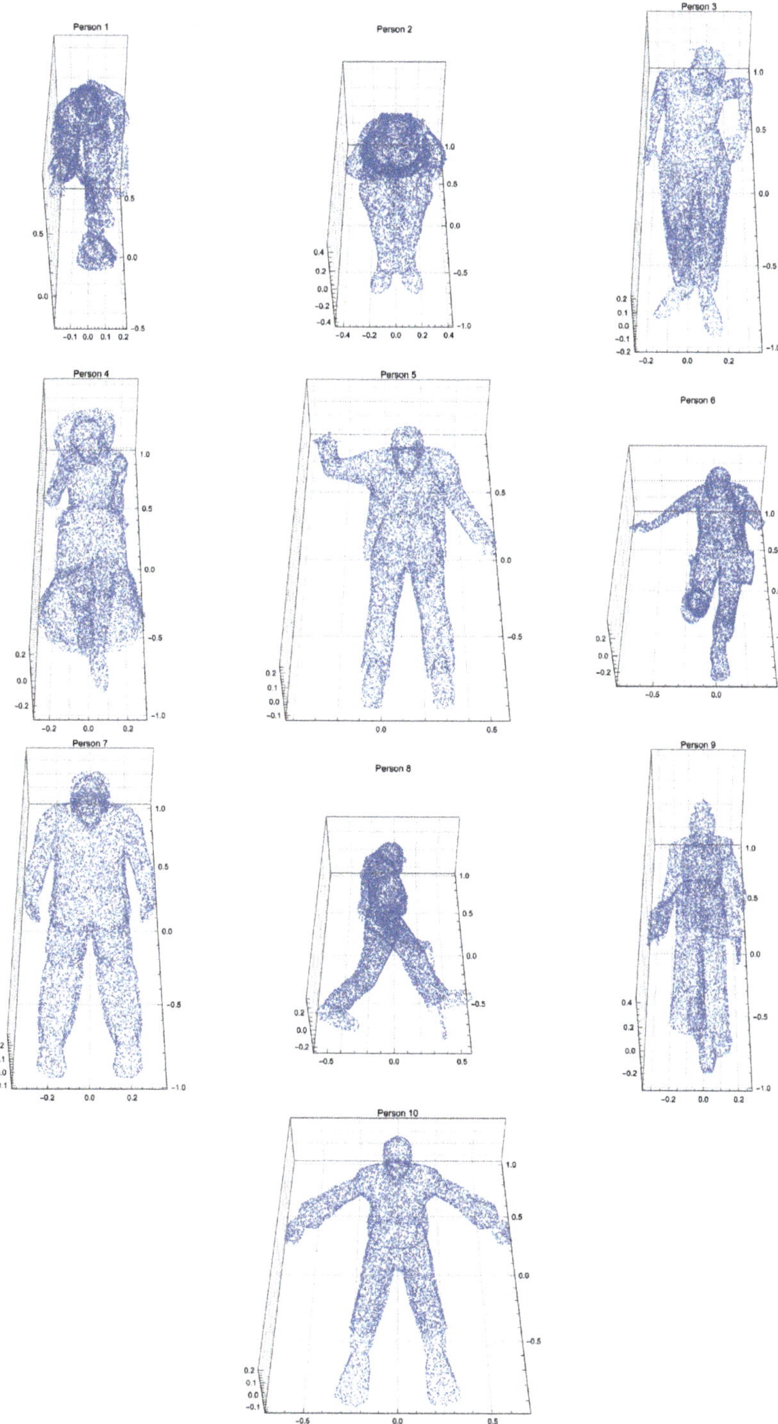

Figure A4. Person Models.

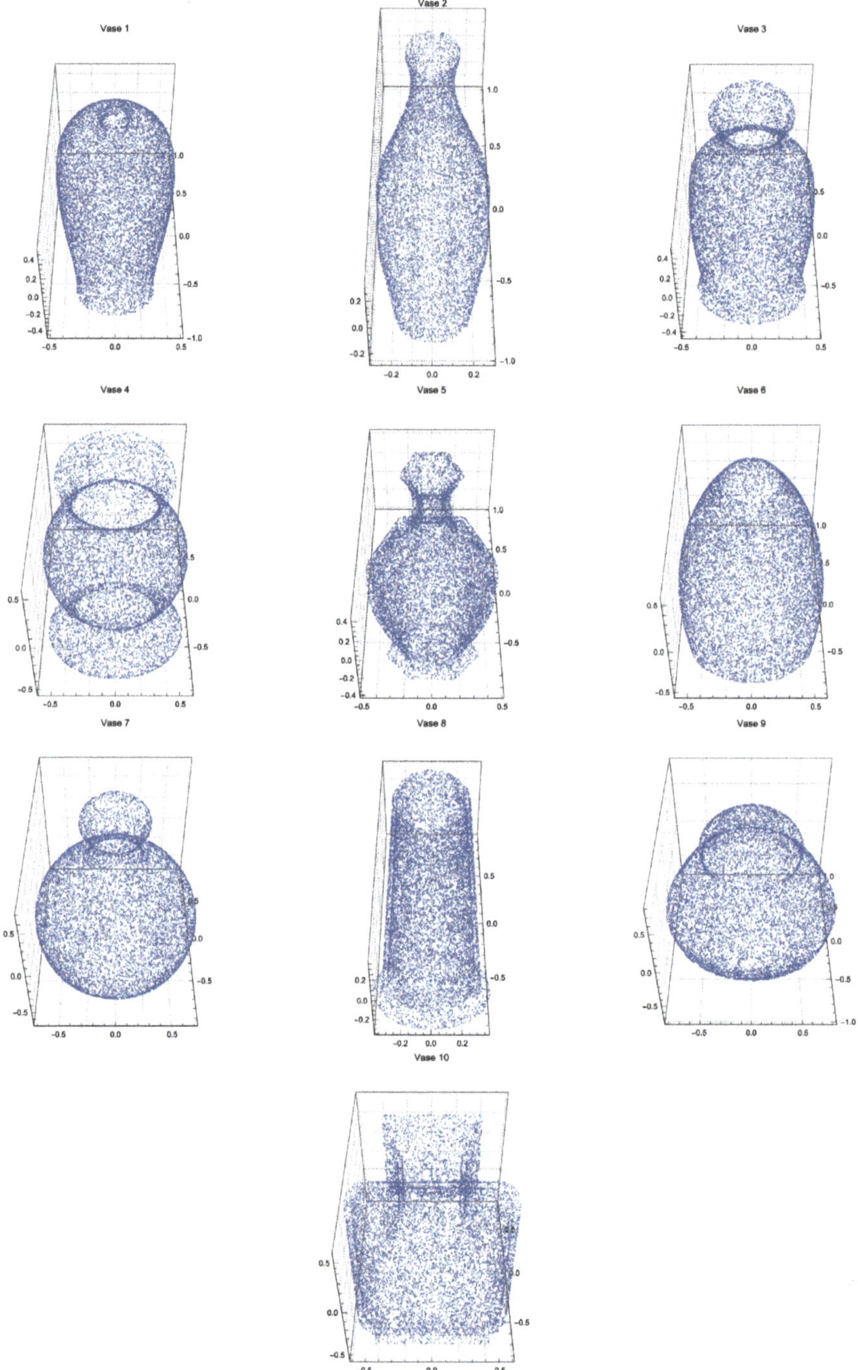

Figure A5. Vase Models.

References

1. Chougule, V.; Mulay, A.; Ahuja, B. Three dimensional point cloud generations from CT scan images for bio-cad modeling. In Proceedings of the International Conference on Additive Manufacturing Technologies—AM2013, Bengaluru, India, 7–8 October 2013 Volume 7, p. 8.
2. Fu, Y.; Lei, Y.; Wang, T.; Patel, P.; Jani, A.B.; Mao, H.; Curran, W.J.; Liu, T.; Yang, X. Biomechanically constrained non-rigid MR-TRUS prostate registration using deep learning based 3D point cloud matching. *Med Image Anal.* **2021**, *67*, 101845. [CrossRef]
3. Gidea, M.; Katz, Y. Topological data analysis of financial time series: Landscapes of crashes. *Phys. A Stat. Mech. Its Appl.* **2018**, *491*, 820–834. [CrossRef]
4. Ismail, M.S.; Hussain, S.I.; Noorani, M.S.M. Detecting early warning signals of major financial crashes in bitcoin using persistent homology. *IEEE Access* **2020**, *8*, 202042–202057. [CrossRef]
5. Liu, J.; Guo, P.; Sun, X. An Automatic 3D Point Cloud Registration Method Based on Biological Vision. *Appl. Sci.* **2021**, *11*, 4538. [CrossRef]
6. Barra, V.; Biasotti, S. 3D shape retrieval and classification using multiple kernel learning on extended Reeb graphs. *Vis. Comput.* **2014**, *30*, 1247–1259. [CrossRef]
7. Chen, L.M. *Digital and Discrete Geometry: Theory and Algorithms*; Springer: Berlin/Heidelberg, Germany, 2014.
8. Song, M. A personalized active method for 3D shape classification. *Vis. Comput.* **2021**, *37*, 497–514. [CrossRef]
9. Wang, F.; Lv, J.; Ying, G.; Chen, S.; Zhang, C. Facial expression recognition from image based on hybrid features understanding. *J. Vis. Commun. Image Represent.* **2019**, *59*, 84–88. [CrossRef]
10. Boulch, A.; Puy, G.; Marlet, R. FKAConv: Feature-kernel alignment for point cloud convolution. In Proceedings of the Asian Conference on Computer Vision, Kyoto, Japan, 30 November–4 December 2020.
11. Gou, J.; Yi, Z.; Zhang, D.; Zhan, Y.; Shen, X.; Du, L. Sparsity and geometry preserving graph embedding for dimensionality reduction. *IEEE Access* **2018**, *6*, 75748–75766. [CrossRef]
12. Meltzer, P.; Mallea, M.D.G.; Bentley, P.J. PiNet: Attention Pooling for Graph Classification. *arXiv* **2020**, arXiv:2008.04575.
13. Wang, T.; Liu, H.; Li, Y.; Jin, Y.; Hou, X.; Ling, H. Learning combinatorial solver for graph matching. In Proceedings of the IEEE/CVF Conference on Computer Vision and Pattern Recognition, Seattle, WA, USA, 13–19 June 2020; pp. 7568–7577.
14. Wu, J.; Pan, S.; Zhu, X.; Zhang, C.; Philip, S.Y. Multiple structure-view learning for graph classification. *IEEE Trans. Neural Netw. Learn. Syst.* **2017**, *29*, 3236–3251. [CrossRef]
15. Mitchell, J.S.; Mount, D.M.; Papadimitriou, C.H. The discrete geodesic problem. *SIAM J. Comput.* **1987**, *16*, 647–668. [CrossRef]
16. Polthier, K.; Schmies, M. Straightest geodesics on polyhedral surfaces. In *ACM SIGGRAPH 2006 Courses*; Association for Computing Machinery: New York, NY, USA, 2006; pp. 30–38.
17. Atici, M.; Vince, A. Geodesics in graphs, an extremal set problem, and perfect hash families. *Graphs Comb.* **2002**, *18*, 403–413. [CrossRef]
18. Bernstein, M.; De Silva, V.; Langford, J.C.; Tenenbaum, J.B. *Graph Approximations to Geodesics on Embedded Manifolds*; Technical Report; Citeseer: University Park, PA, USA, 2000.
19. Alexa, M.; Behr, J.; Cohen-Or, D.; Fleishman, S.; Levin, D.; Silva, C.T. Point set surfaces. In Proceedings of the Visualization, 2001—VIS'01, San Diego, CA, USA, 21–26 October 2001; pp. 21–29.
20. Amenta, N.; Kil, Y.J. Defining point-set surfaces. *ACM Trans. Graph. (TOG)* **2004**, *23*, 264–270. [CrossRef]
21. Levin, D. Mesh-independent surface interpolation. In *Geometric Modeling for Scientific Visualization*; Springer: Berlin/Heidelberg, Germany, 2004; pp. 37–49.
22. Adamson, A.; Alexa, M. Approximating and intersecting surfaces from points. In Proceedings of the 2003 Eurographics/ACM SIGGRAPH Symposium on Geometry Processing, Aachen, Germany, 23–25 June 2003; pp. 230–239.
23. Amenta, N.; Choi, S.; Kolluri, R.K. The power crust, unions of balls, and the medial axis transform. *Comput. Geom.* **2001**, *19*, 127–153. [CrossRef]
24. Dey, T.K.; Goswami, S. Provable surface reconstruction from noisy samples. *Comput. Geom.* **2006**, *35*, 124–141. [CrossRef]
25. Attene, M.; Patanè, G. Hierarchical structure recovery of point-sampled surfaces. *Comput. Graph. Forum* **2010**, *29*, 1905–1920. [CrossRef]
26. Chen, X. Hierarchical rigid registration of femur surface model based on anatomical features. *Mol. Cell. Biomech.* **2020**, *17*, 139. [CrossRef]
27. Nakarmi, U.; Wang, Y.; Lyu, J.; Liang, D.; Ying, L. A kernel-based low-rank (KLR) model for low-dimensional manifold recovery in highly accelerated dynamic MRI. *IEEE Trans. Med. Imaging* **2017**, *36*, 2297–2307. [CrossRef]
28. Zhang, S.; Cui, S.; Ding, Z. Hypergraph spectral analysis and processing in 3D point cloud. *IEEE Trans. Image Process.* **2020**, *30*, 1193–1206. [CrossRef]
29. Connor, M.; Kumar, P. Fast construction of k-nearest neighbor graphs for point clouds. *IEEE Trans. Vis. Comput. Graph.* **2010**, *16*, 599–608. [CrossRef] [PubMed]
30. Natali, M.; Biasotti, S.; Patanè, G.; Falcidieno, B. Graph-based representations of point clouds. *Graph. Model.* **2011**, *73*, 151–164. [CrossRef]
31. Klein, J.; Zachmann, G. Point cloud surfaces using geometric proximity graphs. *Comput. Graph.* **2004**, *28*, 839–850. [CrossRef]
32. Fan, H.; Wang, Y.; Gong, J. Layout graph model for semantic façade reconstruction using laser point clouds. *Geo-Spat. Inf. Sci.* **2021**, *24*, 403–421. [CrossRef]

33. Wang, Y.; Ma, Y.; Zhu, A.x.; Zhao, H.; Liao, L. Accurate facade feature extraction method for buildings from three-dimensional point cloud data considering structural information. *ISPRS J. Photogramm. Remote Sens.* **2018**, *139*, 146–153. [CrossRef]

34. Wang, Y.; Li, H.; Liao, L. DEM Construction Method for Slopes Using Three-Dimensional Point Cloud Data Based on Moving Least Square Theory. *J. Surv. Eng.* **2020**, *146*, 04020013. [CrossRef]

35. Wu, B.; Yang, L.; Wu, Q.; Zhao, Y.; Pan, Z.; Xiao, T.; Zhang, J.; Wu, J.; Yu, B. A Stepwise Minimum Spanning Tree Matching Method for Registering Vehicle-Borne and Backpack LiDAR Point Clouds. *IEEE Trans. Geosci. Remote Sens.* **2022**, *60*, 1–13. [CrossRef]

36. Guo, Q.; Wang, Y.; Yang, S.; Xiang, Z. A method of blasted rock image segmentation based on improved watershed algorithm. *Sci. Rep.* **2022**, *12*, 7143. [CrossRef]

37. Wang, Y.; Tu, W.; Li, H. Fragmentation calculation method for blast muck piles in open-pit copper mines based on three-dimensional laser point cloud data. *Int. J. Appl. Earth Obs. Geoinf.* **2021**, *100*, 102338. [CrossRef]

38. Wang, Y.; Zhou, T.; Li, H.; Tu, W.; Xi, J.; Liao, L. Laser point cloud registration method based on iterative closest point improved by Gaussian mixture model considering corner features. *Int. J. Remote Sens.* **2022**, *43*, 932–960. [CrossRef]

39. Alaba, S.Y.; Ball, J.E. A survey on deep-learning-based lidar 3d object detection for autonomous driving. *Sensors* **2022**, *22*, 9577. [CrossRef] [PubMed]

40. Song, F.; Shao, Y.; Gao, W.; Wang, H.; Li, T. Layer-wise geometry aggregation framework for lossless lidar point cloud compression. *IEEE Trans. Circuits Syst. Video Technol.* **2021**, *31*, 4603–4616. [CrossRef]

41. Wang, S.; Jiao, J.; Cai, P.; Wang, L. R-pcc: A baseline for range image-based point cloud compression. In Proceedings of the 2022 International Conference on Robotics and Automation (ICRA), Philadelphia, PA, USA, 23–27 May 2022; pp. 10055–10061.

42. Xiong, J.; Gao, H.; Wang, M.; Li, H.; Ngan, K.N.; Lin, W. Efficient geometry surface coding in V-PCC. *IEEE Trans. Multimed.* **2022**, *early access*. [CrossRef]

43. Zhao, L.; Ma, K.K.; Liu, Z.; Yin, Q.; Chen, J. Real-time scene-aware LiDAR point cloud compression using semantic prior representation. *IEEE Trans. Circuits Syst. Video Technol.* **2022**, *32*, 5623–5637. [CrossRef]

44. Sun, X.; Wang, S.; Wang, M.; Cheng, S.S.; Liu, M. An advanced LiDAR point cloud sequence coding scheme for autonomous driving. In Proceedings of the 28th ACM International Conference on Multimedia, Seattle, WA, USA 12–16 October 2020; pp. 2793–2801.

45. Sun, X.; Wang, M.; Du, J.; Sun, Y.; Cheng, S.S.; Xie, W. A Task-Driven Scene-Aware LiDAR Point Cloud Coding Framework for Autonomous Vehicles. *IEEE Trans. Ind. Inform.* **2022**, *early access*. [CrossRef]

46. Filippone, M.; Camastra, F.; Masulli, F.; Rovetta, S. A survey of kernel and spectral methods for clustering. *Pattern Recognit.* **2008**, *41*, 176–190. [CrossRef]

47. Kriege, N.M.; Johansson, F.D.; Morris, C. A survey on graph kernels. *Appl. Netw. Sci.* **2020**, *5*, 1–42. [CrossRef]

48. Nikolentzos, G.; Siglidis, G.; Vazirgiannis, M. Graph Kernels: A Survey. *J. Artif. Intell. Res.* **2021**, *72*, 943–1027. [CrossRef]

49. Gaidon, A.; Harchaoui, Z.; Schmid, C. A time series kernel for action recognition. In Proceedings of the BMVC 2011-British Machine Vision Conference, Dundee, UK, 29 August–2 September 2011; pp. 63.1–63.11.

50. Bach, F.R. Graph kernels between point clouds. In Proceedings of the 25th International Conference on Machine Learning, Helsinki, Finland, 5–9 July 2008; pp. 25–32.

51. Wang, L.; Sahbi, H. Directed acyclic graph kernels for action recognition. In Proceedings of the IEEE International Conference on Computer Vision, Sydney, Australia, 1–8 December 2013; pp. 3168–3175.

52. Harchaoui, Z.; Bach, F. Image classification with segmentation graph kernels. In Proceedings of the 2007 IEEE Conference on Computer Vision and Pattern Recognition, Minneapolis, MN, USA, 17–22 June 2007; pp. 1–8.

53. Borgwardt, K.M.; Kriegel, H.P. Shortest-path kernels on graphs. In Proceedings of the Fifth IEEE International Conference on Data Mining (ICDM'05), Houston, TX, USA, 27–30 November 2005; p. 8.

54. Aziz, F.; Wilson, R.C.; Hancock, E.R. Backtrackless walks on a graph. *IEEE Trans. Neural Netw. Learn. Syst.* **2013**, *24*, 977–989. [CrossRef]

55. Johansson, F.; Jethava, V.; Dubhashi, D.; Bhattacharyya, C. Global graph kernels using geometric embeddings. In Proceedings of the International Conference on Machine Learning, Beijing, China, 22–24 June 2014; pp. 694–702.

56. Shervashidze, N.; Schweitzer, P.; Van Leeuwen, E.J.; Mehlhorn, K.; Borgwardt, K.M. Weisfeiler-lehman graph kernels. *J. Mach. Learn. Res.* **2011**, *12*, 2539–2561.

57. Bai, L.; Rossi, L.; Zhang, Z.; Hancock, E. An aligned subtree kernel for weighted graphs. In Proceedings of the International Conference on Machine Learning, Lille, France, 7–9 July 2015; pp. 30–39.

58. Fröhlich, H.; Wegner, J.K.; Sieker, F.; Zell, A. Optimal assignment kernels for attributed molecular graphs. In Proceedings of the 22nd International Conference on Machine Learning, Bonn, Germany, 7–11 August 2005; pp. 225–232.

59. Hatcher, A. *Algebraic Topology*; Cambridge University Press: Cambridge, UK, 2002.

60. Dey, T.K. *Curve and Surface Reconstruction: Algorithms with Mathematical Analysis*; Cambridge University Press: Cambridge, UK, 2006; Volume 23.

61. Shewchuk, J.R. Lecture Notes on Delaunay Mesh Generation. 1999. Available online: https://people.eecs.berkeley.edu/~jrs/meshpapers/delnotes.pdf (accessed on 10 January 2023).

62. Edelsbrunner, H.; Mücke, E.P. Three-dimensional alpha shapes. *ACM Trans. Graph. (TOG)* **1994**, *13*, 43–72. [CrossRef]

63. De Floriani, L.; Hui, A. Data Structures for Simplicial Complexes: An Analysis Furthermore, A Comparison. In Proceedings of the Symposium on Geometry Processing, Vienna, Austria, 4–6 July 2005; pp. 119–128.

64. Edelsbrunner, H.; Harer, J. Persistent homology-a survey. *AMS Contemp. Math.* **2008**, *453*, 257–282.
65. Kahle, M. Topology of random simplicial complexes: A survey. *AMS Contemprory Math.* **2014**, *620*, 201–222.
66. Zomorodian, A. Fast construction of the Vietoris-Rips complex. *Comput. Graph.* **2010**, *34*, 263–271. [CrossRef]
67. Boissonnat, J.D.; Guibas, L.J.; Oudot, S.Y. Manifold reconstruction in arbitrary dimensions using witness complexes. *Discret. Comput. Geom.* **2009**, *42*, 37–70. [CrossRef]
68. Burnham, K.P.; Anderson, D.R. Kullback–Leibler information as a basis for strong inference in ecological studies. *Wildl. Res.* **2001**, *28*, 111–119. [CrossRef]
69. Pérez-Cruz, F. Kullback–Leibler divergence estimation of continuous distributions. In Proceedings of the 2008 IEEE International Symposium on Information Theory, Toronto, ON, Canada, 6–11 July 2008; pp. 1666–1670.
70. Yu, S.; Mehta, P.G. The Kullback–Leibler rate pseudo-metric for comparing dynamical systems. *IEEE Trans. Autom. Control* **2010**, *55*, 1585–1598.
71. Potamias, M.; Bonchi, F.; Castillo, C.; Gionis, A. Fast shortest path distance estimation in large networks. In Proceedings of the 18th ACM Conference on Information and Knowledge Management, Hong Kong, China, 2–6 November 2009; pp. 867–876.
72. Panaretos, V.M.; Zemel, Y. Statistical Aspects of Wasserstein Distances. *Annu. Rev. Stat. Its Appl.* **2019**, *6*, 405–431. [CrossRef]
73. Wu, Z.; Song, S.; Khosla, A.; Yu, F.; Zhang, L.; Tang, X.; Xiao, J. 3d shapenets: A deep representation for volumetric shapes. In Proceedings of the IEEE Conference on Computer Vision and Pattern Recognition, Boston, MA, USA, 7–12 June 2015; pp. 1912–1920.
74. Kriege, N.M.; Giscard, P.L.; Wilson, R. On valid optimal assignment kernels and applications to graph classification. *Adv. Neural Inf. Process. Syst.* **2016**, *29*, 1623–1631.
75. Kudo, T.; Maeda, E.; Matsumoto, Y. An application of boosting to graph classification. *Adv. Neural Inf. Process. Syst.* **2004**, *17*, 1–8.
76. Ma, T.; Shao, W.; Hao, Y.; Cao, J. Graph classification based on graph set reconstruction and graph kernel feature reduction. *Neurocomputing* **2018**, *296*, 33–45. [CrossRef]
77. Wu, J.; He, J.; Xu, J. Net: Degree-specific graph neural networks for node and graph classification. In Proceedings of the 25th ACM SIGKDD International Conference on Knowledge Discovery & Data Mining, Anchorage, AK, USA, 4–8 August 2019; pp. 406–415.
78. Yao, H.R.; Chang, D.C.; Frieder, O.; Huang, W.; Lee, T.S. Graph Kernel prediction of drug prescription. In Proceedings of the 2019 IEEE EMBS International Conference on Biomedical & Health Informatics (BHI), Chicago, IL, USA, 19–22 May 2019; pp. 1–4.
79. Parés, F.; Gasulla, D.G.; Vilalta, A.; Moreno, J.; Ayguadé, E.; Labarta, J.; Cortés, U.; Suzumura, T. Fluid communities: A competitive, scalable and diverse community detection algorithm. In Proceedings of the International conference on Complex Networks and Their Applications, Lyon, France, 29 November–1 December 2017; pp. 229–240.
80. Hua, B.S.; Tran, M.K.; Yeung, S.K. Pointwise convolutional neural networks. In Proceedings of the IEEE Conference on Computer Vision and Pattern Recognition, Salt Lake City, UT, USA, 18–22 June 2018; pp. 984–993.
81. Zaheer, M.; Kottur, S.; Ravanbakhsh, S.; Poczos, B.; Salakhutdinov, R.R.; Smola, A.J. Deep sets. *Adv. Neural Inf. Process. Syst.* **2017**, *30*, 3391–3401.
82. Simonovsky, M.; Komodakis, N. Dynamic edge-conditioned filters in convolutional neural networks on graphs. In Proceedings of the IEEE Conference on Computer Vision and Pattern Recognition, Honolulu, HI, USA, 21–26 July 2017; pp. 3693–3702.
83. Qi, C.R.; Su, H.; Mo, K.; Guibas, L.J. Pointnet: Deep learning on point sets for 3d classification and segmentation. In Proceedings of the IEEE Conference on Computer Vision and Pattern Recognition, Honolulu, HI, USA, 21–26 July 2017; pp. 652–660.
84. Xie, S.; Liu, S.; Chen, Z.; Tu, Z. Attentional shapecontextnet for point cloud recognition. In Proceedings of the IEEE Conference on Computer Vision and Pattern Recognition, Salt Lake City, UT, USA, 18–22 June 2018; pp. 4606–4615.
85. Groh, F.; Wieschollek, P.; Lensch, H. Flex-convolution (million-scale point-cloud learning beyond grid-worlds). *arXiv* **2018**, arXiv:1803.07289.
86. Klokov, R.; Lempitsky, V. Escape from cells: Deep kd-networks for the recognition of 3d point cloud models. In Proceedings of the IEEE International Conference on Computer Vision, Venice, Italy, 22–29 October 2017; pp. 863–872.
87. Qi, C.R.; Yi, L.; Su, H.; Guibas, L.J. Pointnet++: Deep hierarchical feature learning on point sets in a metric space. *Adv. Neural Inf. Process. Syst.* **2017**, *30*, 1–14.
88. Shen, Y.; Feng, C.; Yang, Y.; Tian, D. Mining point cloud local structures by kernel correlation and graph pooling. In Proceedings of the IEEE Conference on Computer Vision and Pattern Recognition, Salt Lake City, UT, USA, 18–22 June 2018; pp. 4548–4557.
89. Wang, Y.; Sun, Y.; Liu, Z.; Sarma, S.E.; Bronstein, M.M.; Solomon, J.M. Dynamic graph cnn for learning on point clouds. *Acm Trans. Graph. (TOG)* **2019**, *38*, 1–12. [CrossRef]
90. Atzmon, M.; Maron, H.; Lipman, Y. Point convolutional neural networks by extension operators. *arXiv* **2018**, arXiv:1803.10091.
91. Liu, Y.; Fan, B.; Xiang, S.; Pan, C. Relation-shape convolutional neural network for point cloud analysis. In Proceedings of the IEEE/CVF Conference on Computer Vision and Pattern Recognition, Long Beach, CA, USA, 15–20 June 2019; pp. 8895–8904.

Article

Point Cloud Instance Segmentation with Inaccurate Bounding-Box Annotations

Yinyin Peng [1], Hui Feng [1], Tao Chen [1] and Bo Hu [1,2,*]

1   Department of Electronic Engineering, Fudan University, Shanghai 200433, China
2   Yiwu Research Institute, Fudan University, Yiwu 322000, China
*   Correspondence: bohu@fudan.edu.cn

Abstract: Most existing point cloud instance segmentation methods require accurate and dense point-level annotations, which are extremely laborious to collect. While incomplete and inexact supervision has been exploited to reduce labeling efforts, inaccurate supervision remains under-explored. This kind of supervision is almost inevitable in practice, especially in complex 3D point clouds, and it severely degrades the generalization performance of deep networks. To this end, we propose the first weakly supervised point cloud instance segmentation framework with inaccurate box-level labels. A novel self-distillation architecture is presented to boost the generalization ability while leveraging the cheap but noisy bounding-box annotations. Specifically, we employ consistency regularization to distill self-knowledge from data perturbation and historical predictions, which prevents the deep network from overfitting the noisy labels. Moreover, we progressively select reliable samples and correct their labels based on the historical consistency. Extensive experiments on the ScanNet-v2 dataset were used to validate the effectiveness and robustness of our method in dealing with inexact and inaccurate annotations.

Keywords: point cloud instance segmentation; learning with noisy labels; weakly supervised learning; self-distillation

Citation: Peng, Y.; Feng, H.; Chen, T.; Hu, B. Point Cloud Instance Segmentation with Inaccurate Bounding-Box Annotations. *Sensors* **2023**, *23*, 2343. https://doi.org/10.3390/s23042343

Academic Editors: Miaohui Wang, Guanghui Yue, Jian Xiong and Sukun Tian

Received: 12 January 2023
Revised: 15 February 2023
Accepted: 17 February 2023
Published: 20 February 2023

1. Introduction

The rapid development of 3D sensors, such as LiDARs and RGB-D cameras, has brought about an increasing amount of 3D data, thus promoting a wide range of applications, including autonomous driving [1], robotics [2], and medical treatment [3]. With the benefit of rich geometric information and the challenge of intrinsic irregularity, more and more attention has been paid to deep learning on 3D point clouds [4–7].

As one of the fundamental tasks in 3D scene understanding, point cloud instance segmentation aims to predict the semantic label of each point and simultaneously distinguish points within the same class but in different instances. Numerous deep learning methods have been proposed to achieve progressively better performance [8–14]. However, the success of most existing segmentation methods depends heavily on accurately and densely annotated training data, which are time-consuming to collect. For example, it takes about 22.3 min to annotate all of the points of one scene in ScanNet [15]. To alleviate the point-level annotation burden of full supervision, a handful of methods have recently taken weak supervision into consideration; they mainly included incomplete [16–18] and inexact supervision [19–22]. The different kinds of weak supervision are illustrated in Figure 1.

For incomplete supervision, current research works perform semi-supervised learning through self-training, self-supervision, label propagation, etc. However, the way of choosing the small fraction of points to annotate is crucial for the segmentation performance. To represent the location of an instance, SegGroup [18] picks the largest segment, and CSC [23] finds exemplary points through active sampling. In other words, additional labeling efforts are required to implement point cloud instance segmentation with incomplete supervision.

For inexact supervision, there exist two leading types, i.e., scene-level (subcloud-level) and box-level supervision. Since it is difficult to extract object localization information from scene-level or subcloud-level tags [19], we focus on box-level supervision, which is of medium granularity and widely available. The key is to identify the foreground points in each bounding box without point-level instance labels. Box2Seg [21] uses attention map modulation and entropy minimization to generate pseudo-labels. SPIB [22] first conducts object detection with partial bounding-box labels and fulfills instance segmentation with three in-box refinement modules. Both of them need multi-stage training, and the box-level annotations are not fully exploited for instance segmentation. Box2Mask [20] allows each point to predict the box in which it belongs and trains the instance segmentation network from end to end with bounding-box annotations.

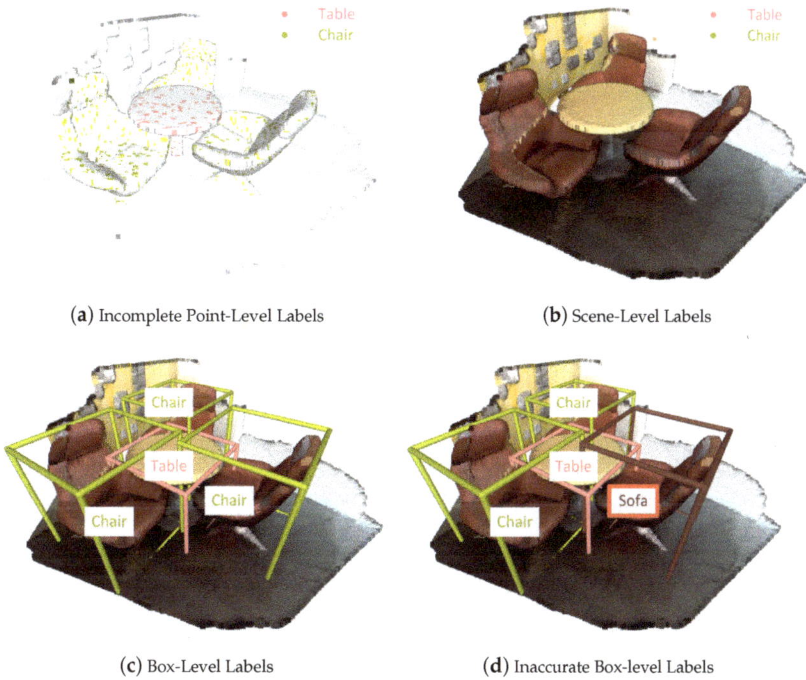

(**a**) Incomplete Point-Level Labels

(**b**) Scene-Level Labels

(**c**) Box-Level Labels

(**d**) Inaccurate Box-level Labels

Figure 1. Illustration of various weak supervision methods for point cloud segmentation. (**a**) Incomplete point-level labels denote the classes to which a small fraction of points belong. (**b**) Scene-level (subcloud-level) labels indicate all of the classes appearing in the scene (subcloud). (**c**) Box-level labels indicate the class and location of each object. (**d**) Inaccurate box-level labels indicate the portion of boxes that are mislabeled. For example, a "chair" is mislabelled as a "sofa".

Nonetheless, the methods presented above implicitly assume that the labels are highly accurate, which may not be guaranteed in practice. Regardless of the granularity at which data are labeled, label noise exists due to the carelessness of annotators and the difficulty of annotating itself. When it comes to box-level label noise, Hu et al. [24] designed a noise-resistant focal loss for 2D object detection. With NLTE [25], it was found that it was essential for domain adaptive objective detection to address noisy box annotations, including miss-annotated boxes and class-corrupted ones. In 3D point cloud instance segmentation, the rough location information of most points is unaffected by slight fluctuations in box coordinates, while mislabeling the box semantics can lead to serious confusion of all of the in-box points. Therefore, we took the semantic label noise of each box into account while leaving the geometric coordinate noise for future work. The inaccurate box-level annotations are shown in Figure 1d. Since deep neural networks are highly capable of learning any complex function, it is easy to overfit inaccurate labels and reduce the generalization performance [26]. Thus, it is necessary to develop a noise-robust point cloud instance segmentation method. A recent work used PNAL [27] to study point noise in semantic segmentation, but it heavily relied on the early memorization effect, which increased the risk of discarding hard samples or those in the minor class. Furthermore, point cloud instance segmentation with inaccurate bounding-box annotations is even more challenging due to the granularity mismatch of given annotations and the target task. There is an urgent need to combat realistic label noise and sufficiently release the potential of box-level supervision in point cloud instance segmentation.

Extensive research has empirically demonstrated the success of knowledge distillation [28] in boosting the generalization ability, which is in great demand when learning with noisy labels. Traditional knowledge distillation transfers knowledge from a large teacher model, while self-distillation efficiently utilizes knowledge from itself and, thus, attracts more and more attention. As for theoretical analysis, there are various opinions that include label smoothing regularization [29], the multi-view hypothesis [30], and loss landscape flattening [31]. Similarly to our method, PS-KD [32] trained a model with soft targets, which were a weighted summation of the hard targets and the last-epoch predictions, and DLB [33] used predictions from the last iteration as soft targets. However, we considered the entire prediction history and maintained an exponential moving average of the predictions.

In this paper, we present a novel self-distillation framework based on perturbation and history (SDPH) to handle the challenge of point cloud instance segmentation with only inaccurate box annotations. Rather than distilling knowledge from a cumbersome teacher model or an extra clean dataset [34], we perform self-distillation by taking full advantage of self-supervision in the data and the learning process. To be specific, we assume that the predictions over the input point cloud are perturbation-invariant. Both geometric and semantic consistency regularization terms are included to provide additional supervision signals. Furthermore, by investigating the consistency of historical predictions, the model is able to locate and correct refurbishable samples with high precision. Finally, we apply temporal consistency regularization to fully utilize the history information and reduce the unstable prediction fluctuations that may hinder the label refurbishment. In a word, we utilize two kinds of consistency regularization to prevent the network from overfitting inaccurate labels and progressively correct the labels during the training process.

Overall, the main contributions of our paper are summarized as follows:

- To the best of our knowledge, this is the first work to simultaneously explore inexact and inaccurate annotations in the point cloud instance segmentation task.
- We propose a novel self-distillation framework for applying consistency regularization and label refurbishment by using data perturbation and history information.
- Extensive experiments were conducted to demonstrate the effectiveness of our method. The results on ScanNet-v2 show that our SDPH achieved comparable performance to that of densely and accurately supervised methods.

The rest of this paper is organized as follows. First, related research is described in Section 2. Next, we present our self-distillation framework in Section 3. Thereafter, the experimental results and analysis are provided in Section 4. Finally, Section 5 concludes the paper and points out future work.

2. Related Works

2.1. Point Cloud Instance Segmentation

Point cloud instance segmentation methods can be roughly divided into two categories: proposal-based methods and proposal-free methods.

2.1.1. Proposal-Based Methods

Proposal-based methods first conduct object detection to generate region proposals and then perform binary classification to separate all of the foreground points in each proposal. GSPN [8] used an analysis-by-synthesis strategy to enforce geometric understanding in generating proposals with high objectness. These object proposals were further processed by Region-Based PointNet (R-PointNet) to obtain the final segmentation results. The method of 3D-SIS [35] first extracted 2D features from multi-view high-resolution RGB images and then projected them back to the associated 3D voxel grids. The geometry and color features were concatenated and fed into a fully convolutional 3D architecture. The method of 3D-BoNet [36] is a single-stage, anchor-free, and end-to-end trainable network. This method directly predicts a fixed number of bounding boxes and fuses the global information into a point mask prediction branch. The method of 3D-MPA [9] generates proposals through center voting, refines them by using a graph convolutional network, and obtains the final instances through proposal aggregation instead of non-maximum suppression.

2.1.2. Proposal-Free Methods

Proposal-free methods focus on discriminative point feature learning and distinguish instances with the same semantic meaning through clustering. SGPN [11] first embeds all of the input points into feature space and then groups the points into instances based on the pairwise feature similarity, which is not scalable. JSIS3D [12] utilizes a multi-value conditional random field model to jointly optimize semantic labels and instance embeddings predicted by a multi-task point-wise network. ASIS [37] utilizes discriminative loss to pull embeddings of the same instance to its center and push those of different instances apart. Moreover, the association of instance segmentation and semantic segmentation further benefits each. PointGroup [38] predicts point offsets towards their respective instance centers and considers both the original point coordinates and the offset-shifted ones in the clustering stage. OccuSeg [13] introduces the occupancy signal to take part in multi-task learning and guide graph-based clustering. PE [14] encodes each point as a tri-variate normal distribution in the probabilistic embedding space, and a novel loss function that benefits both semantic segmentation and subsequent clustering was proposed. HAIS [10] performs point aggregation and set aggregation to progressively generate instance proposals. SoftGroup [39] groups points based on soft semantic scores to avoid error propagation and suppresses false positive instances by learning to categorize them as the background.

We follow the proposal-free approach because of its superior performance and flexible architecture. Nevertheless, we utilize inaccurate box-level supervision to learn point-level instance segmentation, which greatly alleviates the labeling cost.

2.2. Weakly Supervised Point Cloud Segmentation

Generally speaking, there are three typical types of weak supervision in machine learning: incomplete supervision, inexact supervision, and inaccurate supervision [40].

Most point cloud segmentation methods are concerned with incomplete supervision, where only a small subset of training data are given with labels [16,17,41–43]. This setting is also known as semi-supervised learning. Xu et al. [16] combined multi-instance learning, self-supervision, and smoothness constraints to achieve semantic segmentation with only 10 times fewer labels. Zhang et al. [41] constructed a self-supervised pre-training task through point cloud colorization and proposed an efficient sparse label propagation mechanism to improve the effectiveness of the weakly supervised semantic segmentation task. PSD [42] enforced the prediction consistency between the perturbed branch and the original branch, and a context-aware module for regularizing the affinity correlation of labeled points was presented. Liu et al. [17] adopted a self-training approach with a super-voxel graph propagation module. Similarly, SSPC-Net [43] built super-point graphs for dynamic label propagation and the coupled attention mechanism to extract discriminative contextual features.

Inexact supervision means that the training data are given with only coarse-grained labels, such as scene-level tags [19,44] and box-level annotations [20–22] in the segmentation context. MPRM [19] applied various attention mechanisms to acquire point class activation maps (PCAMs). After generating pseudo-point-level labels from PCAMs, a segmentation network could be trained in a fully supervised manner. WyPR [44] jointly performed semantic segmentation and object detection through a series of self- and cross-task consistency losses with multi-instance learning objectives. SPIB [22] first leveraged partially labeled bounding boxes to train a proposal generation network with perturbation consistency regularization and then predicted the instance mask inside each target box with three smoothness regularization and refinement modules. Box2Seg [21] learned pseudo-labels from bounding-box-level foreground annotations and subcloud-level background tags, and it achieved semantic segmentation through fully supervised retraining. Box2Mask [20] directly voted for bounding boxes and obtained instance masks via non-maximum clustering.

Inaccurate supervision means that the given labels are not always the ground truth. Although learning from noisy labels with deep neural networks has been explored very much, especially in image classification [45], few researchers have investigated noisy labels with increasing amounts of point cloud data. As the pioneering work in noise-robust point cloud semantic segmentation, PNAL [27] selected reliable points based on their consistency among historical predictions, and it corrected locally similar points with the most likely label, which was voted on in each cluster.

Most of the above weakly supervised methods focused merely on one type of weak supervision. However, the circumstances are usually more complicated in reality, where the label noise in particular is almost inevitable but often ignored. Thus, we consider both inexact and inaccurate supervision and develop a robust point cloud instance segmentation framework with inaccurate box annotations.

3. Our Method

3.1. Overview

The pipeline of our SDPH is depicted in Figure 2. Given a point cloud $\mathcal{P}$ with inaccurate bounding-box annotations, we first assign point-level pseudo-labels based on the spatial inclusion relations between points and boxes. This simple association process allows for a fully supervised training manner. After the label preparation, the backbone network takes a voxelized point cloud as input and produces embeddings for each voxel. To lessen the computational cost, we perform over-segmentation to group voxels into super-voxels. This basic training process will be introduced in Section 3.3. The final instances are obtained through super-voxel-level non-maximum clustering and backward projection.

Apart from the whole forward inference procedure, our self-distillation training framework consists of two main parts that leverage data perturbation and historical information. First, we construct a perturbed branch and keep the prediction consistency between the original branch and the perturbed one. Furthermore, the past predictions are fully exploited to select refurbishable samples and provide soft targets.

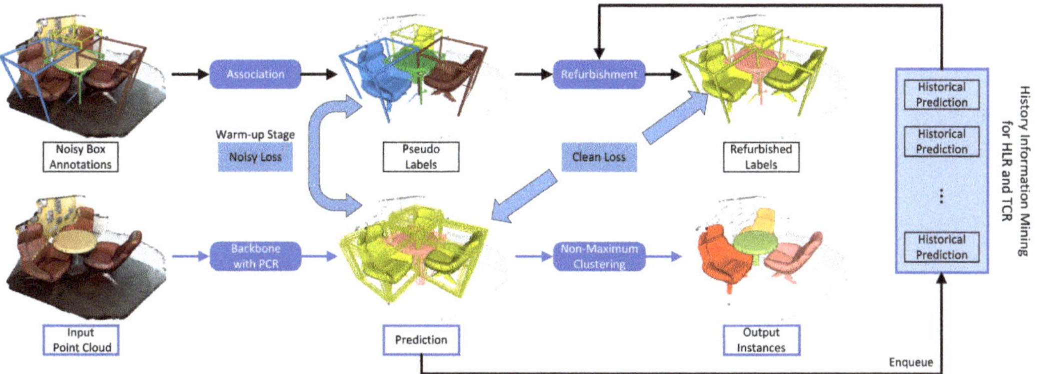

Figure 2. The training framework of self-distillation based on perturbation and history. We first generate pseudo-labels according to the point–box association (c.f. Section 3.2) and train a 3D sparse convolutional network with two types of consistency regularization, namely, PCR (c.f. Section 3.4.1) and TCR (c.f. Section 3.4.3). With the help of regularization, the model is able to perform label refurbishment (HLR, c.f. Section 3.4.2) with higher precision. Note that the noisy loss is used only in the warm-up stage, and afterward, it is replaced by the clean loss, since the cleaned (i.e., refurbished) labels are available.

3.2. Pseudo-Label Generation

Since ground-truth point-level labels are not available, directly training a segmentation network with only box-level labels is infeasible. Therefore, we need to establish the box–point association first. Specifically, we categorize points according to the numbers of boxes containing them. If a point is contained in only one box, it is simply labeled as the unique box, which is represented by both the geometric coordinates and the semantic category. If a point is inside more than one box, the smallest one is associated with it. A point is treated as background if it is outside all of the boxes.

Let $\mathcal{B}$ denote a set of box annotations, with each box $\mathbf{b} \in \mathbb{R}^7$ representing its three-dimensional center, three-dimensional size, and one-dimensional semantic label. For clarity, we use $\mathbf{p}_i \in \mathbf{b}_j$ ($\mathbf{p}_i \notin \mathbf{b}_j$) to show that the i-th point is (not) contained by the j-th box. The pseudo-labels are generated through the following mapping function.

$$\phi(\mathbf{p}_i) = \begin{cases} \mathbf{b}_j, & j = \underset{j \in \{k | \mathbf{p}_i \in \mathbf{b}_k\}}{\arg\min} \ \mathrm{sizeof}(\mathbf{b}_j), \\ \mathrm{background}, & \forall j, \mathbf{p}_i \notin \mathbf{b}_j. \end{cases} \tag{1}$$

Although this mapping function seems plausible, the generated point-level pseudo-labels inevitably suffer from inaccurate associations, as do the super-voxel-level pseudo-labels. The label quality will further degrade due to inaccurate box annotations, which motivated us to design a noise-robust self-distillation training framework.

3.3. Point Cloud Instance Segmentation Network

Before self-distillation, we introduce the basic point cloud instance segmentation network, where the labels are regarded noise-free. As a common choice, we adopted a UNet-like sparse convolutional network as the backbone [10,46,47]. The input point cloud is converted into volumetric grids and then fed into the backbone to extract voxel features, which are pooled into super-voxel features by using the over-segmentation results. Next, multiple output heads are applied to predict the semantic label, the associated box coordinates (offset and size), and the intersection-over-union (IoU) score of the predicted box with the ground-truth box. The basic network is trained with the following multi-task loss.

$$\mathcal{L}_{basic} = \mathcal{L}_{sem} + \mathcal{L}_{offset} + \mathcal{L}_{size} + \mathcal{L}_{score}. \tag{2}$$

Here, $\mathcal{L}_{sem}$ is a normal cross-entropy loss for learning the semantics, which are formulated as

$$\mathcal{L}_{sem} = -\frac{1}{N}\sum_{i=1}^{N}\sum_{c=1}^{C} y_{ic}\log p_{ic}, \tag{3}$$

where y_{ic} represents the one-hot semantic label of the i-th super-voxel, $p_{ic} = \frac{\exp(z_{ic})}{\sum_{k=1}^{C}\exp(z_{ik})}$ denotes the probability of being predicted as the c-th category, N is the number of super-voxels, and C represents the number of semantic categories. Note that the background is also included in the categories concerned with $\mathcal{L}_{sem}$.

As for the box regression, we use the L_1 loss.

$$\mathcal{L}_{offset} = \frac{1}{M}\sum_{i=1}^{M}\|\mathbf{d}_i - \hat{\mathbf{d}}_i\|_1,$$
$$\mathcal{L}_{size} = \frac{1}{M}\sum_{i=1}^{M}\|\mathbf{s}_i - \hat{\mathbf{s}}_i\|_1, \tag{4}$$

where M is the number of foreground super-voxels. $\mathbf{d}_i$ and $\hat{\mathbf{d}}_i$ represent the ground-truth and predicted offsets of the i-th super-voxel with respect to the associated box center, respectively. $\mathbf{s}_i$ and $\hat{\mathbf{s}}_i$ represent the corresponding box sizes.

To assist in the later non-maximum clustering and average precision calculation, the IoU score loss is defined as

$$\mathcal{L}_{score} = -\frac{1}{M}\sum_{i=1}^{M}[u_i\log v_i + (1-u_i)\log(1-v_i)], \tag{5}$$

where u_i and v_i represent the true and predicted IoUs between the predicted box and the associated ground-truth box, respectively.

At the inference stage, we follow Box2Mask [20] in performing non-maximum clustering (NMC), which follows exactly the same procedure of non-maximum suppression (NMS) in object detection. Instead of dropping redundant boxes, in NMC, they are collected to form clusters with the corresponding representative boxes. The semantic category of each cluster is assigned through a majority vote. Finally, the clustering structure of super-voxels is projected back to points, which completes the instance segmentation.

3.4. Self-Distillation Based on Perturbation and History

3.4.1. Perturbation-Based Consistency Regularization

Since the original supervision method is inaccurate and untrustworthy, we turn to self-supervision, which has shown great power in deep learning. To provide additional supervision, we construct a perturbed branch and constrain the predictions of the perturbed and original branches to be consistent.

We adopt three kinds of perturbation strategies: scaling, flipping, and rotation. For scaling, we sample a scaling factor ζ from a uniform distribution $\mathcal{U}(0.8, 1.2)$. The origin-centered scaling process is represented as $\tilde{P} = \zeta \cdot P$, where $P \in \mathbb{R}^{N_p \times 3}$ is the coordinate matrix of the input point cloud and $\tilde{P}$ is the transformed one. For flipping, we randomly sample the flipping indicators f_x, f_y from $\{-1, 1\}$, where -1 means flipping over the corresponding axis. Thus, the flipping can be expressed as $\tilde{P} = P \cdot \mathrm{diag}(f_x, f_y, 1)$. For rotation, the rotation angle θ around z-axis is denoted as θ_z and sampled from the uniform distribution $\mathcal{U}(0, 2\pi)$. Rotating the point cloud means multiplying its coordinates with a rotation matrix as follows:

$$\tilde{P} = P \cdot R(\theta_z) = P \cdot \begin{bmatrix} \cos\theta_z & \sin\theta_z & 0 \\ -\sin\theta_z & \cos\theta_z & 0 \\ 0 & 0 & 1 \end{bmatrix}. \tag{6}$$

Obviously, both semantic and geometric predictions should be consistent between the two branches, i.e., the perturbation-based consistency regularization loss ("PCR loss" in Figure 3) is defined as

$$\mathcal{L}_{pcr} = \mathcal{L}_{pcr}^{sem} + \mathcal{L}_{pcr}^{geo}. \tag{7}$$

The KL-divergence and MSE losses are used as consistency regularization terms. To be specific, we formulate the semantic consistency loss as

$$\mathcal{L}_{pcr}^{sem} = \frac{1}{N} \sum_{i=1}^{N} D_{KL}(\mathbf{p}_i \| \tilde{\mathbf{p}}_i)$$
$$= \frac{1}{N} \sum_{i=1}^{N} \sum_{c=1}^{C} p_{ic} \log \frac{p_{ic}}{\tilde{p}_{ic}}. \tag{8}$$

The geometric consistency loss is defined as

$$\mathcal{L}_{pcr}^{geo} = \frac{1}{N} \sum_{i=1}^{N} \left[\| \tilde{\mathbf{o}}_i - \hat{\mathbf{o}}_i \|_2^2 + \| \tilde{\mathbf{s}}_i - \hat{\mathbf{s}}_i \|_2^2 \right], \tag{9}$$

where $\mathbf{o}_i$ represents the center of the i-th super-voxel's associated box. In addition, $\hat{\ }$ indicates the predicted value, and $\tilde{\ }$ means perturbation. To ensure valid consistency regularization, the same perturbation should be applied to the geometric predictions of the original branch.

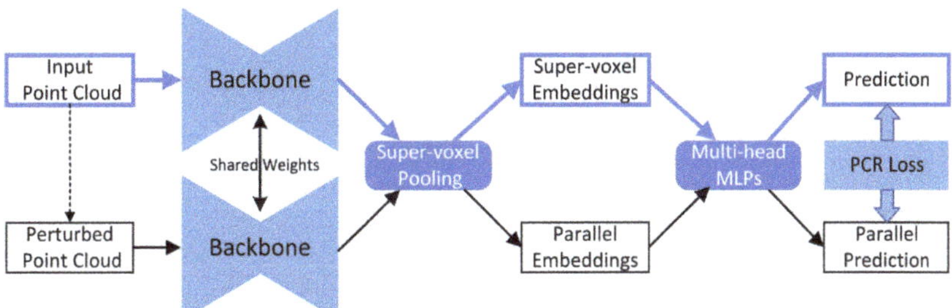

Figure 3. Illustration of the perturbation-based consistency regularization (PCR) module. We construct a parallel branch through data perturbation and force the output predictions of the two branches to be consistent. Note that the predictions include both semantics and geometry.

3.4.2. History-Guided Label Refurbishment

In light of the memorization effect, in which deep networks first learn simple patterns in clean data before memorizing noise by brute force [48], the model is able to identify and correct inaccurate labels by itself during training. Specifically, the consistency of predictions is widely used as a confidence criterion [27,49–51]. Along this line, we consider samples with consistent historical predictions as refurbishable. The refurbishment process is illustrated in Figure 4.

Let $\Psi(q) = \{\hat{y}_{t_1}, \hat{y}_{t_2}, \cdots, \hat{y}_{t_q}\}$ denote the label prediction history of a super-voxel sample, where q is the length of the historical queue. The frequency of the super-voxel being predicted as the c-th category is calculated as

$$F(c|q) = \sum_{i=1}^{q} \frac{[\hat{y}_{t_i} = c]}{q}, \tag{10}$$

where $[\cdot]$ is the Iverson bracket. With the frequency–probability approximation, we apply the following normalized information entropy as the consistency metric:

$$H(q) = \frac{1}{Z} \sum_{c=1}^{C} -F(c|q) \log F(c|q), \tag{11}$$

where $Z = \sum_{c=1}^{C} -\frac{1}{C} \log(\frac{1}{C}) = \log(C)$ is the normalization term representing the maximum entropy. A smaller entropy indicates more consistent predictions. To be concrete, we treat the super-voxel that satisfies $H(q) \leq \epsilon$ $(0 \leq \epsilon \leq 1)$ as the refurbishable sample. The refurbished label is defined as

$$y^* = \underset{1 \leq c \leq C}{\operatorname{argmax}} F(c|q). \tag{12}$$

Apparently, the refurbishment will be applied after an appropriate number of warm-up epochs, which is longer than the historical queue. The refurbishable samples are relocated at each new epoch to avoid the accumulation of correction errors. Instead of dropping the remaining samples, we leave them unaffected to enable full exploration of the dataset. In addition, it is noteworthy that we do not impose any restrictions on the label noise, which makes our refurbishment robust to different noise types and different noise rates.

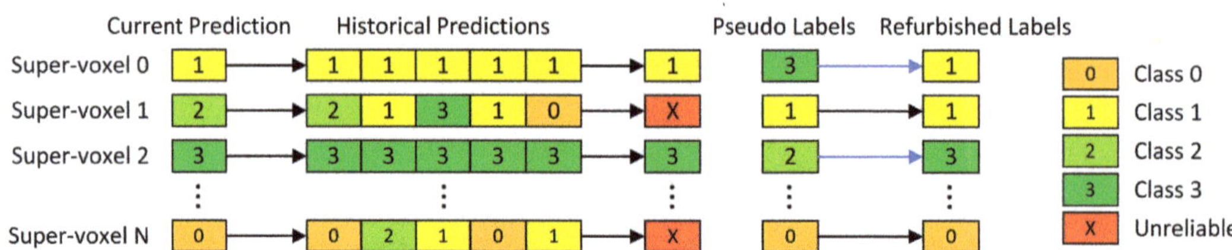

Figure 4. Illustration of the history-guided label refurbishment (HLR) module. We use a historical queue to store the past predictions and correct the previously generated pseudo-labels with consistently predicted classes while keeping the unreliable samples unchanged instead of directly dropping them. Compared with other methods, we take a more conservative strategy, as regularization decreases the overfitting risk.

3.4.3. Temporal Consistency Regularization

The label refurbishment in Section 3.4.2 only utilizes discrete hard labels, overlooking the rich information in the continuous soft distributions. Here, we record the exponential moving average (EMA) of historical logits to impose temporal consistency regularization [32,33].

Let $\mathbf{z}_e$ be the model's output logits at epoch e. After the first trivial epoch, the moving-average logits can be normally updated as

$$\bar{\mathbf{z}}_e = (1 - \alpha)\bar{\mathbf{z}}_{e-1} + \alpha\mathbf{z}_e, \tag{13}$$

where α is the weight of the current epoch. In accordance with the conventional practice, we add the temperature τ to further soften the distribution:

$$p_{ic}^{\tau} = \frac{\exp(\mathbf{z}_{ic}/\tau)}{\sum_{k=1}^{C} \exp(\mathbf{z}_{ik}/\tau)}. \tag{14}$$

The temporal consistency regularization term ("TCR loss" in Figure 5) is then defined as

$$\mathcal{L}_{tcr} = \frac{1}{N} \sum_{i=1}^{N} \tau^2 D_{KL}(\bar{\mathbf{p}}_i^{\tau} \| \mathbf{p}_i^{\tau}), \tag{15}$$

where $\bar{\cdot}$ denotes the corresponding EMA version.

With the temporal consistency regularization, the network tries to learn from itself and make comparatively stable predictions, which is important for correcting mislabeled hard samples and promoting the generalization performance.

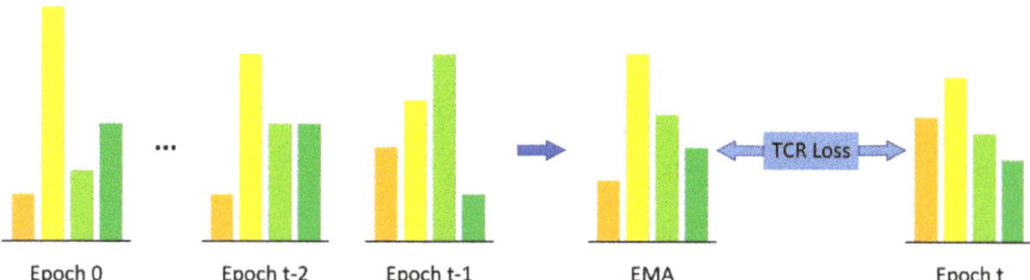

Figure 5. Illustration of the temporal consistency regularization (TCR) module. We record the exponential moving average of the past predicted distributions (logits), which serve as the soft targets for the current prediction.

3.5. Total Loss

Our SDPH can be trained in an end-to-end manner with the total loss $\mathcal{L}$, which contains three parts: the basic loss $\mathcal{L}_{basic}$, the perturbation-based consistency regularization loss $\mathcal{L}_{pcr}$, and the temporal consistency regularization loss $\mathcal{L}_{tcr}$.

$$\mathcal{L} = \mathcal{L}_{basic} + \mathcal{L}_{pcr} + \mathcal{L}_{tcr}, \tag{16}$$

where $\mathcal{L}_{basic}$ is given in Equation (2), $\mathcal{L}_{pcr}$ is given in Equation (7), and $\mathcal{L}_{tcr}$ is given in Equation (15). As we mentioned before, the label refurbishment needs a warm-up stage in which the noisy labels are unchanged in $\mathcal{L}_{basic}$. That is why we call it the "noisy loss" in Figure 2. After the warm-up stage, $\mathcal{L}_{basic}$ is referred to as the "clean loss", since the labels have been cleaned.

4. Experiments
4.1. Experimental Settings
4.1.1. Dataset

We conducted experiments on the widely used ScanNet-v2 [15] dataset. This challenging large-scale indoor point cloud dataset consists of 1201 training scenes, 312 validation scenes, and 100 hidden testing scenes. Each scene of the training and validation sets is richly annotated with point-level semantic-instance labels that are used in the densely supervised methods. However, we created axis-aligned bounding boxes from the point-level annotations to validate our weakly supervised learning framework. To simulate inaccurate annotations, we artificially injected symmetric noise into the training set. Specifically, the semantic labels of the corrupted instance boxes were changed to other labels with equal probability. We used the noise rate, i.e., the probability of each box being mislabeled, to represent the severity of inaccurate supervision. The effects of different noise rates are visualized in Figure 6.

4.1.2. Evaluation Metrics

As with existing methods, we used the mean average precision over 18 foreground object categories as our evaluation metric. To be specific, AP_{25} and AP_{50} denote the scores with IoU thresholds set to 0.25 and 0.5, respectively. In addition, we also report the AP, which averages scores with thresholds varying from 0.5 to 0.95, with a step size of 0.05.

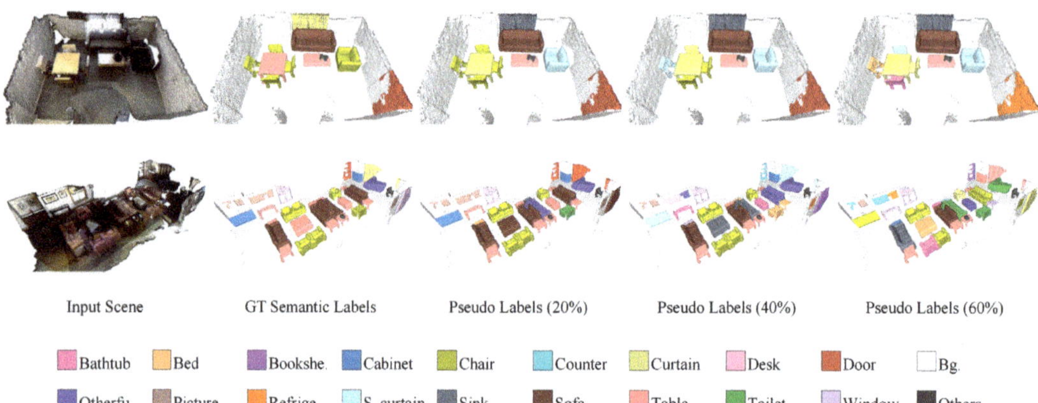

Figure 6. Visualization of different noise rates affecting the semantic labels. From left to right are the input scene, the ground-truth semantics, and the pseudo-labels of noise rates of 20%, 40%, and 60%. The higher the noise rate, the more chaotic the semantics.

4.1.3. Implementation Details

All experiments were performed on a PC with two NVIDIA GeForce RTX 3090 Ti GPUs and an Intel Core i7-12700K CPU. We used two GPUs for distributed training and one for inference. The main software configuration included Python 3.8.13, Pytorch 1.10.2, CUDA 11.3, and MinkowskiEngine 0.5.4. Following the pioneering work of Box2Mask [20], we adopted a six-layer UNet-like sparse convolutional network as our backbone, and the multi-head MLPs were implemented with three layers and 96 hidden units. We set the voxel size to 0.02 m. For history-guided label refurbishment, we set the number of warm-up epochs, the length of the historical queue, and the threshold ϵ to 40, 10, and 0.001, respectively. For temporal consistency regularization, the temperature τ and the EMA coefficient α were empirically set to 3 and 0.9. We trained our network from scratch with a batch size of 4 for 200 epochs in total while using the Adam optimizer with an initial learning rate of 0.001. A cosine annealing scheduler was applied after 100 epochs.

4.2. Instance Segmentation Results

First of all, we conducted comparative experiments with different noise rates to demonstrate the effectiveness of our noise-tolerant learning framework, SDPH. As listed in Table 1, our SDPH achieved consistently better performance than that of Box2Mask (the baseline) under all of the noise rate settings with respect to all of the evaluation metrics. From the overall trend, we observed that higher noise rates were related to larger improvements. When the noise rate is set to 40%, our SDPH still outperformed noise-free Box2Mask in terms of AP. The performance was comparable or even better in terms of AP_{25} and AP_{50} when the noise rate was 20%. These results demonstrate our method's robustness to label noise. Qualitative comparisons of instance and semantic segmentation are shown in Figures 7 and 8, respectively. When training with a noise rate of 40%, our SDPH predicted the semantics more accurately than Box2Mask did, which usually led to better instance segmentation performance.

Table 1. Quantitative comparison of different noise rates on ScanNet-v2.

Method	Metric	0%	10%	20%	30%	40%	50%	60%
Box2Mask [20]	AP	39.1	37.5	36.3	36.3	35.2	33.6	32.0
	AP_{50}	59.7	57.5	55.8	55.4	53.3	50.4	46.7
	AP_{25}	71.8	69.8	68.8	67.3	65.8	62.6	58.2
SDPH	AP	40.1	41.2	40.8	40.0	40.4	37.6	36.5
	AP_{50}	60.4	60.4	60.3	58.7	58.6	55.1	52.5
	AP_{25}	73.0	72.1	71.7	70.7	69.0	65.4	61.9
Improvements	AP	1.0	3.7	4.5	3.7	5.2	4.0	4.5
	AP_{50}	0.7	2.9	4.5	3.3	5.3	4.7	5.8
	AP_{25}	1.2	2.3	2.9	3.4	3.2	2.8	3.7

Even though our method was designed especially for learning with label noise, we acquired a little performance gain in the "noise-free" setting. The possible reasons are two-fold. Firstly, Box2Mask trained the network with associated super-voxel-level pseudo-labels that were not inaccurate. Hence, the label refurbishment worked even without additional noise injection. Secondly, our SDPH benefited from the regularization terms that distilled knowledge from the data and the model itself.

In Table 2, we provide a detailed comparison with state-of-the-art methods that do not explicitly consider label noise. It can be seen that our method performed well in the noise-free setting, which demonstrated the effectiveness of our SDPH. However, instead of attaining consistent performance boosts over different categories, there were some significant declines and increases, especially between SDPH and 3D-MPA [9]. This was probably because 3D-MPA and SDPH adopted different supervision types and instance segmentation routines. The former is a proposal-based method with point-level supervision, while our SDPH is proposal-free and uses the more challenging box-level supervision. As proposal-free methods, PointGroup [38], Box2Mask [20], and SDPH exhibited similar trends when compared with 3D-MPA. For example, their performance greatly declined for refrigerators and shower curtains, and it increases for chairs, desks, sinks, sofas, and other furniture. As shown in Figure 9, the refrigerators had various shapes and sizes and were sometimes surrounded by cabinets. Moreover, curtains were usually beside windows. Even in the case of full point-level supervision—let alone weak box-level supervision—it was difficult to segment them clearly. Furthermore, compared with chairs and desks, there were fewer instances of these categories, which could lead to SDPH's false refurbishment and lower performance.

Table 2. Quantitative comparison with state-of-the-art methods on ScanNet-v2. The highest performance in each column is marked in bold.

Setting	Method	AP_{25}	Bathtub	Bed	Bookshe.	Cabinet	Chair	Counter	Curtain	Desk	Door	Otherfu.	Picture	Refrige.	S. Curtain	Sink	Sofa	Table	Toilet	Window
Full	SegCluster [35]	13.4	16.4	13.5	11.7	11.8	18.9	13.7	12.4	12.2	11.1	12.0	0.0	11.2	18.0	18.9	14.6	13.8	19.5	11.5
	SGPN [11]	22.2	0.0	31.5	13.6	20.7	31.6	17.4	22.2	14.1	16.6	18.6	0.0	0.0	0.0	52.4	40.6	31.9	72.9	15.3
	3D-SIS [35]	35.7	57.6	66.3	16.9	32.0	65.3	22.1	22.6	35.1	26.7	21.1	0.0	28.6	37.2	39.6	56.4	29.4	74.9	10.1
	MTML [52]	55.4	79.4	80.6	45.3	34.6	87.7	9.7	54.2	49.9	45.8	33.5	19.8	44.1	74.9	44.5	80.3	67.4	98.0	47.2
	PointGroup [38]	71.3	86.5	79.5	74.4	67.3	92.5	**64.8**	61.6	74.1	54.8	65.4	**48.2**	38.3	71.1	82.8	85.1	74.2	**100**	**63.6**
	3D-MPA [9]	72.4	**90.3**	83.4	**78.3**	**69.9**	87.6	62.5	**66.0**	69.2	56.6	48.6	48.0	**61.4**	**93.1**	75.2	76.1	74.8	99.2	62.2
Weak	SPIB [22]	61.4	87.4	**86.8**	48.8	45.4	89.0	49.6	47.8	52.3	49.2	45.5	9.9	48.3	82.6	63.2	**88.1**	66.2	95.9	41.9
	Box2Mask [20]	71.8	87.1	83.8	68.2	59.5	94.5	58.5	65.1	78.6	59.8	**67.1**	45.6	46.9	77.4	79.5	87.0	75.5	96.9	61.4
	SDPH	**73.0**	87.1	82.6	73.6	62.1	**95.2**	63.0	61.5	**85.5**	**61.1**	63.1	43.5	46.7	82.0	**85.4**	86.3	**78.2**	98.3	59.3

Figure 7. Qualitative comparison at a noise rate of 40% on ScanNet-v2. The legend is employed to distinguish among different semantic meanings, while the individual instances are randomly colored. The key differences are marked out with red dashed rectangles.

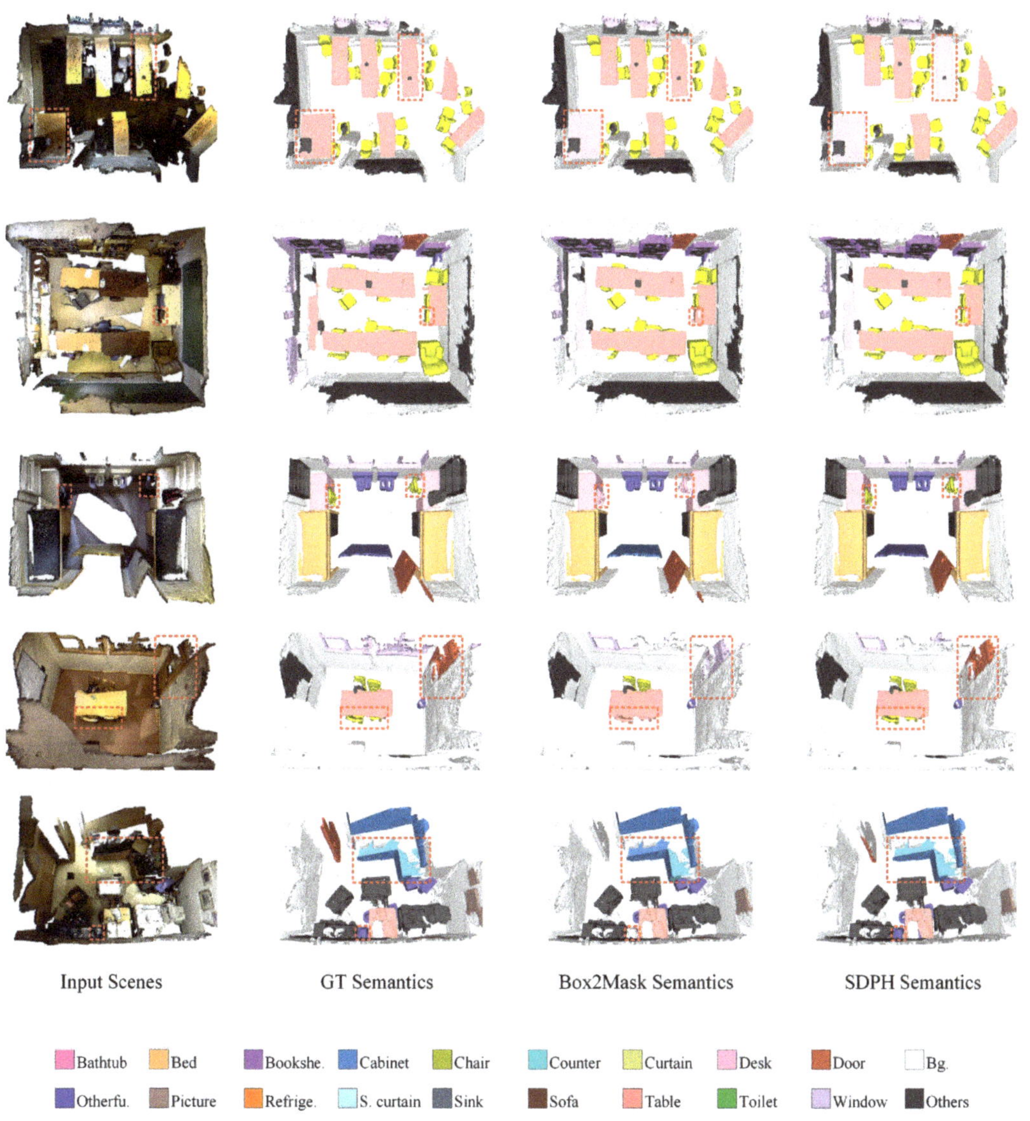

Figure 8. Qualitative comparison at a noise rate of 40% on ScanNet-v2. The legend is employed to distinguish among different semantic meanings, and the key differences are marked out with red dashed rectangles.

4.3. Ablation Study

We analyzed the contribution of each component in our learning framework, including perturbation-based consistency regularization (PCR), history-based label refurbishment (HLR), and temporal consistency regularization (TCR). It should be noted that the models in the ablation study were all trained with a noise rate of 40%. The complete ablation results are shown in Table 3. We found that every single component was able to improve the performance by itself. In particular, TCR alone obtained 2.6, 3.4, and 2.0 percent improvements in terms of AP, AP_{50}, and AP_{25}, respectively. The performance could be further boosted through their combination, and the largest increases in AP, AP_{50}, and AP_{25}

reached 5.2, 5.3, and 3.2 by combining all three components. This thorough ablation study demonstrated that each module plays an important role in our framework.

Table 3. Ablation study on ScanNet-v2. The highest performance in each column is marked in bold.

PCR	HLR	TCR	AP	AP_{50}	AP_{25}
			35.2	53.3	65.8
✓			37.1	53.7	65.1
	✓		37.6	55.4	66.6
		✓	37.8	56.7	67.8
✓	✓		39.5	58.1	67.9
✓		✓	37.1	54.8	65.6
	✓	✓	39.5	57.4	68.8
✓	✓	✓	**40.4**	**58.6**	**69.0**

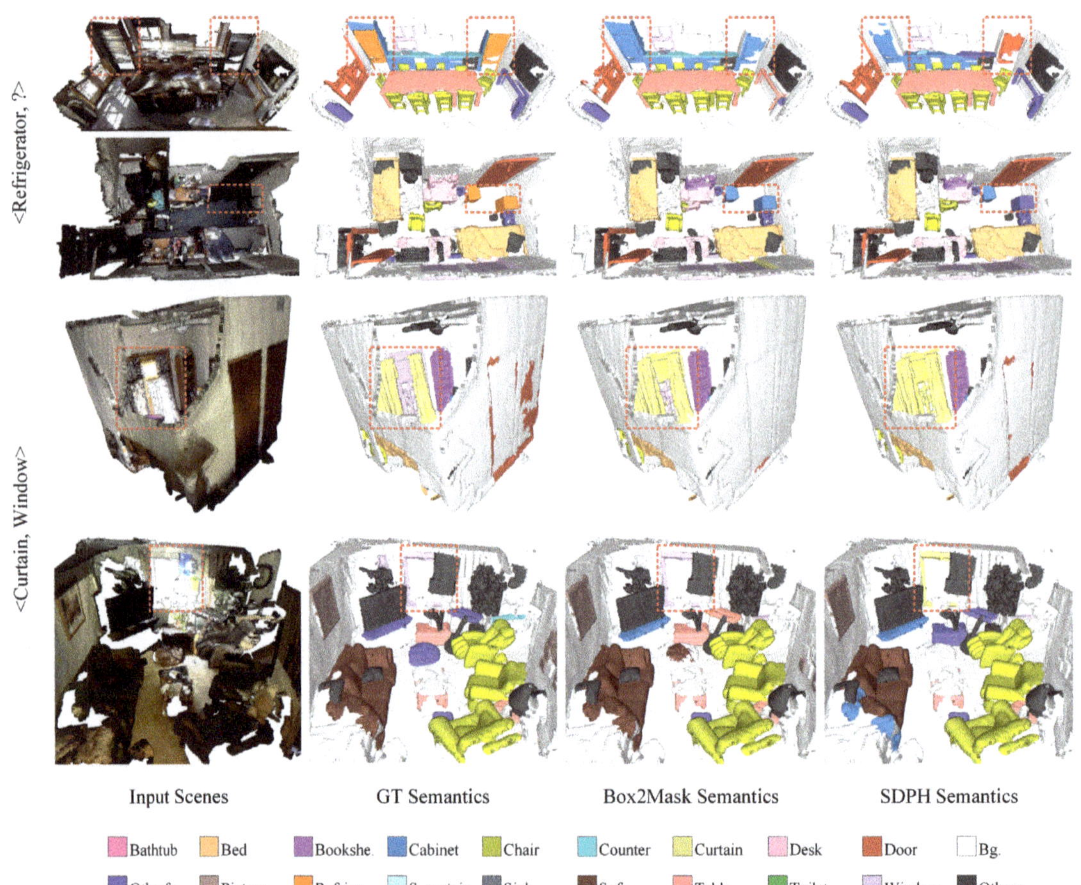

Figure 9. Bad cases on ScanNet-v2 in the noise-free setting. The first two rows show that refrigerators could be misclassified as cabinets, doors, and other furniture. We use "?" to represent this complicated situation. The last two rows show that windows could be misclassified as curtains, which lowered both categories' performance. The legend is employed to distinguish among different semantic meanings, and the key differences are marked out with red dashed rectangles.

4.4. Analysis of Label Refurbishment

To demonstrate the process of label refurbishment, we further recorded two related statistics, as shown in Figure 10. The first was the ratio of refurbishable super-voxel samples, which was defined as

$$\eta = \frac{number\ of\ refurbishable\ samples}{number\ of\ total\ samples}. \tag{17}$$

The second was the correction error, which could be computed as

$$\delta = \frac{number\ of\ mistakenly\ corrected\ samples}{number\ of\ refurbishable\ samples}. \tag{18}$$

Note that both statistics took the entire training set into account. We set the noise rate to 40%.

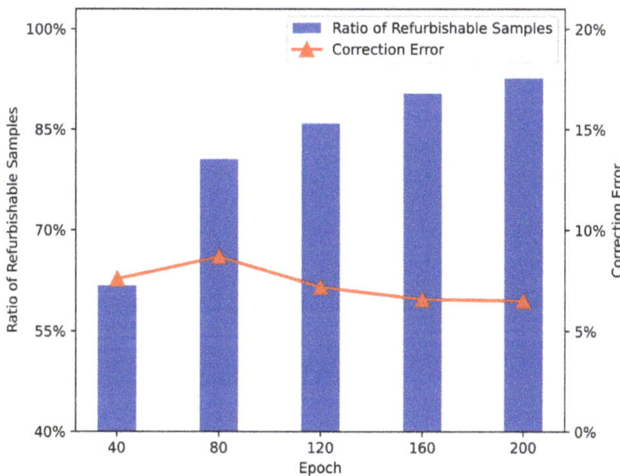

Figure 10. Trend of statistics in history-guided label refurbishment.

The ratio of refurbishable super-voxel samples gradually increased from 61.8% to 92.6%, finally covering the majority of the whole training set. Moreover, the correction error stayed relatively low throughout the training process because we adopted a conservative refurbishment strategy. On the one hand, the refurbishable threshold was quite strict to reduce false correction. On the other hand, we kept the unrefurbishable samples instead of dropping them, which lowered the risk of error accumulation. Therefore, the label quality was steadily improved as the training proceeded, as shown in Figure 11. However, we observed that it was easier to correct the labels of isolated objects with clear boundaries, such as chairs, sofas, and tables. On the contrary, flat objects that were often attached to walls, such as pictures and curtains, were harder to distinguish from the background.

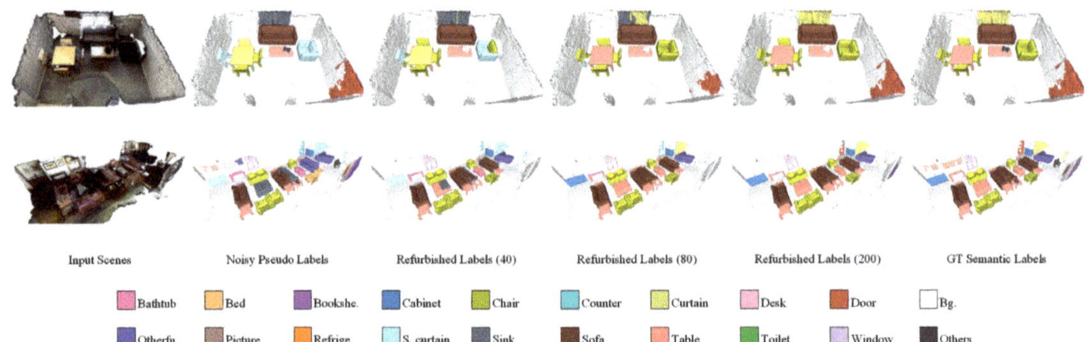

Figure 11. Qualitative demonstration of history-guided label refurbishment. From left to right are the input point clouds, the corresponding noisy pseudo-labels, the refurbished labels in epochs 40, 80, and 200, and the ground-truth semantic labels.

4.5. Complexity Analysis

Apart from the mean average precision, we also compared the time costs to give a full picture of the performance. As shown in Table 4, the inference time of our SDPH was comparable to that of the state-of-the-art weakly supervised method Box2Mask [20], though SDPH required a longer time for training. In fact, our approach mainly focused on the design of loss functions that only affected the training cost. Without extra network parameters, the majority of the additional cost came from perturbation-based consistency regularization (PCR), since it constructed a perturbed network branch. PCR did not affect the inference time, as only the main branch was used in inference.

Table 4. Comparison of the average computation time in milliseconds per scan on ScanNet-v2. The running time was measured in the same environment. Note that a post-processing step was implemented to cluster points into instances in inference.

Method	Training Time (ms)	Inference Time (ms)
Box2Mask [20]	444	1044
SDPH	722	1026

5. Conclusions

In this work, we proposed a novel self-distillation architecture for weakly supervised point cloud instance segmentation with inaccurate bounding boxes as annotations. We employed consistency regularization based on data perturbation and historical records to prevent the network from overfitting noisy labels. Moreover, the noisy labels were refurbished according to the predictions' temporal consistency without knowing the noise rate. An extensive ablation study and analysis verified the importance of each module in SDPH. Our method achieved comparable performance to that of fully supervised methods, and it outperformed recent weakly supervised methods by at least 1.2 percentage points in terms of AP_{25}, which demonstrated the effectiveness and robustness of our framework.

In the future, we plan to extend the noise types to asymmetric semantic noise and geometric coordinate noise, which may require a new confidence criterion. In addition, inspired by the mutual promotion between semantic segmentation and instance segmentation, semantic classification and geometric regression could be associated through smoothness regularization to reduce discontinuity and messy "over-segmentation".

Author Contributions: Conceptualization, Y.P. and H.F.; methodology, Y.P.; software, Y.P.; validation, Y.P.; formal analysis, Y.P.; investigation, Y.P.; resources, H.F.; writing—original draft preparation, Y.P.; writing—review and editing, H.F. and T.C.; visualization, Y.P.; supervision, B.H.; project administration, B.H. All authors have read and agreed to the published version of the manuscript.

Funding: This research received no external funding.

Institutional Review Board Statement: Not applicable.

Informed Consent Statement: Not applicable.

Data Availability Statement: Publicly available datasets were analyzed in this study. This data was accessed at http://www.scan-net.org/ on 27 August 2022.

Acknowledgments: We are particularly grateful to those who provided useful suggestions and kind help with programming during the study.

Conflicts of Interest: The authors declare no conflict of interest.

References

1. El Madawi, K.; Rashed, H.; El Sallab, A.; Nasr, O.; Kamel, H.; Yogamani, S. RGB and LiDAR fusion based 3D Semantic Segmentation for Autonomous Driving. In Proceedings of the 2019 IEEE Intelligent Transportation Systems Conference (ITSC), Auckland, New Zealand, 27–30 October 2019; pp. 7–12. [CrossRef]
2. Yan, Z.; Duckett, T.; Bellotto, N. Online learning for 3D LiDAR-based human detection: Experimental analysis of point cloud clustering and classification methods. *Auton. Robot.* **2020**, *44*, 147–164. [CrossRef]
3. Zhao, Y.; Zhang, L.; Liu, Y.; Meng, D.; Cui, Z.; Gao, C.; Gao, X.; Lian, C.; Shen, D. Two-Stream Graph Convolutional Network for Intra-Oral Scanner Image Segmentation. *IEEE Trans. Med. Imaging* **2022**, *41*, 826–835. [CrossRef] [PubMed]
4. Guo, Y.; Wang, H.; Hu, Q.; Liu, H.; Liu, L.; Bennamoun, M. Deep Learning for 3D Point Clouds: A Survey. *IEEE Trans. Pattern Anal. Mach. Intell.* **2021**, *43*, 4338–4364. [CrossRef] [PubMed]
5. Lu, H.; Shi, H. Deep Learning for 3D Point Cloud Understanding: A Survey. *arXiv* **2020**, arXiv:2009.08920.
6. Bello, S.A.; Yu, S.; Wang, C.; Adam, J.M.; Li, J. Review: Deep Learning on 3D Point Clouds. *Remote Sens.* **2020**, *12*, 1729. [CrossRef]
7. Liu, W.; Sun, J.; Li, W.; Hu, T.; Wang, P. Deep Learning on Point Clouds and Its Application: A Survey. *Sensors* **2019**, *19*, 4188. [CrossRef] [PubMed]
8. Yi, L.; Zhao, W.; Wang, H.; Sung, M.; Guibas, L.J. GSPN: Generative Shape Proposal Network for 3D Instance Segmentation in Point Cloud. In Proceedings of the IEEE/CVF Conference on Computer Vision and Pattern Recognition (CVPR), Long Beach, CA, USA, 15–20 June 2019.
9. Engelmann, F.; Bokeloh, M.; Fathi, A.; Leibe, B.; Niessner, M. 3D-MPA: Multi-Proposal Aggregation for 3D Semantic Instance Segmentation. In Proceedings of the IEEE/CVF Conference on Computer Vision and Pattern Recognition (CVPR), Virtual, 13–19 June 2020.
10. Chen, S.; Fang, J.; Zhang, Q.; Liu, W.; Wang, X. Hierarchical aggregation for 3d instance segmentation. In Proceedings of the IEEE/CVF International Conference on Computer Vision, Virtual, 10–17 October 2021; pp. 15467–15476.
11. Wang, W.; Yu, R.; Huang, Q.; Neumann, U. SGPN: Similarity Group Proposal Network for 3D Point Cloud Instance Segmentation. In Proceedings of the IEEE Conference on Computer Vision and Pattern Recognition (CVPR), Salt Lake City, UT, USA, 18–22 June 2018.
12. Pham, Q.H.; Nguyen, T.; Hua, B.S.; Roig, G.; Yeung, S.K. JSIS3D: Joint Semantic-Instance Segmentation of 3D Point Clouds with Multi-Task Pointwise Networks and Multi-Value Conditional Random Fields. In Proceedings of the IEEE/CVF Conference on Computer Vision and Pattern Recognition (CVPR), Long Beach, CA, USA, 16–20 June 2019.
13. Han, L.; Zheng, T.; Xu, L.; Fang, L. OccuSeg: Occupancy-Aware 3D Instance Segmentation. In Proceedings of the IEEE/CVF Conference on Computer Vision and Pattern Recognition (CVPR), Virtual, 13–19 June 2020.
14. Zhang, B.; Wonka, P. Point Cloud Instance Segmentation Using Probabilistic Embeddings. In Proceedings of the IEEE/CVF Conference on Computer Vision and Pattern Recognition (CVPR), Virtual, 19–25 June 2021; pp. 8883–8892.
15. Dai, A.; Chang, A.X.; Savva, M.; Halber, M.; Funkhouser, T.; Niessner, M. ScanNet: Richly-Annotated 3D Reconstructions of Indoor Scenes. In Proceedings of the IEEE Conference on Computer Vision and Pattern Recognition (CVPR), Honolulu, HI, USA, 21–26 July 2017.
16. Xu, X.; Lee, G.H. Weakly Supervised Semantic Point Cloud Segmentation: Towards 10x Fewer Labels. In Proceedings of the IEEE/CVF Conference on Computer Vision and Pattern Recognition (CVPR), Virtual, 13–19 June 2020.
17. Liu, Z.; Qi, X.; Fu, C.W. One Thing One Click: A Self-Training Approach for Weakly Supervised 3D Semantic Segmentation. In Proceedings of the IEEE/CVF Conference on Computer Vision and Pattern Recognition (CVPR), Virtual, 19–25 June 2021; pp. 1726–1736.
18. Tao, A.; Duan, Y.; Wei, Y.; Lu, J.; Zhou, J. SegGroup: Seg-Level Supervision for 3D Instance and Semantic Segmentation. *IEEE Trans. Image Process.* **2022**, *31*, 4952–4965. [CrossRef] [PubMed]
19. Wei, J.; Lin, G.; Yap, K.H.; Hung, T.Y.; Xie, L. Multi-Path Region Mining for Weakly Supervised 3D Semantic Segmentation on Point Clouds. In Proceedings of the IEEE/CVF Conference on Computer Vision and Pattern Recognition (CVPR), Virtual, 13–19 June 2020.

20. Chibane, J.; Engelmann, F.; Anh Tran, T.; Pons-Moll, G. Box2Mask: Weakly Supervised 3D Semantic Instance Segmentation using Bounding Boxes. In Proceedings of the Computer Vision—ECCV 2022, Tel Aviv, Israel, 23–27 October 2022; Avidan, S., Brostow, G., Cissé, M., Farinella, G.M., Hassner, T., Eds.; Springer Nature: Cham, Switzerland, 2022; pp. 681–699.
21. Liu, Y.; Hu, Q.; Lei, Y.; Xu, K.; Li, J.; Guo, Y. Box2Seg: Learning Semantics of 3D Point Clouds with Box-Level Supervision. *arXiv* **2022**, arXiv:2201.02963.
22. Liao, Y.; Zhu, H.; Zhang, Y.; Ye, C.; Chen, T.; Fan, J. Point Cloud Instance Segmentation with Semi-Supervised Bounding-Box Mining. *IEEE Trans. Pattern Anal. Mach. Intell.* **2022**, *44*, 10159–10170. [CrossRef] [PubMed]
23. Hou, J.; Graham, B.; Niessner, M.; Xie, S. Exploring Data-Efficient 3D Scene Understanding with Contrastive Scene Contexts. In Proceedings of the IEEE/CVF Conference on Computer Vision and Pattern Recognition (CVPR), Virtual, 19–25 June 2021; pp. 15587–15597.
24. Hu, Z.; Gao, K.; Zhang, X.; Dou, Z. Noise resistant focal loss for object detection. In Proceedings of the Chinese Conference on Pattern Recognition and Computer Vision (PRCV), Nanjing, China, 16–18 October 2020; Springer: Cham, Switzerland, 2020; pp. 114–125.
25. Liu, X.; Li, W.; Yang, Q.; Li, B.; Yuan, Y. Towards Robust Adaptive Object Detection Under Noisy Annotations. In Proceedings of the IEEE/CVF Conference on Computer Vision and Pattern Recognition (CVPR), New Orleans, LA, USA, 19–20 June 2022; pp. 14207–14216.
26. Zhang, C.; Bengio, S.; Hardt, M.; Recht, B.; Vinyals, O. Understanding Deep Learning (Still) Requires Rethinking Generalization. *Commun. ACM* **2021**, *64*, 107–115. [CrossRef]
27. Ye, S.; Chen, D.; Han, S.; Liao, J. Learning with Noisy Labels for Robust Point Cloud Segmentation. In Proceedings of the IEEE/CVF International Conference on Computer Vision (ICCV), Virtual, 11–17 October 2021; pp. 6443–6452.
28. Hinton, G.; Vinyals, O.; Dean, J. Distilling the Knowledge in a Neural Network. *arXiv* **2015**, arXiv:1503.02531.
29. Yuan, L.; Tay, F.E.; Li, G.; Wang, T.; Feng, J. Revisiting Knowledge Distillation via Label Smoothing Regularization. In Proceedings of the IEEE/CVF Conference on Computer Vision and Pattern Recognition (CVPR), Virtual, 13–19 June 2020.
30. Allen-Zhu, Z.; Li, Y. Towards Understanding Ensemble, Knowledge Distillation and Self-Distillation in Deep Learning. *arXiv* **2020**, arXiv:2012.09816.
31. Pham, M.; Cho, M.; Joshi, A.; Hegde, C. Revisiting Self-Distillation. *arXiv* **2022**, arXiv:2206.08491.
32. Kim, K.; Ji, B.; Yoon, D.; Hwang, S. Self-Knowledge Distillation with Progressive Refinement of Targets. In Proceedings of the IEEE/CVF International Conference on Computer Vision (ICCV), Virtual, 10–17 October 2021; pp. 6567–6576.
33. Shen, Y.; Xu, L.; Yang, Y.; Li, Y.; Guo, Y. Self-Distillation From the Last Mini-Batch for Consistency Regularization. In Proceedings of the IEEE/CVF Conference on Computer Vision and Pattern Recognition (CVPR), New Orleans, LA, USA, 19–20 June 2022; pp. 11943–11952.
34. Li, Y.; Yang, J.; Song, Y.; Cao, L.; Luo, J.; Li, L.J. Learning From Noisy Labels with Distillation. In Proceedings of the IEEE International Conference on Computer Vision (ICCV), Venice, Italy, 22–29 October 2017.
35. Hou, J.; Dai, A.; Niessner, M. 3D-SIS: 3D Semantic Instance Segmentation of RGB-D Scans. In Proceedings of the IEEE/CVF Conference on Computer Vision and Pattern Recognition (CVPR), Long Beach, CA, USA, 15–20 June 2019.
36. Yang, B.; Wang, J.; Clark, R.; Hu, Q.; Wang, S.; Markham, A.; Trigoni, N. Learning Object Bounding Boxes for 3D Instance Segmentation on Point Clouds. In Proceedings of the Annual Conference on Neural Information Processing Systems 2019, NeurIPS 2019, Vancouver, BC, Canada, 8–14 December 2019; Wallach, H., Larochelle, H., Beygelzimer, A., d'Alché-Buc, F., Fox, E., Garnett, R., Eds.; NIPS: La Jolla, CA, USA, 2019; Volume 32.
37. Wang, X.; Liu, S.; Shen, X.; Shen, C.; Jia, J. Associatively Segmenting Instances and Semantics in Point Clouds. In Proceedings of the IEEE/CVF Conference on Computer Vision and Pattern Recognition (CVPR), Long Beach, CA, USA, 15–20 June 2019.
38. Jiang, L.; Zhao, H.; Shi, S.; Liu, S.; Fu, C.W.; Jia, J. PointGroup: Dual-Set Point Grouping for 3D Instance Segmentation. In Proceedings of the IEEE/CVF Conference on Computer Vision and Pattern Recognition (CVPR), Virtual, 13–19 June 2020.
39. Vu, T.; Kim, K.; Luu, T.M.; Nguyen, T.; Yoo, C.D. SoftGroup for 3D Instance Segmentation on Point Clouds. In Proceedings of the IEEE/CVF Conference on Computer Vision and Pattern Recognition (CVPR), New Orleans, LA, USA, 19–20 June 2022; pp. 2708–2717.
40. Zhou, Z.H. A brief introduction to weakly supervised learning. *Natl. Sci. Rev.* **2018**, *5*, 44–53. [CrossRef]
41. Zhang, Y.; Li, Z.; Xie, Y.; Qu, Y.; Li, C.; Mei, T. Weakly Supervised Semantic Segmentation for Large-Scale Point Cloud. *Proc. AAAI Conf. Artif. Intell.* **2021**, *35*, 3421–3429. [CrossRef]
42. Zhang, Y.; Qu, Y.; Xie, Y.; Li, Z.; Zheng, S.; Li, C. Perturbed Self-Distillation: Weakly Supervised Large-Scale Point Cloud Semantic Segmentation. In Proceedings of the IEEE/CVF International Conference on Computer Vision (ICCV), Virtual, 10–17 October 2021; pp. 15520–15528.
43. Cheng, M.; Hui, L.; Xie, J.; Yang, J. SSPC-Net: Semi-supervised Semantic 3D Point Cloud Segmentation Network. *Proc. AAAI Conf. Artif. Intell.* **2021**, *35*, 1140–1147. [CrossRef]
44. Ren, Z.; Misra, I.; Schwing, A.G.; Girdhar, R. 3D Spatial Recognition Without Spatially Labeled 3D. In Proceedings of the IEEE/CVF Conference on Computer Vision and Pattern Recognition (CVPR), Virtual, 19–25 June 2021; pp. 13204–13213.
45. Song, H.; Kim, M.; Park, D.; Shin, Y.; Lee, J.G. Learning From Noisy Labels with Deep Neural Networks: A Survey. In *IEEE Transactions on Neural Networks and Learning Systems*; IEEE: Piscataway, NJ, USA, 2022; pp. 1–19. [CrossRef]

46. Graham, B.; Engelcke, M.; van der Maaten, L. 3D Semantic Segmentation with Submanifold Sparse Convolutional Networks. In Proceedings of the IEEE Conference on Computer Vision and Pattern Recognition (CVPR), Salt Lake City, UT, USA, 18–22 June 2018.
47. Choy, C.; Gwak, J.; Savarese, S. 4D Spatio-Temporal ConvNets: Minkowski Convolutional Neural Networks. In Proceedings of the IEEE/CVF Conference on Computer Vision and Pattern Recognition (CVPR), Virtual, 19–25 June 2021.
48. Arpit, D.; Jastrzębski, S.; Ballas, N.; Krueger, D.; Bengio, E.; Kanwal, M.S.; Maharaj, T.; Fischer, A.; Courville, A.; Bengio, Y.; et al. A Closer Look at Memorization in Deep Networks. In *International Conference on Machine Learning, Proceedings of the 34th International Conference on Machine Learning, Sydney, NSW, Australia, 6–11 August 2017*; Precup, D., Teh, Y.W., Eds.; PMLR: San Diego, CA, USA, 2017; Volume 70, pp. 233–242.
49. Reed, S.; Lee, H.; Anguelov, D.; Szegedy, C.; Erhan, D.; Rabinovich, A. Training Deep Neural Networks on Noisy Labels with Bootstrapping. *arXiv* **2014**, arXiv:1412.6596.
50. Song, H.; Kim, M.; Lee, J.G. SELFIE: Refurbishing Unclean Samples for Robust Deep Learning. In *Machine Learning Research, Proceedings of the 36th International Conference on Machine Learning, Long Beach, CA, USA, 9–15 June 2019*; Chaudhuri, K., Salakhutdinov, R., Eds.; PMLR: San Diego, CA, USA, 2019; Volume 97, pp. 5907–5915.
51. Nguyen, T.; Mummadi, C.K.; Ngo, T.P.N.; Nguyen, T.H.P.; Beggel, L.; Brox, T. SELF: Learning to filter noisy labels with self-ensembling. In Proceedings of the International Conference on Learning Representations (ICLR), Virtual, 26–30 April 2020.
52. Lahoud, J.; Ghanem, B.; Pollefeys, M.; Oswald, M.R. 3D Instance Segmentation via Multi-Task Metric Learning. In Proceedings of the IEEE/CVF International Conference on Computer Vision (ICCV), Seoul, Republic of Korea, 27 October–2 November 2019.

Article

Automatic Registration of Homogeneous and Cross-Source TomoSAR Point Clouds in Urban Areas

Lei Pang [1], Dayuan Liu [1,*], Conghua Li [1] and Fengli Zhang [2]

[1] School of Geomatics and Urban Spatial Informatics, Beijing University of Civil Engineering and Architecture, Beijing 102616, China

[2] Aerospace Information Research Institute, Chinese Academy of Sciences, Beijing 100094, China

* Correspondence: 2108570020094@stu.bucea.edu.cn

Abstract: Building reconstruction using high-resolution satellite-based synthetic SAR tomography (TomoSAR) is of great importance in urban planning and city modeling applications. However, since the imaging mode of SAR is side-by-side, the TomoSAR point cloud of a single orbit cannot achieve a complete observation of buildings. It is difficult for existing methods to extract the same features, as well as to use the overlap rate to achieve the alignment of the homologous TomoSAR point cloud and the cross-source TomoSAR point cloud. Therefore, this paper proposes a robust alignment method for TomoSAR point clouds in urban areas. First, noise points and outlier points are filtered by statistical filtering, and density of projection point (DoPP)-based projection is used to extract TomoSAR building point clouds and obtain the facade points for subsequent calculations based on density clustering. Subsequently, coarse alignment of source and target point clouds was performed using principal component analysis (PCA). Lastly, the rotation and translation coefficients were calculated using the angle of the normal vector of the opposite facade of the building and the distance of the outer end of the facade projection. The experimental results verify the feasibility and robustness of the proposed method. For the homologous TomoSAR point cloud, the experimental results show that the average rotation error of the proposed method was less than $0.1°$, and the average translation error was less than 0.25 m. The alignment accuracy of the cross-source TomoSAR point cloud was evaluated for the defined angle and distance, whose values were less than $0.2°$ and 0.25 m.

Keywords: homologous TomoSAR point cloud; cross-source TomoSAR point cloud; the normal vector of the opposite facade; the facade projection

Citation: Pang, L.; Liu, D.; Li, C.; Zhang, F. Automatic Registration of Homogeneous and Cross-Source TomoSAR Point Clouds in Urban Areas. *Sensors* **2023**, *23*, 852. https://doi.org/10.3390/s23020852

Academic Editors: Miaohui Wang, Guanghui Yue, Jian Xiong and Sukun Tian

Received: 14 December 2022
Revised: 5 January 2023
Accepted: 9 January 2023
Published: 11 January 2023

1. Introduction

As an extension of the interferometric synthetic aperture radar (InSAR) technology, the synthetic aperture principle is extended to the elevation direction, solving the overlay mask problem caused by the SAR imaging geometry, and realizing the three-dimensional imaging of the distance direction, azimuth direction, and elevation direction [1,2]. In recent years, with the improvement of airborne SAR and satellite-based SAR systems and the advancement of technology, the resolution, signal-to-noise ratio, and other indices have been improved, and high-precision 3D building point clouds in the observation area can now be generated using airborne or satellite-based SAR tomography technology, while even higher-dimensional information such as building deformation can be obtained using differential tomography technology [3–5].

The three-dimensional visualization of urban buildings plays an extremely important role in the process of urban digital construction. Since the synthetic aperture radar is side-imaging, the TomoSAR point cloud generated by a single track only shows the structure of one side of the building, and the TomoSAR point cloud generated by SAR images of at least two tracks is needed to show the complete structure of the building in the target area. The team of Zhu Xiaoxiang [6] fused the TomoSAR point cloud of the ascending and descending orbit of the Berlin urban area for the first time, and the fused TomoSAR point cloud could realize the construction of urban dynamic models and 3D visualization. However, in the environment of urban expansion, the low coherence and noise of buildings in the observation area lead to a reduction in the total amount of SAR images and the quality of TomoSAR point clouds, thus limiting the application of spaceborne SAR 3D imaging. Due to the different number of SAR images in different orbits and the error of geocoding, there are some rotation and translation errors in multi-view TomoSAR point clouds. The 3D visualization of a complete building structure based on the TomoSAR point cloud can not only rely on the TomoSAR point cloud generated by the SAR image of rising and falling tracks but also be realized by combining TomoSAR and LiDAR point clouds. The backpack mobile 3D laser scanner uses the laser SLAM principle, and the operation is very simple [7,8]. It restores the spatial 3D data through the algorithm as a function of its attitude data and laser point cloud. The detection distance of the backpack mobile 3D laser scanner is 50–120 m. Scanning a high-rise building of more than 100 m can easily cause the loss of the facade and top point cloud of the high-rise building; in contrast, SAR is prone to missing point clouds at the bottom of buildings in complex environments, but it can detect the upper floors and top areas of buildings. Compared with TomoSAR point clouds, mobile laser scanning (MLS) point clouds have higher density and a very low overlap rate. However, how to extract the same features to fuse ascending and descending TomoSAR point clouds with single-track TomoSAR and MLS point clouds is the main research problem addressed in this paper.

Point cloud automatic registration mainly adopts the registration strategy from coarse to fine. Firstly, the rough point cloud registration algorithm is used to roughly estimate the altitude conversion parameters between the two-point cloud data, i.e., the initial rotation and translation parameters. Then, the initial conversion parameters are used as the input parameters of the point cloud precision registration algorithm to further accurately register the two-point clouds, and a higher precision point cloud registration result is obtained [9–11]. The main features used for point cloud automatic coarse configuration are point [12,13], line [10,14], and face [15,16]. Point features such as SIFT [17,18], Harris [19,20], and FPFH [21,22] are extracted for automatic registration of airborne laser scanned (ALS) and terrestrial laser scanned(TLS) building point cloud data [10,23]. Extracting line and surface features for automatic registration leads to higher robustness than point features, and it can effectively reduce the interference of point cloud noise [11]. The 2-D contours of buildings were extracted to automatically register ground and airborne point clouds in [24]. A parameterization based on complex numbers was used to determine the corresponding relationship between planes, which was effectively applied to the ground laser scanning data with a certain degree of overlap.

At present, the commonly used methods of point cloud precise registration are the iterative nearest neighbor algorithm (ICP), random sampling consistency (RANSAC), normal distribution transformation algorithm (NDT), etc. Among them, the ICP algorithm [25] iteratively corrects the rigid body transformation (translation and rotation) of two original point clouds to minimize the distance between all point sets. The RANSAC algorithm [26] achieves this goal by iteratively selecting a set of random subsets of point cloud data with a certain probability to get a reasonable alignment result, and the number of iterations must be increased to improve the probability. The NDT algorithm [27] uses the statistical information of the point cloud data, whereby the probability density of the transformed points is maximized if the transformed parameters are the best alignment result of the two point clouds.

The above methods are commonly used in the coalignment of LiDAR point clouds and point clouds derived from optical images, but they are not applicable to point clouds derived from SAR images because the overlap rate of TomoSAR point clouds generated from cross-directional orbits is extremely low, and the density and accuracy of point clouds are low compared with LiDAR point clouds; hence, the automatic alignment of TomoSAR point clouds faces greater difficulties. Gernhardt et al. [28] used the fused PSI for detailed monitoring of individual buildings. In [6], the automatic alignment of the TomoSAR point clouds of ascending and descending orbits was achieved by extracting the L-shaped endpoints of the TomoSAR building facade point clouds, and the L-shaped endpoints of the TomoSAR point clouds of the two orbits could not be accurately corresponded when there were fewer point clouds of one of the building L-shaped facades. Hence, this method was limited to buildings with L-shaped facades. A robust alignment method for urban area array InSAR point clouds was proposed in [29], using the concave and convex facades of buildings for rotation correction and fine displacement. The TomoSAR and MLS point clouds in urban areas also have rich facade information. In this paper, the geometric features of the TomoSAR and MLS facade point clouds were used to derive the optimal alignment parameters, i.e., rotation matrix and translation matrix, to achieve automatic alignment between homologous and cross-source TomoSAR point clouds. Thus, the main contributions of the work in this paper are as follows:

- A method is proposed for aligning TomoSAR point clouds for both ascending and descending orbits, and TomoSAR point clouds with MLS point clouds;
- Rotation correction is performed using the normal vector angle of the opposing facades of the building;
- Fine translation correction of the spatial position of the opposing facades of the building is achieved using previous information.

2. Materials and Methods

Three-dimensional point clouds were acquired from two different views of the urban area, as shown in Figure 1. For the ascending and descending TomoSAR point clouds, the homologous TomoSAR point clouds had offset and rotation errors perpendicular to the line of sight due to the satellite platform position and geocoding errors, and the overlap rate of the two point clouds was extremely low. For TomoSAR point clouds and MLS point clouds, the point cloud densities were different, and the two point clouds obtained from different viewpoints belonged to the opposite facades of buildings; therefore, the overlap information could not be used for alignment fusion. Accordingly, we propose an alignment method using the point cloud characteristics of building facades.

Figure 1. Schematic map of multi-view urban building point cloud acquisition (tomographic synthetic aperture radar system and backpack mobile 3D laser scanner).

Our method was implemented in C++ based on the existing functions of PCL. All the experiments are conducted on a computer with an Intel i7-11700 and 32-GB RAM. The flow chart of the method is shown in Figure 2. The source and target point clouds were composed of the TomoSAR point clouds of the ascending and descending orbits, or the source and target point clouds were composed of the single-orbit TomoSAR point cloud and the MLS point cloud, and the two point clouds provided the front and reverse sides of the building. Firstly, statistical filtering was used for filtering, and most of the noise and outlier points were eliminated. The filtered point cloud still had some of the denser outlier block point clouds. According to the DoPP algorithm, to extract the building facade points, the extracted building facade points were clustered by density to obtain the building facade blocks, thus further eliminating the outlier points. To ensure that the source and target point clouds had good initial positions, the PCA-based initial coarse alignment method was used, which mainly used the principal axis direction of the point cloud data to align the two sets of point clouds with good initial positions after alignment. Using RANSAC to fit the building facade points to get the plane of the building facade, the normal vector of the plane, angle of the normal vector according to the topological relationship with the building facade, and rotation coefficient were sequentially calculated. After projecting the fitted plane point cloud onto the xy-axis, the least squares method was used to fit a two-dimensional straight line, and the translation coefficient was calculated according to the spatial position of the building facade.

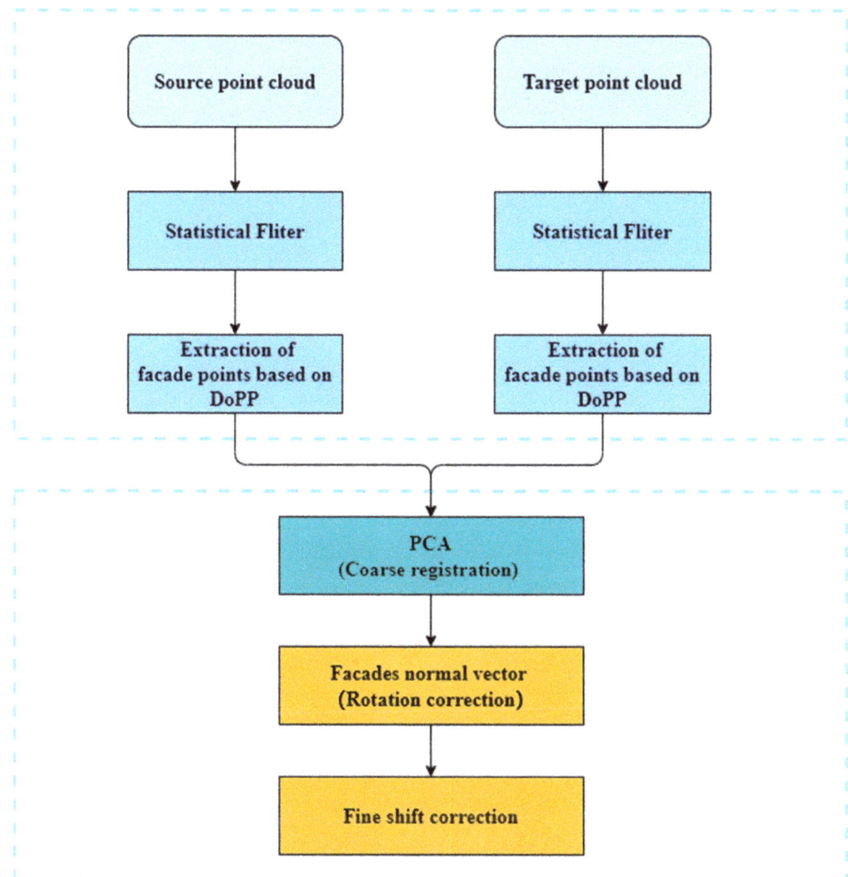

Figure 2. Flowchart of the proposed method. The rectangular box in the upper half represents the preprocessing of the data: filtering and extraction of building elevation points. The rectangular box in the lower half represents the step of data alignment: coarse alignment followed by rotation and fine translation correction.

The method of data pre-processing described above involves several artificially given parameters, including statistical filtering, building point extraction, and RANSAC-based façade extraction. We set these parameters in combination with building spacing, number of building storys and point cloud density:

- Statistical filtering: the number of close points analyzed for each point is set to 50 and the multiple of the standard deviation is set to 1. This means that a point is marked as an outlier if it exceeds the mean distance by more than one standard deviation;
- Building elevation point extraction: Set the grid size of the DoPP projection to 0.5 m × 0.5 m, and set the number of points of a single grid to 15 for TomoSAR facade point cloud estimation;
- Facade extraction based on RANSAC: As the remaining thickness of the building facade points is about 1~2 m, the facade plane fitting tolerance is set to 0.5 m, resulting in an average facade thickness of 1.5 m.

2.1. TomoSAR System Model

TomoSAR, which originated from medical CT imaging technology, extends the two-dimensional imaging principle of SAR to three dimensions. TomoSAR uses multiple aligned two-dimensional SAR images obtained from observations of the same target feature to invert its scattering values at different heights in the oblique distance direction, thus restoring the real three-dimensional scene [30,31]. The geometric model of the TomoSAR imaging principle is shown in Figure 3. One of the M + 1 view aerial pass SAR single-view complex images of the same target area was selected as the main image, and the complex value gm of each resolution unit in the m-th aerial pass image except the main image could be regarded as the superposition of N scattered target signals in the same orientation at the same oblique distance in the laminar direction s. This can be expressed as follows:

$$g_n = \int_{\Delta s} \gamma(s) exp(-j2\pi\xi_n s)ds, n = 1, 2, 3, \ldots, N, \tag{1}$$

where $\gamma(s)$ is the backward reflectivity function along the elevation direction of the imaging area, and the spatial sampling interval ξ_n can be calculated as $\xi_n = -2b_{\perp n}/(\lambda r)$, $b_{\perp n}$ is the vertical baseline distance, λ is the incident wavelength, Δs is the range of elevation angles depending on the width of the antenna diffraction pattern, and r is the central slope distance. After discretization, Equation (1) can be simply approximated as follows:

$$g = R\gamma + \varepsilon \tag{2}$$

where g is the measurement vector of length N, R is the dictionary matrix with size N × L, L is the number of grid cells divided on the s-axis, $A_{ik} = exp(-j2\pi\xi_k s_k)$ is the element of the i-th row and k-th column of the matrix, and ε is the noise vector.

According to linear algebra theory, Equation (2) becomes an underdetermined equation with a nonunique solution space when the number of samples in the elevation direction is much larger than the actual number of coherent trajectories. A common solution is to use compressed sensing methods [32]. The objective function with a sparse constraint term is as follows:

$$\hat{\gamma} = arg \min_x \left\{ \| R\gamma - g \|_2^2 + \lambda \| \gamma \|_1 \right\} \tag{3}$$

where λ denotes the sparsity factor. A larger value indicates a sparser solution. $\| \gamma \|_1$ is the sparsity constraint term that limits the solution space.

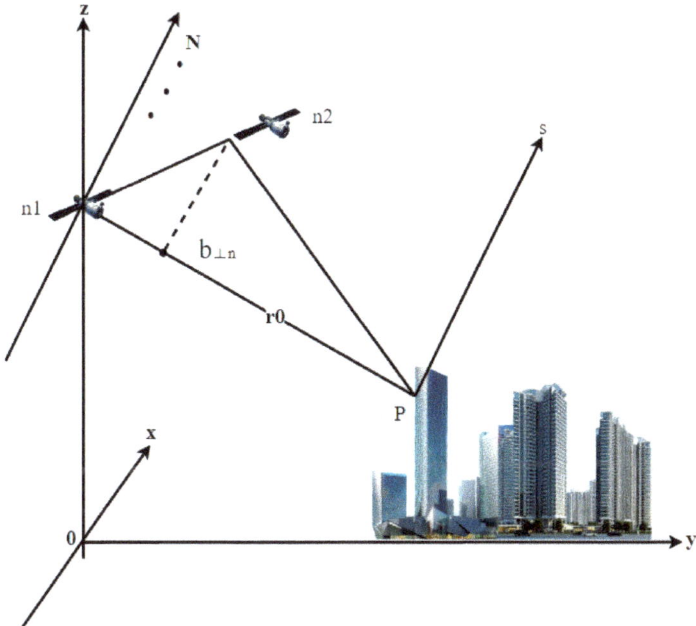

Figure 3. The imaging mode of TomoSAR.

2.2. Filtering and Facade Point Extraction

The TomoSAR point cloud of urban scenes had more outliers; in order to extract the building facade points effectively, statistical filtering was first used to remove obvious outliers. The outliers were sparsely distributed, and the distances of all points in the point cloud formed a Gaussian distribution. The average distance of each point to its nearest k points was calculated, and the mean and variance were designed to eliminate the outliers smaller than the set value.

For both TomoSAR and MLS point clouds, the building facade points could be extracted on the basis of the density of the projection point, whereby the point cloud is divided using a horizontal grid, and the number of projection points falling on each grid cell is counted. For the characteristics of TomoSAR point clouds, the DoPP values were much larger in the building facade than in other areas. The DoPP values of noise points caused by multiple scattering and noise points on the ground were uniform and small; for the point clouds on top of buildings and ground features, the DoPP values were locally larger. Using the above characteristics, a reasonable threshold value could be selected to classify the TomoSAR point clouds, with DoPP greater than T1 for the building facade point clouds, DoPP less than T2 for the noise points, and the remaining DoPP for the point clouds of the top of buildings and ground features [33].

Figure 4b shows the statistically filtered point cloud with most of the outlier points removed. Figure 4c shows the results of extracting the elevation points according to the DoPP projection point density, where the blue point cloud is the building elevation point and the green point cloud is composed of the small-scale building points and the top point cloud. In Figure 4d, the observations in the $n \times p$ data matrix X are divided into clusters according to the DBSCAN algorithm, and the extracted building facade points are partitioned into point cloud blocks by clusters.

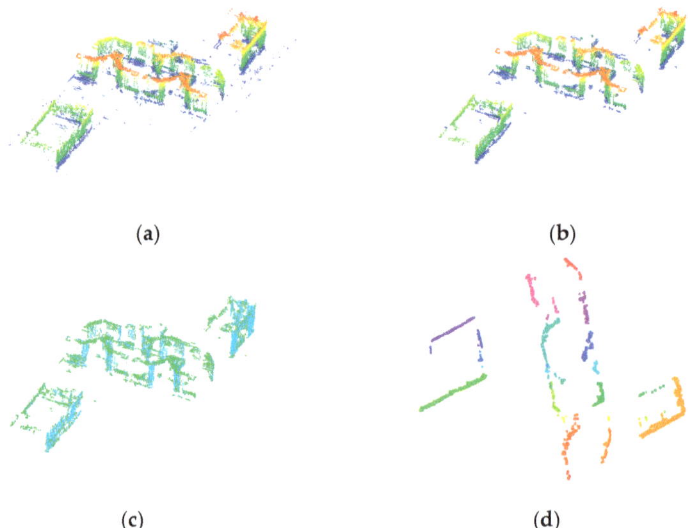

(a)

(b)

(c)

(d)

Figure 4. Results of filtering and façade point extraction: (**a**) original point cloud; (**b**) point cloud based on statistical filtering; (**c**) building points extracted according to DoPP projection, with blue indicating building façade points and green indicating building planes or other structural points; (**d**) results of density clustering.

2.3. Coarse Alignment

Point cloud alignment was divided into two steps: coarse alignment and fine alignment. Coarse alignment referred to when the transformation between two point clouds was unknown, aimed at providing a better initial value of transformation for the fine alignment; the fine alignment criterion was given an initial transformation and further optimized to obtain a more accurate transformation.

For rigidly transformed point cloud alignment, the transformation factor T can be expressed as follows [34]:

$$
T = \begin{vmatrix} R_{11} & R_{12} & R_{13} & x' \\ R_{21} & R_{22} & R_{23} & y' \\ R_{31} & R_{32} & R_{33} & z' \\ 0 & 0 & 0 & 1 \end{vmatrix}, R^* = \begin{vmatrix} R_{11} & R_{12} & R_{13} \\ R_{21} & R_{22} & R_{23} \\ R_{31} & R_{32} & R_{33} \end{vmatrix}, t^* = \begin{vmatrix} x' \\ y' \\ z' \end{vmatrix} \tag{4}
$$

where R^* and t^* are the rotation and translation coefficients.

The PCA-based initial alignment method mainly uses the principal axis direction of the extracted façade point cloud data for alignment [35]. Firstly, the covariance matrix of the two sets of point clouds is calculated, and the main feature components, i.e., the principal axis directions of the point cloud data, are calculated according to the covariance matrix. Then, the rotation matrix is derived from the principal axis direction, and the translation vector is directly derived by calculating the translational shift of the center coordinates of the two sets of point clouds. As shown in Figure 5a, the source and target point clouds were not parallel and had rotation and translation errors. Figure 5b shows the results after coarse alignment based on PCA, where the two point clouds had good initial spatial positions after coarse alignment but still have some rotation and translation errors. Therefore, a rotation and translation correction using the characteristics of the point cloud of the building façade is proposed below in order to recover the correct spatial position between the building facades.

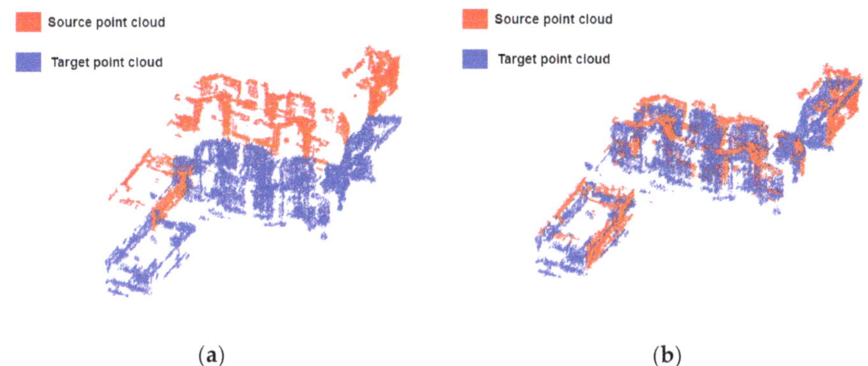

Figure 5. (**a**) The initial positions of the source and target point clouds. (**b**) The two point clouds after coarse alignment based on PCA.

2.4. Rotation Correction

The easiest way to fit the plane is least squares fitting, but the accuracy of least squares fitting is easily affected by noise, while the random sample consensus (RANSAC) algorithm is a method to calculate mathematical model parameters from a series of data containing outliers. By fitting the plane with RANSAC, the effect of noise can be excluded, and the fitting accuracy can be greatly improved [36]. As shown in Figure 6a, the facades of the same color were the opposite facades of the same building, and the point clouds of the opposite facades were fitted using RANSAC after coarse alignment.

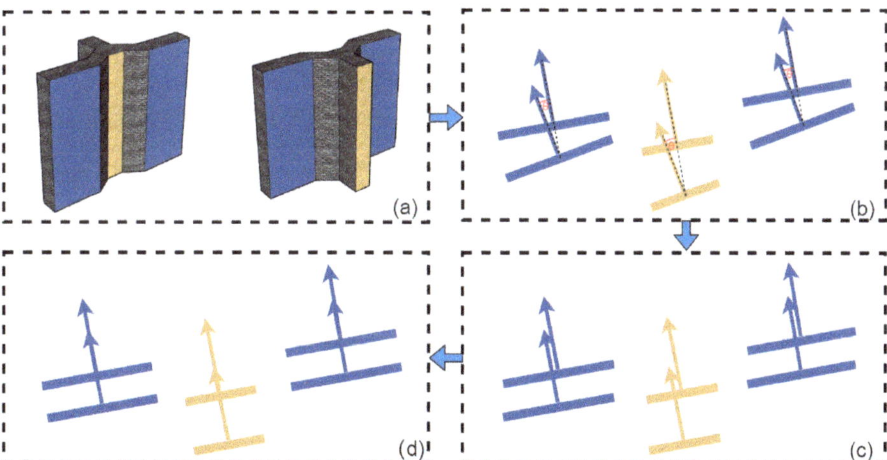

Figure 6. Schematic diagram of the spatially positioned fusion point cloud process using pairs of building elevations. (**a**) Ideal paired target and source facades with the same color facade spatial positions parallel. (**b**) The arrows indicate the corresponding normal vectors of the facades, and the normal vectors are not parallel with obvious pinch angles. (**c**) The normal vectors of the opposing elevations are parallel after rotation correction of the facade positions. (**d**) Exact translation correction of the facade position.

When the height or width of the fitted elevation is close to that of its opposite elevation, the normal vector of the plane should be calculated to perform the rotation correction. The eigenvector corresponding to the minimum eigenvalue of the covariance matrix calculated by PCA is the normal vector of the plane. Since the eigenvectors calculated by PCA are

3. Results and Discussion

In this section, three sets of experimental data with different scenes are used to evaluate the performance of the proposed method in this paper. The TomoSAR point clouds in the experiments were generated by 3D imaging of the ascending and descending orbits of TerraSAR-X spotlight data in Baoan District, Shenzhen, with 18 images of the ascending orbit and 37 images of the descending orbit, both of which had a time span greater than 800 days. Three urban scenes with different complexity were selected for experiments, as shown in Figure 9. Experiment 1 and Experiment 2 verified the robustness of this paper's method to align the ascending- and descending-orbit TomoSAR point clouds, while Experiment 3 verified the robustness of this paper's method to align the single-orbit TomoSAR and MLS point clouds.

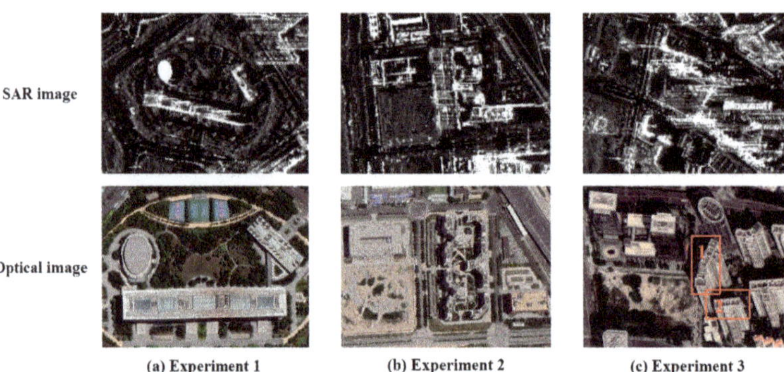

Figure 9. SAR images and optical images of three experimental scenes. The experimental area is red wireframe building 1 and building 2.

The alignment accuracy evaluation measured the angular rotation deviation and translation deviation between the aligned point cloud and the true point cloud. Due to the lack of real point clouds, we selected the TomoSAR point clouds generated when the geocoded distance direction and azimuth direction fitting error were less than 1 as the validation data for Experiments 1 and 2. The low sparse density of the TomoSAR point cloud and the small amount of information for extracting conjugate features did not allow the introduction of line, surface, and body-based feature elements for accuracy evaluation. Therefore, the alignment accuracy was evaluated by the difference between the calculated transformation parameters E^r and E^t and the validation data, according to the RMSE. In addition, for Experiment 3, which lacked validation data, we evaluated the difference between the angle θ of the two normal vectors between each facade shown in Figure 6, as well as the difference between the vertical distance of the outer endpoints of the aligned facade shown in Figure 8 and the true value d.

3.1. Homologous TomoSAR Point Cloud Alignment Experiment 1

In Experiment 1, an open urban area with no high-rise buildings around was selected; the main building in the area had 10 floors, and the building height was about 50 m. The descending-orbit TomoSAR and ascending-orbit TomoSAR point clouds were selected as the source and target point clouds, respectively. The two point clouds were extracted by statistical filtering and DoPP-based building points, and then, PCA-based coarse alignment was used; the coarse-aligned point clouds had good initial positions. Then, the point clouds were finely aligned using the method of this paper, ICP algorithm and FPFH algorithm, after which the results of the alignment were compared and evaluated in terms of accuracy.

In our experiments, the elevation information of the source and target point clouds were used for the fine alignment of the whole building point clouds; therefore, in the data preprocessing step, we extracted and filtered out the available elevation information and showed it with red point clouds in Figure 10a.The ICP algorithm was more sensitive to the initial position and rotation error of the point clouds, and the algorithm combined the two point clouds through the nearest neighbor search; The FPFH algorithm aligned two point clouds together by calculating the neighborhood features of the points. Since the TomoSAR point cloud is a sparse point cloud with an uneven distribution, the algorithm is limited by the process of feature extraction; however, due to the overlap of the two point clouds being extremely low, the ICP algorithm tended to obtain a local optimal solution, which led to unstable results of the alignment. Table 1 presents the quantitative evaluation results, revealing that the two point clouds still had large rotation and translation errors after coarse alignment. Although the ICP algorithm reduced the translation error, the rotation error increased due to its instability. The FPFH algorithm reduced the translation error and rotation error, but their values are still large. Our method could achieve a rotation error of $0.019°$ and a translation error of 0.1242 m, which achieved a good alignment.

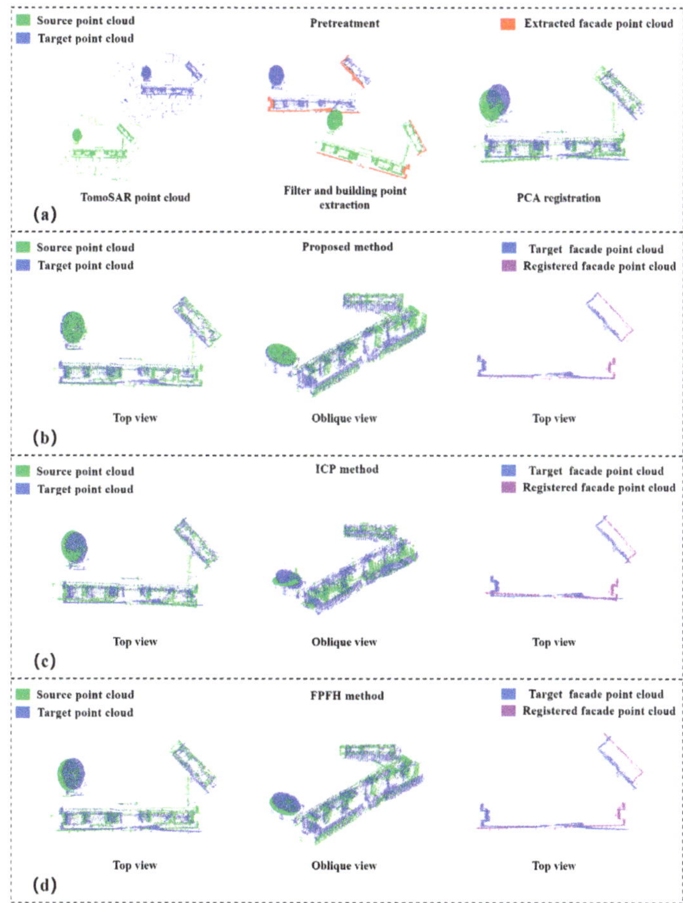

Figure 10. Alignment results of homologous TomoSAR point cloud of Experiment 1: (**a**) raw data and preprocessing of experimental data, including statistical filtering, extraction of DoPP-based facade points, density clustering, and PCA-based coarse alignment; (**b**) top and oblique views of the alignment results of the method in this paper and top view of the aligned facade points; (**c**) top and oblique views of the alignment results of ICP algorithm and top view of the aligned facade points; (**d**) top and oblique views of the alignment results of FPFH algorithm and top view of the aligned facade points.

Table 1. Alignment accuracy parameters of Experiment 1.

Method	Rotation Error E^r (°)	Translation Error E^t (m)	RMSE (m)
PCA	15.0321	9.2350	8.9866
ICP	16.1078	6.7562	6.9740
FPFH	9.2041	2.4132	4.5473
Proposed method	0.0189	0.1242	0.1913

For the alignment of the ascending and descending TomoSAR point clouds, we also calculated the root-mean-square error to evaluate the alignment results, and the RMSE was used to measure the deviation between the observed and real values. The distance between the aligned point cloud and the real value was greater than 6 m. Although the FPFH algorithm reduced the registration error, its RMSE is still 4.5473 m. As shown in Figure 10c,d, the aligned point cloud of the building facade still had a large deviation from the real building. The ICP algorithm was weakly applicable to the lift-track TomoSAR point cloud. The FPFH algorithm has some applicability to the homogenous TomoSAR point cloud, but it was limited by the quality of the point cloud itself, and its applicability was reduced for TomoSAR point cloud with fractures and uneven distribution. The main reason for its non-applicability was that the TomoSAR point cloud was not homogeneous, and the overlap rate of the two points was too low. The method in this paper does not use the overlap information for alignment but uses the elevation information for alignment; the results of the alignment are shown in Figure 10b. From the top and oblique views, the aligned point cloud and the target point cloud showed the structure of the building accurately, and the elevation point cloud also had a precise position. Since the TomoSAR point cloud contains many noise points and the TerraSAR-X image has a 3D resolution of 0.25 m, the alignment error of about 0.25 m is within the controllable range because the method in this paper has a high alignment accuracy and strong robustness for aligning the ascending and descending TomoSAR point clouds.

3.2. Homologous TomoSAR Point Cloud Alignment Experiment 2

A denser building complex was selected for Experiment 2. The main building in this area had 18 floors and was about 65 m high. The descending-orbit TomoSAR and ascending-orbit TomoSAR point clouds were selected as the source and target point clouds, respectively. Since the region was more complex compared with the scene of Experiment 1, the filtered point cloud still had a small number of dense outliers. Due to the limitation of the onboard TerraSAR-X incidence angle, the two TomoSAR elevation point clouds extracted using DoPP could not represent the facades of the six buildings in the scene, and the point cloud on the facade of the rightmost rectangular building in the scene has been removed because they were too sparse, but this did not affect the calculation of the method in this paper; accordingly, the extracted building facades were sufficient to complete the calculation of the rotation matrix and translation matrix.

The coarse alignment based on PCA roughly aligned the two point clouds together according to their principal axes, and the two point clouds after the coarse alignment also had certain rotation and translation errors. As in Experiment 1, we used the method of this paper, ICP and FPFH, to finely align them. From the top and oblique views in Figure 11b–d, our method achieved more accurate alignment results than the ICP and FPFH algorithm, and the purple building elevation points were parallel to the source building facade points and had precise spatial positions after alignment. As shown in Table 2, the alignment results of the ICP and FPFH algorithm also had large rotation and translation errors, while the errors of the methods in this paper were less than 0.25 m, demonstrating the high accuracy and robustness of the TomoSAR point cloud alignment.

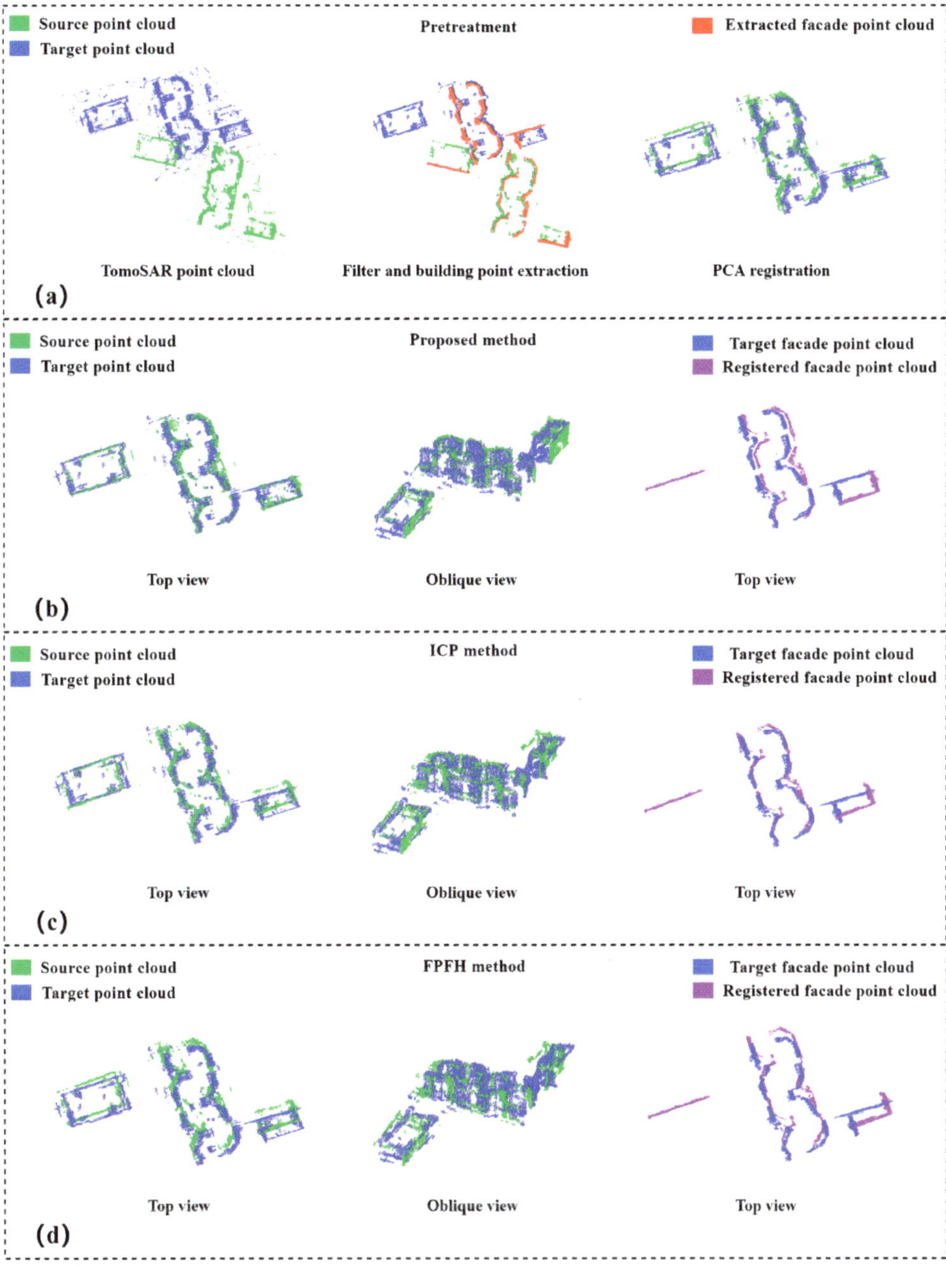

Figure 11. Alignment results of homologous TomoSAR point cloud in Experiment 2: (**a**) preprocessing of raw and experimental data, including statistical filtering, DoPP-based extraction of elevation points, density clustering, and PCA-based coarse alignment; (**b**) top and oblique views of the alignment results of the method in this paper and top view of the aligned facade points; (**c**) top and oblique views of the alignment results of ICP algorithm and top view of the aligned facade points; (**d**) top and oblique views of the alignment results of FPFH algorithm and top view of the aligned facade points.

Table 2. Alignment accuracy parameters of Experiment 2.

Method	Rotation Error E^r (°)	Translation Error E^t (m)	RMSE (m)
PCA	12.0547	8.9407	8.6295
ICP	8.1752	6.8431	7.0945
FPFH	6.0982	9.8456	8.7361
Proposed method	0.0107	0.1584	0.1802

The results of Experiments 1 and 2 demonstrate the high accuracy and strong adaptability of the method in this paper for aligning homologous TomoSAR point clouds, as well as the good robustness of the alignment using the characteristics of the building facade for the very low overlap rate of the two point clouds. The limitations of the image data and building environment led to the point cloud of a particular track not being able to support it for alignment. Our experimental solution was to align the TomoSAR point cloud on one side with the point cloud on one side scanned by other sensors.

3.3. Cross-Source TomoSAR Point Cloud Alignment Experiment

Experiment 3 selected the TomoSAR point cloud obtained from 38-view downlinked TerraSAR-X data after 3D imaging and the point cloud of one side of the building scanned by a ZEB-REVO RTT portable laser scanner of CHC NAVIGATION. The experimental area was a high-rise building with a complex building environment, and the main building had 32 floors and was about 100 m high. The downlinked TomoSAR and MLS point clouds were selected as the source and target point clouds, respectively. Since the density of the MLS point cloud was denser than that of TomoSAR point cloud, the threshold value for setting filtering and building facade point extraction needed to be increased.

The results based on PCA coarse alignment are shown in Figure 12b. From the top and oblique views, the two point clouds after aggregation had obvious rotation and translation errors, and the top view of the building facade points showed the spatial position relationship between the aligned facade point cloud and the target point cloud facade point cloud, whereas the coarsely aligned facade points did not show their spatial position correctly and could not express the building structure in the top view. In Experiments 1 and 2, the method of this paper and the ICP algorithm were used to finely align the point clouds. The alignment results of the ICP and FPFH algorithm are shown in Figure 12d,e, where the two point clouds showed visually obvious rotation errors, and the two facade points did not have correct spatial positions. The alignment results of the method in this paper are shown in Figure 12c. From the top and oblique views of the two points, the two point clouds correctly represented visually the buildings for which the accuracy was evaluated, and the facade points also had the correct spatial positions.

In the absence of real validation data, we evaluated the alignment results by the angular difference θ of the designed facade normal vectors and the outer endpoint Δd of the facade. We calculated θ and Δd for planar buildings 1 and 2 within the scene in Figure 9c after alignment by the ICP algorithm, FPFH algorithm and the method in this paper, and the results are shown in Table 3. The ICP method could not correctly rotate the two point clouds, and the angle between the opposite elevations of buildings 1 and 2 after alignment was about 24°, while the value of Δd was greater than 1.5 m. The FPFH method also could not rotate the two point clouds correctly, after registration, the relative elevation angle of building 1 and 2 is about-19°, and the Δd value is more than 3.0 m. From the experimental results and accuracy analysis, the ICP and FPFH algorithm could not calculate the exact correspondence between the TomoSAR point cloud and the MLS point cloud, and although the translation error was close to the meter level, it had a more obvious rotation error. However, our proposed method had significantly improved accuracy compared with the ICP and FPFH algorithm; the translation error reached 0.25 m, and the normal vectors of the rotated building facade point clouds were parallel with minimal rotation error. Since the TomoSAR and MLS point clouds contained many noise points and the difference between the two point clouds was too large, we believe that the error after alignment is within the acceptable range, but the accuracy and efficiency of the alignment still have room for improvement; in particular, the efficiency of the algorithm needs to be further optimized.

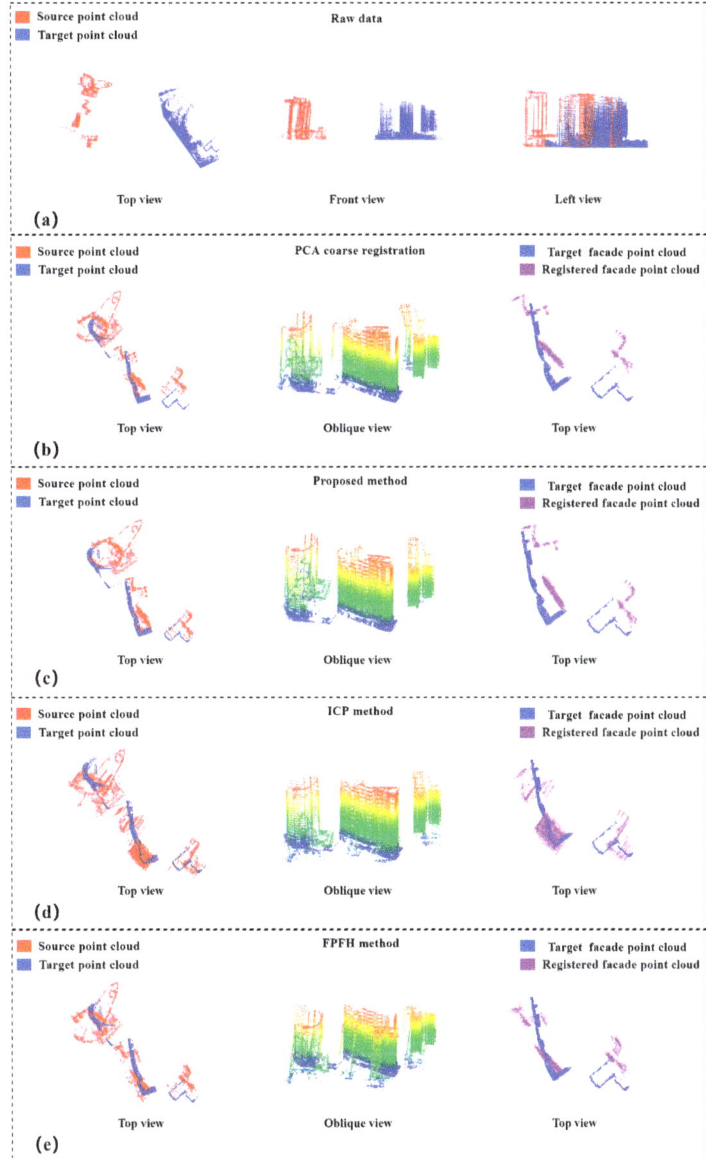

Figure 12. Experiment 3 alignment results of cross-source TomoSAR point cloud with MLS point cloud: (**a**) top, front, and left views of the original data; (**b**) top and oblique views of the coarse alignment results based on PCA and top view of the aligned facade points; (**c**) top and oblique views of the alignment results of this paper and top view of the aligned facade points; (**d**) top and oblique views of the alignment result of ICP algorithm and top view of the facade point after alignment; (**e**) top and oblique views of the alignment results of FPFH algorithm and top view of the aligned facade points.

Table 3. Alignment accuracy parameters of Experiment 3. Processing times are obtained by a regular desktop PC (Intel i7-11700).

Building	Method	Θ (°)	Δd (m)	Time (s)
	ICP	24.5624	1.5898	4.1847
Building one	FPFH	−19.4571	3.2489	64.4830
	Proposed	0.1681	0.2259	24.4830
	ICP	24.2873	1.5788	3.7857
Building two	FPFH	−19.6317	3.3115	47.6954
	Proposed	0.1487	0.2314	22.8704

3.4. Discussion

In this paper, we proposed an alignment method to align homologous and cross-source TomoSAR point clouds using the normal vectors and outer endpoints of building facades. The above experimental results verified the effectiveness of the proposed method. Compared with the famous PCA, ICP, and FPFH algorithms, the proposed method has the following advantages and disadvantages:

(1) The most important significance of our method is that it could be applied to both homologous and cross-source TomoSAR point cloud registration, helps to accurately correct rotation and translation errors, and could realize the complete observation of 3D buildings based on TomoSAR point clouds. In contrast, PCA algorithm could only achieve rough registration of two point clouds, and subsequent fine alignment was required to obtain more accurate alignment results. Although ICP algorithm can reduce the translation error, the registration result had a large rotation error and could not correctly display the building structure. The FPFH algorithm was applicable to homologous TomoSAR point clouds in simple environments, but it was not used in homologous TomoSAR point clouds with complex environments and cross-source TomoSAR point clouds.

(2) Our method does not depend on the overlap between the two point clouds but obtains the architectural points in the experimental scene and extracts the facade points through statistical filtering and DoPP projection filtering and calculates the rotation matrix by using the angle of the normal vector of the opposite side of the building and then uses the outer endpoint of the building facade projection to estimate the fine translation. The experimental results and actual data show that the method proposed in this paper had higher accuracy than other algorithms.

(3) However, the engineering process of this paper is more complex complicated and less time-efficient, especially in the extraction of building facade information. Furthermore, the calculation of facade information consumes most of the time. In addition, we need to measure the vertical distance between opposite building elevations from high-precision remote sensing images, cadastral information, or in the field.

In summary, for satellite-based synthetic SAR tomography point clouds, the method in this paper can achieve the alignment of their homologous or cross-source urban multi-view point clouds using building facade information. Moreover, we will continue to refine the method and apply it to the alignment of point clouds acquired by other sensors of different quality.

4. Conclusions

The TomoSAR point cloud of a single track cannot show the complete building structure. In order to solve this problem, this paper proposed a robust homologous and cross-source TomoSAR point cloud registration method. Under the condition of many noise points and a low overlap rate, a complete TomoSAR point cloud registration process was designed and implemented. The experimental process includes statistical filtering, building facade point extraction based on DoPP, density clustering, and rough registration based on PCA. The final rotation and translation coefficients are calculated from the angle of the normal vector of the building facade and the distance between the outer endpoints.

Experimental results showed that, compared with the ICP algorithm, the proposed method is more robust in registering homologous and cross-source TomoSAR point clouds.

However, there are several aspects of our work that can be improved. First of all, for the facade of building facade points, the method of this paper depends on the selection of parameters, which greatly reduces the efficiency of registration. When there are enough spaceborne TomoSAR point cloud data, we can use some deep learning methods to classify and segment them. Secondly, the method in this paper can be applied to urban point cloud registration collected by sensors of different quality. In future work, we will optimize the efficiency of this method and further analyze and evaluate its performance of this method.

Author Contributions: Conceptualization, L.P., D.L., C.L. and F.Z.; methodology, D.L. and C.L.; software, D.L.; writing—original draft preparation, L.P. and D.L.; writing—review and editing, L.P., D.L., C.L. and F.Z.; visualization, L.P.; supervision, D.L.; project administration, D.L.; funding acquisition, L.P. and F.Z. All authors have read and agreed to the published version of the manuscript.

Funding: This research was funded by the National Natural Science Foundation of China (No. 41671359), the Common Application Support Platform for Land Observation Satellites of China's Civil Space Infrastructure (CASPLOS_CCSI), and the China high-resolution Earth observation system (21-Y20B01-9003-19/22).

Institutional Review Board Statement: Not applicable.

Informed Consent Statement: Not applicable.

Data Availability Statement: Not applicable.

Acknowledgments: The authors would like to thank Deutsches Zentrum für Luft-und Raumfahrt (DLR) for providing the SAR tomographic experiment datasets (TerraSAR-X proposal RES3668).

Conflicts of Interest: The authors declare no conflict of interest.

References

1. Fornaro, G.; Serafino, F.; Soldovieri, F. Three-dimensional focusing with multipass SAR data. *IEEE Trans. Geosci. Remote Sens.* **2003**, *41*, 507–517. [CrossRef]
2. Budillon, A.; Johnsy, A.C.; Schirinzi, G. Contextual information based SAR tomography of urban areas. In Proceedings of the 2019 Joint Urban Remote Sensing Event (JURSE), Vannes, France, 22–24 May 2019; pp. 1–4.
3. Zhu, X.X.; Bamler, R. Tomographic SAR inversion by L_1-norm regularization—The compressive sensing approach. *IEEE Trans. Geosci. Remote Sens.* **2010**, *48*, 3839–3846. [CrossRef]
4. Fornaro, G.; Reale, D.; Serafino, F. Four-dimensional SAR imaging for height estimation and monitoring of single and double scatterers. *IEEE Trans. Geosci. Remote Sens.* **2008**, *47*, 224–237. [CrossRef]
5. Chai, H.; Lv, X.; Yao, J.; Xue, F. Off-grid differential tomographic SAR and its application to railway monitoring. *IEEE J. Sel. Top. Appl. Earth Obs. Remote Sens.* **2019**, *12*, 3999–4013. [CrossRef]
6. Wang, Y.; Zhu, X.X. Automatic feature-based geometric fusion of multiview TomoSAR point clouds in urban area. *IEEE J. Sel. Top. Appl. Earth Obs. Remote Sens.* **2014**, *8*, 953–965. [CrossRef]
7. Deschaud, J.-E. IMLS-SLAM: Scan-to-model matching based on 3D data. In Proceedings of the 2018 IEEE International Conference on Robotics and Automation (ICRA), Brisbane, QLD, Australia, 21–25 May 2018; pp. 2480–2485.
8. Guo, S.; Rong, Z.; Wang, S.; Wu, Y. A LiDAR SLAM With PCA-Based Feature Extraction and Two-Stage Matching. *IEEE Trans. Instrum. Meas.* **2022**, *71*, 8501711. [CrossRef]
9. Guo, Y.; Bennamoun, M.; Sohel, F.; Lu, M.; Wan, J. 3D object recognition in cluttered scenes with local surface features: A survey. *IEEE Trans. Pattern Anal. Mach. Intell.* **2014**, *36*, 2270–2287. [CrossRef] [PubMed]
10. Cheng, L.; Tong, L.; Li, M.; Liu, Y. Semi-automatic registration of airborne and terrestrial laser scanning data using building corner matching with boundaries as reliability check. *Remote Sens.* **2013**, *5*, 6260–6283. [CrossRef]
11. Cheng, X.; Cheng, X.; Li, Q.; Ma, L. Automatic registration of terrestrial and airborne point clouds using building outline features. *IEEE J. Sel. Top. Appl. Earth Obs. Remote Sens.* **2018**, *11*, 628–638. [CrossRef]
12. Hänsch, R.; Weber, T.; Hellwich, O. Comparison of 3D interest point detectors and descriptors for point cloud fusion. *ISPRS Ann. Photogramm. Remote Sens. Spat. Inf. Sci.* **2014**, *2*, 57. [CrossRef]
13. Kuçak, R.A.; Erol, S.; Erol, B. An experimental study of a new keypoint matching algorithm for automatic point cloud registration. *ISPRS Int. J. Geo-Inf.* **2021**, *10*, 204. [CrossRef]
14. Cheng, J.; Cheng, M.; Lin, Y.; Wang, C. A line segment based registration method for Terrestrial Laser Scanning point cloud data. In Proceedings of the 2nd ISPRS International Conference on Computer Vision in Remote Sensing (CVRS 2015), Xiamen, China, 28–30 April 2015; pp. 206–211.

15. Xu, Y.; Boerner, R.; Yao, W.; Hoegner, L.; Stilla, U. Automated coarse registration of point clouds in 3d urban scenes using voxel based plane constraint. *ISPRS Ann. Photogramm. Remote Sens. Spat. Inf. Sci.* **2017**, *4*, 185–191. [CrossRef]
16. Chen, S.; Nan, L.; Xia, R.; Zhao, J.; Wonka, P. PLADE: A plane-based descriptor for point cloud registration with small overlap. *IEEE Trans. Geosci. Remote Sens.* **2019**, *58*, 2530–2540. [CrossRef]
17. Böhm, J.; Becker, S. Automatic marker-free registration of terrestrial laser scans using reflectance. In Proceedings of the 8th conference on optical 3D measurement techniques, Zurich, Switzerland, 9–12 July 2007; pp. 9–12.
18. Jiao, Z.; Liu, R.; Yi, P.; Zhou, D. A Point Cloud Registration Algorithm Based on 3d-Sift. In *Transactions on Edutainment XV*; Springer: Berlin/Heidelberg, Germany, 2019; pp. 24–31.
19. Sipiran, I.; Bustos, B. Harris 3D: A robust extension of the Harris operator for interest point detection on 3D meshes. *Vis. Comput.* **2011**, *27*, 963–976. [CrossRef]
20. Zhong, Y.; Bai, F.; Liu, Y.; Huang, L.; Yuan, X.; Zhang, Y.; Zhong, J. Point Cloud Splicing Based on 3D-Harris Operator. In Proceedings of the 2021 3rd International Symposium on Smart and Healthy Cities (ISHC), Toronto, ON, Canada, 28–29 December 2021; pp. 61–66.
21. Rusu, R.B.; Blodow, N.; Beetz, M. Fast point feature histograms (FPFH) for 3D registration. In Proceedings of the 2009 IEEE international conference on robotics and automation, Kobe, Japan, 12–17 May 2009; pp. 3212–3217.
22. Zheng, L.; Li, Z. Virtual Namesake Point Multi-Source Point Cloud Data Fusion Based on FPFH Feature Difference. *Sensors* **2021**, *21*, 5441. [CrossRef] [PubMed]
23. Zheng, L.; Yu, M.; Song, M.; Stefanidis, A.; Ji, Z.; Yang, C. Registration of long-strip terrestrial laser scanning point clouds using ransac and closed constraint adjustment. *Remote Sens.* **2016**, *8*, 278. [CrossRef]
24. Pavan, N.L.; dos Santos, D.R.; Khoshelham, K. Global registration of terrestrial laser scanner point clouds using plane-to-plane correspondences. *Remote Sens.* **2020**, *12*, 1127. [CrossRef]
25. Li, P.; Wang, R.; Wang, Y.; Tao, W. Evaluation of the ICP algorithm in 3D point cloud registration. *IEEE Access* **2020**, *8*, 68030–68048. [CrossRef]
26. Buch, A.G.; Kraft, D.; Kamarainen, J.-K.; Petersen, H.G.; Krüger, N. Pose estimation using local structure-specific shape and appearance context. In Proceedings of the 2013 IEEE International Conference on Robotics and Automation, Karlsruhe, Germany, 6–10 May 2013; pp. 2080–2087.
27. Stoyanov, T.; Magnusson, M.; Lilienthal, A.J. Point set registration through minimization of the L2 distance between 3d-ndt models. In Proceedings of the 2012 IEEE International Conference on Robotics and Automation, Saint Paul, MN, USA, 14–18 May 2012; pp. 5196–5201.
28. Gernhardt, S.; Bamler, R. Deformation monitoring of single buildings using meter-resolution SAR data in PSI. *ISPRS J. Photogramm. Remote Sens.* **2012**, *73*, 68–79. [CrossRef]
29. Tong, X.; Zhang, X.; Liu, S.; Ye, Z.; Feng, Y.; Xie, H.; Chen, L.; Zhang, F.; Han, J.; Jin, Y. Automatic Registration of Very Low Overlapping Array InSAR Point Clouds in Urban Scenes. *IEEE Trans. Geosci. Remote Sens.* **2022**, *60*, 5224125. [CrossRef]
30. Ge, N.; Gonzalez, F.R.; Wang, Y.; Shi, Y.; Zhu, X.X. Spaceborne staring spotlight SAR tomography—A first demonstration with TerraSAR-X. *IEEE J. Sel. Top. Appl. Earth Obs. Remote Sens.* **2018**, *11*, 3743–3756. [CrossRef]
31. Fornaro, G.; Lombardini, F.; Serafino, F. Three-dimensional multipass SAR focusing: Experiments with long-term spaceborne data. *IEEE Trans. Geosci. Remote Sens.* **2005**, *43*, 702–714. [CrossRef]
32. Pang, L.; Gai, Y.; Zhang, T. Joint Sparsity for TomoSAR Imaging in Urban Areas Using Building POI and TerraSAR-X Staring Spotlight Data. *Sensors* **2021**, *21*, 6888. [CrossRef] [PubMed]
33. Wang, G.; Wang, Q.; Zhao, R.; Chen, C.; Lu, Y. Building Segmentation of UAV-based Oblique Photography Point Cloud Using DoPP and DBSCAN. In Proceedings of the 2022 3rd International Conference on Geology, Mapping and Remote Sensing (ICGMRS), Zhoushan, China, 22–24 April 2022; pp. 233–236.
34. Zhou, Q.-Y.; Park, J.; Koltun, V. Fast Global Registration. In *Proceedings of European Conference on Computer Vision*; Springer: Berlin/Heidelberg, Germany, 2016; pp. 766–782.
35. Bellekens, B.; Spruyt, V.; Berkvens, R.; Weyn, M. A survey of rigid 3d pointcloud registration algorithms. In Proceedings of the AMBIENT 2014: The Fourth International Conference on Ambient Computing, Applications, Services and Technologies, Rome, Italy, 24–28 August 2014; pp. 8–13.
36. Zeineldin, R.A.; El-Fishawy, N.A. A survey of RANSAC enhancements for plane detection in 3D point clouds. *Menoufia J. Electron. Eng. Res* **2017**, *26*, 519–537. [CrossRef]
37. Zefran, M.; Kumar, V.; Croke, C.B. On the generation of smooth three-dimensional rigid body motions. *IEEE Trans. Robot. Autom.* **1998**, *14*, 576–589. [CrossRef]
38. Liang, K.K. Efficient conversion from rotating matrix to rotation axis and angle by extending Rodrigues' formula. *arXiv* **2018**, arXiv:1810.02999.

Article

A Sequential Color Correction Approach for Texture Mapping of 3D Meshes

Lucas Dal'Col [1,2], Daniel Coelho [1,2], Tiago Madeira [1,3], Paulo Dias [1,3,*] and Miguel Oliveira [1,2]

[1]   Intelligent System Associate Laboratory (LASI), Institute of Electronics and Informatics Engineering of Aveiro (IEETA), University of Aveiro, 3810-193 Aveiro, Portugal
[2]   Department of Mechanical Engineering (DEM), University of Aveiro, 3810-193 Aveiro, Portugal
[3]   Department of Electronics, Telecommunications and Informatics (DETI), University of Aveiro, 3810-193 Aveiro, Portugal
*   Correspondence: paulo.dias@ua.pt

Abstract: Texture mapping can be defined as the colorization of a 3D mesh using one or multiple images. In the case of multiple images, this process often results in textured meshes with unappealing visual artifacts, known as texture seams, caused by the lack of color similarity between the images. The main goal of this work is to create textured meshes free of texture seams by color correcting all the images used. We propose a novel color-correction approach, called sequential pairwise color correction, capable of color correcting multiple images from the same scene, using a pairwise-based method. This approach consists of sequentially color correcting each image of the set with respect to a reference image, following color-correction paths computed from a weighted graph. The color-correction algorithm is integrated with a texture-mapping pipeline that receives uncorrected images, a 3D mesh, and point clouds as inputs, producing color-corrected images and a textured mesh as outputs. Results show that the proposed approach outperforms several state-of-the-art color-correction algorithms, both in qualitative and quantitative evaluations. The approach eliminates most texture seams, significantly increasing the visual quality of the textured meshes.

Keywords: pairwise-based color correction; texture mapping; joint image histogram; color mapping function; weighted graphs

Citation: Dal'Col, L.; Coelho, D.; Madeira, T.; Dias, P.; Oliveira, M. A Sequential Color Correction Approach for Texture Mapping of 3D Meshes. *Sensors* **2023**, *23*, 607. https://doi.org/10.3390/s23020607

Academic Editors: Miaohui Wang, Guanghui Yue, Jian Xiong and Sukun Tian

Received: 5 December 2022
Revised: 29 December 2022
Accepted: 3 January 2023
Published: 5 January 2023

1. Introduction

Three-dimensional reconstruction can be summarized as the creation of digital 3D models from the acquisition of real-world objects. It is an extensively researched topic in areas such as robotics, autonomous driving, cultural heritage, agriculture and medical imaging. In robotics and autonomous driving, 3D reconstruction can be used to obtain a 3D perception of the car's surroundings and is crucial to accomplish tasks such as localization and navigation in a fully autonomous approach [1–3]. In cultural heritage, 3D reconstruction aids in the restoration of historical constructions that have deteriorated over time [4]. In agriculture, 3D reconstruction facilitates the improvement of vehicle navigation, crop, and animal husbandry [5]. In medical imaging, reconstruction produces enhanced visualizations which are used to assist diagnostics [6].

The technologies used to reconstruct a 3D model are typically based on RGB cameras [7], RGB-D cameras [8,9], and light detection and ranging (LiDAR) [10,11]. Point clouds from LiDAR sensors contain quite precise depth information, but suffer from problems such as occlusions, sparsity, and noise, and images from RGB cameras provide color and high-resolution data, but no depth information [12]. Nonetheless, the fusion of LiDAR point clouds and RGB images can achieve better 3D reconstruction results than a single data modality [13] and, for this reason, is often used to create textured 3D models.

Texture mapping is the colorization of a 3D mesh using one or more images [14–17]. In the latest applications, usually several overlapping images are available to texture the

3D mesh, and a technique to manage the redundant photometric information is required. Those techniques are often referred to as multiview texture mapping [18–21]. The winner-take-all approach is usually the technique used to handle the redundant photometric information. For each face of the 3D mesh, this approach selects a single image from the set of available images to colorize it. The selection of the image raises the problem of how to select the most suitable one from the set of available images and how to consistently select the images for all faces of the 3D mesh. To address those problems, [22] employ a Markov random field to select the images. The authors of [23] use graph optimization mechanisms. Others use optimization procedures to minimize the discontinuities between adjacent faces [24]. Another approach that is commonly used is the average-based approach, which fuses the texture from all available images to color each face. The authors of [25,26] propose the fusion of the textures based on different forms of weighted average of the contributions of textures in the image space. However, these approaches are highly affected by the geometric registration error between the images, causing blurring and ghost artifacts, which are not visually appealing [27]. The winner-take-all approach has the advantage of solving the problem of blurring and ghost artifacts. However, particularly in the transitions between selected images, texture seams are often perceptible. These texture seams are caused by the color dissimilarities between the images. To tackle this additional problem, robust color correction techniques are often used to color correct all images, in an attempt to create the most seamless textured 3D meshes possible. Color correction can be defined as the general problem of compensating the photometrical disparities between two coarsely geometrically registered images. In other words, color correction consists of transferring the color palette of a source image S to a target image T [28].

There are several color-correction approaches in the literature; however, the majority of them focus on correcting a single pair of images, whereas our objective is to increase the level of similarity between multiple images from a 3D model. Nonetheless, in the past few years, the color correction across image sets from the same scene has become a widely researched subject [29–31]. Apart from the different techniques to ensure the color consistency between multi-view images, color correction for both image pairs and for image sets can be defined as the same problem of adjusting the color of two or more images in order to obtain photometric consistency [31]. The methods of color correction across image sets are usually formulated as a global optimization problem to minimize the color discrepancy between all images [30]. On the other hand, the methods of color correction for image pairs are usually designed to transfer the color palette of a source image S to a target image T and are only used to color correct a single pair of images [28]. Nevertheless, we believe that pairwise-based methods can be extended to address the problem of color correcting multiple images of a scene, which is usually the case for texture mapping applications. Pairwise-based methods are naturally less complex and easier to implement than optimization methods. However, the extension of these methods to color correct several overlapping images is not straightforward.

In this light, we propose a novel pairwise-based approach that uses a clever sequence of steps, allowing for a pairwise method that was originally designed to color correct a single pair of images, to color correct multiple images from a scene. The manner in which we carry out the sequence of pairwise steps is the key to achieving an accurate color correction of multiple images, as will be detailed in Section 3.

The remainder of the paper is structured as follows: Section 2 presents the related work, concerning the main color correction approaches available in the literature and how we can contribute to the state-of-the-art; Section 3 describes the details of the proposed color correction approach; Section 4 discusses the results achieved by the proposed approach, and provides a comparison with other state-of-the-art approaches; and finally, Section 5 presents the conclusions and future work.

2. Related Work

There are two main color correction approaches: based on image pairs, or using a larger set of images. In this section, we start by reviewing image pairs color-correction techniques. Then we present techniques of color correction for image sets. At the end, a critical analysis of the state-of-the-art and a contextualization of our work is presented. Regarding color correction based on image pairs, existing techniques are divided in two classes: model-based parametric and model-less non-parametric. Model-based uses the statistical distribution of the images to guide the correction process. The authors of [32] were the pioneers with a linear transformation to model the global color distribution from a source image **S** to a target image **T**. They used Gaussian distributions (mean and standard) to correct the source image color space according to the target image. This approach is commonly used as a baseline for comparison [33–35]. Another approach [34] used the mean and covariance matrix to model the global color distribution of the images in the RGB color space. To estimate the color distribution more accurately, the authors of [36] proposed the usage of Gaussian mixture models (GMMs) and expectation maximization (EM) rather than simple Gaussian. They also proposed a local approach by modeling the color distribution of matched regions of the images. This local correction is also used in [37], where the mean shift algorithm was used to create spatially connected regions modeled as a collection of truncated Gaussians using a maximum likelihood estimation procedure. Local color correction approaches can better handle several reasons for differences between images: different color clusters in the same image, differing optics, sensor characteristics, among others [38]. A problem that may appear when correcting the three image channels independently are cross-channel artifacts. To avoid these artifacts, 3D GMMs can be used to model the color distribution in all three channels [35] or multichannel blending [39,40] since they model the three color channels simultaneously.

Model-less non-parametric approaches are usually based on joint image histogram (JIH) from the overlapped regions of a pair of images. A JIH is a 2D histogram that shows the relationship of color intensities at the exact position between a target image **T**, and a source image **S**. Then, a JIH is used to estimate a color mapping function (CMF): a function that maps the colors of a source image **S** to a target image **T** [28]. However, CMFs are highly affected by outliers that may appear due to camera exposure variation [41], vignetting effect [42], different illumination, occlusions, reflection properties of certain objects, and capturing angles, among others [28]. The radiometric response function of the camera may also affect the effectiveness of the CMF in color correcting a pair of images. To reduce this problem, the monotonicity of the CMF has to be ensured using, for example, dynamic programming [43]. In [44,45], the CMF estimation is performed with a 2D tensor voting approach followed by a heuristic local adjustment method to force monotonicity. The same authors estimated the CMF using a Bayesian framework in [46]. Other possibilities for CMF estimation are energy minimization methods [47], high-dimensional Bezier patches [48] or the use of a Vandermonde matrix to model the CMF as a polynomial [49]. A root-polynomial regression, invariant to scene irradiance and to camera exposure, was proposed in [41]. A different approach corrects images in the RGB color space with a moving least squares framework with spatial constraints [50]. Color-correction approaches for image pairs are not straightforward to extend to image set that uses three or more images for the correction.

Most color-correction approaches for image sets are based on global optimization algorithms that minimize the color difference of the overlapping regions to obtain a set of transformation parameters for each image. These approaches are common to several applications, such as image stitching, and 3D reconstruction, among others. The authors of [39] addressed the problem with a global gain compensation to minimize the color difference in the overlapped regions. This approach was proposed for image stitching and has is implemented in the *OpenCV stitching model* https://docs.opencv.org/4.x/d1 /d46/group__stitching.html (accessed on 19 November 2022) and in the panorama software *Autostitch* http://matthewalunbrown.com/autostitch/autostitch.html (accessed on

19 November 2022). In [51], a linear model was used to estimate independently three global transformations for each color channel, minimizing the color difference through histogram matching. The overlapped regions were computed using image-matching techniques, such as scale-invariant feature transform (SIFT) key points [52]. This method was implemented in *OpenMVG* https://github.com/openMVG/openMVG (accessed on 19 November 2022) [53]. Linear models might not be flexible enough to minimize high color differences between regions of the images. In this context, the authors of [54] proposed a gamma model rather than a linear model to avoid the oversaturation of luminance by applying it in the luminance channel in the YCrCb color space. In the other two chromatic channels, the authors maintained the linear model through gain compensation.

The approaches described in the previous paragraph apply the least-square loss function as the optimization solver, but this method is very sensitive to outliers and missing data. An alternative proposed in [29] estimated the gamma model parameters through low-rank matrix factorization [55]. Although the linear and gamma models achieved good results, they present some limitations with challenging datasets with high color differences, resulting in the use of other models. A quadratic spline curve, more flexible to correct significant color differences, was proposed in [56]. The same quadratic spline curves were used in [30] with additional constraints concerning image properties, such as gradient, contrast, and dynamic range. In [31], the authors estimated a global quadratic spline curve by minimizing the color variance of all points generated by the structure from motion (SfM) technique [57]. The robustness of this method was improved with the adoption of strong geometric constraints across multiview images. Moreover, to improve efficiency through large-scale image sets, the authors proposed a parallelizable hierarchical image color correction strategy based on the m-ray tree structure.

The previous paragraphs have detailed several approaches to deal with the problem of color correcting a pair of images or multi-view image sets. Many approaches have been proposed that aim to color correct image sets through global optimization processes. However, only a few of them are directed to the problems of texture mapping applications. For the case of indoor scenarios, these problems include the following: (i) overlapped regions with a very different size because of the varying distance from the cameras to the captured objects; (ii) considerable amounts of occlusions; and (iii) lack of overlapped regions. In recent papers for color correction in texture mapping applications, [31,58] tackle the problem of texture mapping for outdoor scenarios, which do not have many of the issues discussed above, such as overlapped regions with very different sizes and occlusions. On the other hand, color-correction approaches for image pairs are simpler methods, but are designed to color correct only a single pair of images, making it difficult to generalize these approaches to color correct multiple images of a scene. However, we believe that color-correction algorithms for image pairs can be adapted to color correct multiple images for two reasons: (i) these algorithms are simpler and more efficient methods because only two images at each step are handled by the algorithm, and in contrast, the optimization-based algorithms try to deal with the color inconsistencies of all images at the same time; and (ii) these algorithms could be executed in an intelligent sequence of steps across the set of images in order to increase the color consistency between all images. By using a pairwise method, we take advantage of all the years of research and all the approaches developed to color correct an image pair.

Our previous work [59] has focused on the usage of 3D information from the scene, namely point clouds and 3D mesh, to improve the effectiveness of the color correction procedure, by filtering incorrect correspondences between images from an indoor scenario. These incorrect correspondences are usually due to occlusions and different sizes of the overlapped regions and may lead to poor performance of the color correction algorithm.

In this paper, we propose a novel pairwise-based color correction approach adapted to solve the problem of color correcting multiple images from the same scene. Our approach aims to use color-correction algorithms for image pairs since we believe that, in some aspects, optimization-based methods may become overly complicated to color correct several images from large and complex scenes. The proposed approach uses a robust pairwise-based method in a clever sequence of steps, taking into consideration the amount of photometric information shared between the pairs of images. This sequential approach color corrects each image with respect to a selected reference image increasing the color consistency across the entire set of images.

3. Proposed Approach

The architecture of the proposed approach is depicted in Figure 1. As input, the system receives RGB images, point clouds from different scans, and a 3D mesh reconstructed from those point clouds. All these are geometrically registered with respect to each other. Our approach consists of 8 steps. In the first step, the faces of the 3D mesh are projected onto the images, to **compute the pairwise mappings**. Pairwise mappings are corresponding pixel coordinates from the same projected face vertices, in a pair of images. Then, several techniques are applied to **filter incorrect pairwise mappings** that would undermine both color-correction and texture-mapping processes. Once the incorrect pairwise mappings have been removed, a **color correction weighted graph is created**, containing all the pairs of images available to use in the color correction procedure. A **color correction path is computed** for each image using a path selection algorithm between each of those images and a reference image that must be selected by the user. Subsequently, we **compute a joint image histogram (JIH)** for each pair of images within the color correction paths and **estimate a color mapping function (CMF)** that best fits each JIH data. To finish the color correction procedure, we perform what we call **sequential pairwise color correction**, by using the color correction paths and the CMFs, to effectively color correct all images with respect to the reference image. At the end of this stage, the corrected RGB images are produced and then used in the last step to colorize each face of the 3D mesh through a **texture mapping image selection** algorithm, resulting in the textured 3D mesh. The information from the pairwise mappings filter is also used to increase the robustness of the image selection technique.

This paper focuses on three components of the pipeline: color correction weighted graph creation (Section 3.3), color correction paths computation (Section 3.4), and sequential pairwise color correction (Section 3.7). Other components, such as pairwise mappings computation and pairwise mappings filtering, have been addressed in detail in our previous work [59]. In this context, we consider it important for the reader to have an overview of the complete framework, which is why we offer a less detailed explanation of these procedures in Sections 3.1 and 3.2. Sections 3.3, 3.4 and 3.7 describe the core contributions of this paper in detail.

To produce the results and show the entire pipeline, we used a dataset from a laboratory at IEETA — Institute of Electronics and Informatics Engineering of Aveiro in University of Aveiro. This dataset represents an entire room and contains 24 images from different viewpoints, 9 point clouds from different scans, and a 3D mesh reconstructed from those point clouds with 41,044 faces and 21,530 vertices representing the entire room.

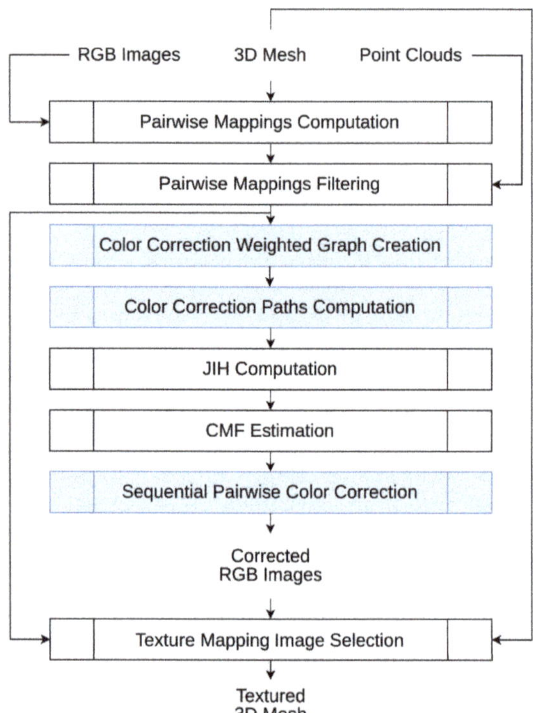

Figure 1. Architecture of the proposed approach. As input, our system receives RGB images, a 3D mesh, and point clouds from different scans, all geometrically registered to each other. In the first part, seven steps are performed to produce the corrected RGB images. Subsequently, we execute one last step to produce the textured 3D mesh. The core contributions of this paper are highlighted in blue. The non-squared elements represent data. The squared elements represent processes.

3.1. Computation of Pairwise Mappings

The geometric registration between a pair of images is usually determined based on overlapping areas, which contain photometrical information between regions of pixels that may or may not correspond to the same object. The usage of 3D information, namely 3D meshes and point clouds, to compute the correspondences between images, can be more accurate and reliable than image-based techniques, especially in indoor scenarios, which are more cluttered and contain more occlusions.

As discussed above, our previous work [59] initially proposed to use 3D meshes to compute correspondences between image pairs. For this reason, the proposed approach starts with the computation of the pairwise mappings that takes as input the RGB images and the 3D mesh. Firstly, the faces of the 3D mesh are projected onto the images. We make use of the pinhole camera model [60] to compute the projections of all vertices of the faces, resulting in pixel coordinates of the image for each projected vertex. The projection of a face onto an image is valid when two conditions are met: (i) all vertices of the face must be projected inside the width and the height of the image; and (ii) the z component of the 3D coordinates of the face vertices, transformed to the camera's coordinate reference system, must be greater than 0 to avoid the projection of vertices that are behind the camera.

After the computation of the valid projections onto all images, we compute the pairwise mappings. Pairwise mappings are pixel coordinates from a pair of images that correspond to the same projected vertices. For every pair of images, we evaluate, for each face, if the projection of that face is valid in both images. For each valid face in those conditions, we create a pairwise mapping for each vertex of that face. One pairwise mapping is represented by the pixel coordinates of the vertex projection onto the first image and onto the second image. In Figure 2, the pairwise mappings are illustrated using the same color to showcase corresponding mappings between a pair of images.

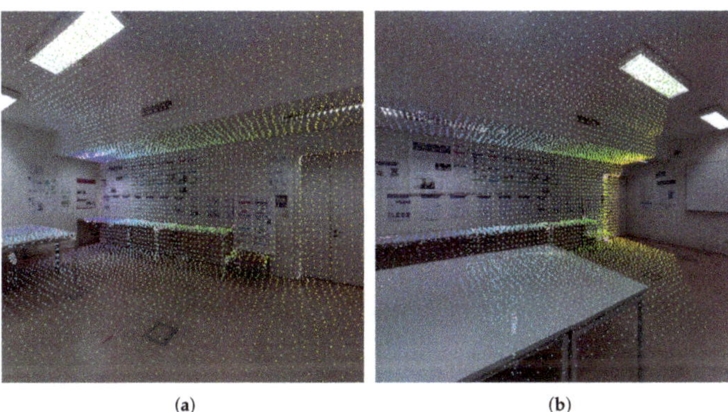

(a)   (b)

Figure 2. Pairwise mappings between images (**a**,**b**). The projections of the vertices are colored to identify the pairwise mappings between images. For example, in both images, the projections in orange represent the entrance door of the room.

3.2. Filtering of Pairwise Mappings

It is noticeable that the pairwise mappings at this stage of the pipeline contain noisy data (see Figure 2) caused by the problems mentioned in Section 2: occlusions, reflection properties of certain objects, and capturing angles, among others. For example, in Figure 2a, there are mappings represented in blue and green colors that represent the ground at the center of the room. In Figure 2b, the same mappings represent the center table. This example demonstrates inaccurate pairwise mappings between two different 3D objects. These inaccurate associations introduce incorrect photometrical correspondences between a pair of images, which would disrupt the color-correction procedure. Occluded faces and registration errors are normally the cause of inaccurate pairwise mappings, especially in indoor scenarios. A face is considered occluded from a camera point of view in the following cases: (a) the occluding face intersects the line of sight between the camera and the occluded face; and (b) the occluding face is closer to the camera. Regarding registration errors, we used a high-precision laser scanner that guarantees a very low average registration error. However, in situations where the angle between the camera viewpoint (focal axis of the camera) and the normal vector of the face is excessively oblique, or in other words, close to 90°, the impact of the registration error on the accuracy of the pairwise mappings is amplified, increasing the possibility of inaccurate pairwise mappings. In this paper, we leverage previous work [59], which uses a filtering procedure composed of a combination of 3 filters: z-buffering filtering, depth consistency filtering and camera viewpoint filtering. Each pairwise mapping is only considered valid if it passes the three filtering steps.

The z-buffering filtering is used to discard the occluded faces from an image point of view. This filter evaluates all faces to assess if they are occluded or not by other faces, considering that point of view. A face is not occluded by another face when considering an image point of view, when one of two conditions are satisfied: (i) the projection of the vertices of the evaluated face does not intersect the projection of the vertices of the other face, i.e., the faces do not overlap each other; (ii) or, in case that there is an intersection between the faces, the maximum Euclidean distance of the three vertices of the evaluated face must be less than the minimum Euclidean distance of the three vertices of the other face, i.e., the evaluated face is in front of the other face and, therefore, is not occluded.

This filter is computed for each viewpoint, i.e., for each image. Figure 3 depicts the entire filtering procedure applied in the same images shown in Figure 2 from the original pairwise mappings to the filtered pairwise mappings, illustrating the impact of each filter to eliminate the noisy data. Figure 3c,d show the pairwise mappings with only the z-buffering filter applied. It is possible to observe that in Figure 3c, the mappings on the ground at the center of the room were discarded because in Figure 3d, they are occluded by the center table.

The depth consistency filtering aims to discard the remaining occlusions due to discrepancies between the mesh and the point cloud. These discrepancies can have multiple sources, such as mesh irregularities, non-defined mesh, and registration issues, among others. This filter estimates the distance from the camera pose to the center of the face, or in other words, the depth of the face, with two different methods. The first method computes the depth of the face based on the 3D mesh. This is done by computing the L_2 norm of the vertices coordinates with respect to the camera, and then selecting the minimum value. The second method computes the depth of the face based on the point cloud. This is done by using the partial point clouds from the setups where the images were taken to project the depth information and then create depth images. However, pixels without information occur due to the sparse points of the point cloud. An image inpainting process based on the Navier–Stokes equations for fluid dynamics is used to fill the missing parts of each depth image using information from the surrounding area [61]. The depth can be directly extracted from the depth images using the coordinates obtained from the vertex projection. Finally, the algorithm compares the difference between those depth values to find depth inconsistencies and discard the inaccurate pairwise mappings. This is computed for all faces from each camera viewpoint. Figure 3e,f show the pairwise mappings with both the z-buffering filter and the depth consistency filter applied. By comparing them with Figure 3c,d, this filter discarded the mappings on the legs of the tables with respect to the wall under the table. The reason for this is that the legs of the tables are not defined on the mesh due to their reduced thickness, and yet the filter was able to discard those incorrect pairwise mappings. Additionally, the pairwise mappings on the table against the wall were discarded since the tables are not well-defined on the mesh due to difficulties in correctly acquiring both the top and bottom surfaces of the table.

The last filter is the camera viewpoint filtering that aims to remove pairwise mappings in which the angle between the camera viewpoint (focal axis of the camera) and the normal vector of the face is excessively oblique. When this occurs, the impact of the registration error on the accuracy of those pairwise mappings is amplified. For this reason, these pairwise mappings are more likely to be incorrect, and therefore should be discarded. Figure 3g,h show the pairwise mappings with all the three filtering methods applied. Regarding the camera viewpoint filtering, we can observe that the remaining pairwise mappings on the center table were discarded due to its excessive oblique angle relative to the camera viewpoint in Figure 3g. Additionally, the discarded mappings on the ground have an excessive oblique angle relative to the camera viewpoint in Figure 3h. Figure 3h clearly shows the extensive amount of noise in the dataset used. This reinforces the importance of using 3D information not only to compute the correspondences between images, but to discard the incorrect ones.

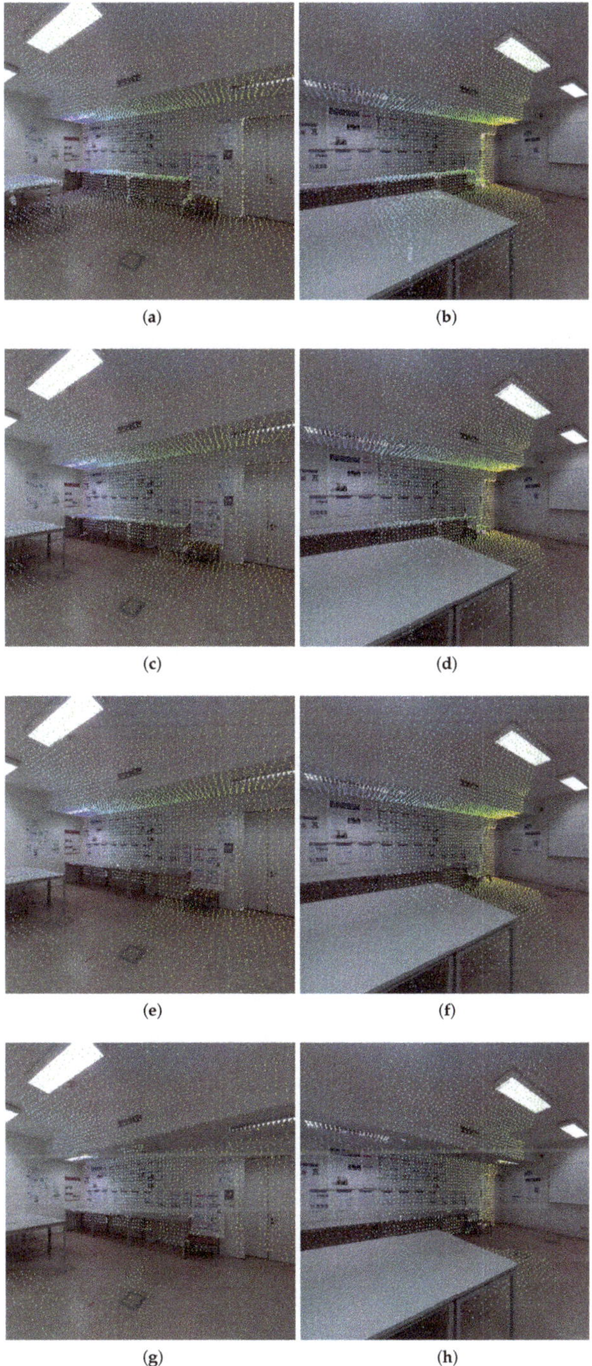

Figure 3. Pairwise mappings filtering procedure: (**a**,**b**) pairwise mappings with no filter applied; (**c**,**d**) pairwise mappings with z-buffering applied; (**e**,**f**) pairwise mappings with z-buffering filter and depth consistency applied; and (**g**,**h**) pairwise mappings with the entire filtering procedure applied.

3.3. Creation of Color-Correction Weighted Graph

For texture mapping applications, especially in indoor scenarios, there are many pairs of images that do not overlap or have minimal overlap, as described in Section 2. Those pairs share none or insufficient photometrical information, to support an accurate color correction. Whenever possible, these pairs should be avoided, and not used in the color-correction procedure. The number of pairwise mappings computed previously can be used as an indicator of this shared photometrical information between each pair of images, as detailed in Sections 3.1 and 3.2. Therefore, a threshold using the number of pairwise mappings t_{npm} can be used to discard the unfavorable pairs from the color correction procedure. By discarding these, the algorithm only carries out color correction between pairs that share enough photometrical information, leading to a more accurate color correction procedure. Furthermore, the higher the number of pairwise mappings, the higher the amount of shared photometrical information between a pair of images.

The color-correction procedure in this paper is pairwise-based. Hence, the information about all the pairs of images available to carry out the color-correction procedure, according to the criterion described above, should be represented in a data structure that supports the creation of color-correction paths between images. In this light, we propose the creation of a color correction graph G, where the nodes represent the individual images, and the edges represent all the pairs of images, which were not filtered, according to the threshold t_{npm}. G is represented as follows:

$$G = \{V, E, w\}, \tag{1}$$

where $V = \{I_1, I_2, ..., I_n\}$ are the graph nodes representing the individual images, $E \subseteq \{\{i, k\} \mid i, k \in V \ and \ i \neq k \ and \ n(\mathcal{M}_{\langle i,k \rangle}) \geq t_{npm}\}$ are the graph undirected edges representing the pairs of images available to the color-correction procedure and $w : E \to \mathbb{Q}$ are the edge weight functions. $n(\mathcal{M}_{\langle i,k \rangle})$ is the cardinality of the set that contains the pairwise mappings associated with the *i-th* and *k-th* images. It would be helpful that each pair of images contains a color-correction quality indicator that represents how suitable that pair is to contribute to a successful color-correction procedure. For example, a path-selection algorithm can use this color correction quality indicator, to decide the sequence of images to transverse, in order to produce an accurate color correction. The number of pairwise mappings seems to be an appropriate indicator of an accurate color correction for each pair of images. As such, we use it as the weights of the graph. The weights are normalized according to the total number of pairwise mappings as follows:

$$w_{ik} = 1 - \frac{n(\mathcal{M}_{\langle i,k \rangle})}{\sum n(\mathcal{M}_{\langle i,k \rangle})}, \tag{2}$$

where w_{ik} is the edge weight associated with the *i-th* and *k-th* images, and $\sum n(\mathcal{M}_{\langle i,k \rangle})$ is the sum of the cardinality of all sets of pairwise mappings.

Figure 4 shows an example of the color correction weighted graph for a subset containing six images as the nodes of the graph. The arrows represent the edges of the graph and the thickness of the edge lines represents their cost, which means that a better path will transverse thinner arrows. For example, the edge that connects the pair of images (**F**, **B**) has a low cost (thinnest arrow), and it is possible to observe that this pair contains a high amount of pairwise mappings, because most of the objects are visible in both images, such as the posters, the wall, and the table. On the other hand, between the pair of images (**B**, **A**) there are only small common regions, resulting in a high cost (thickest arrow).

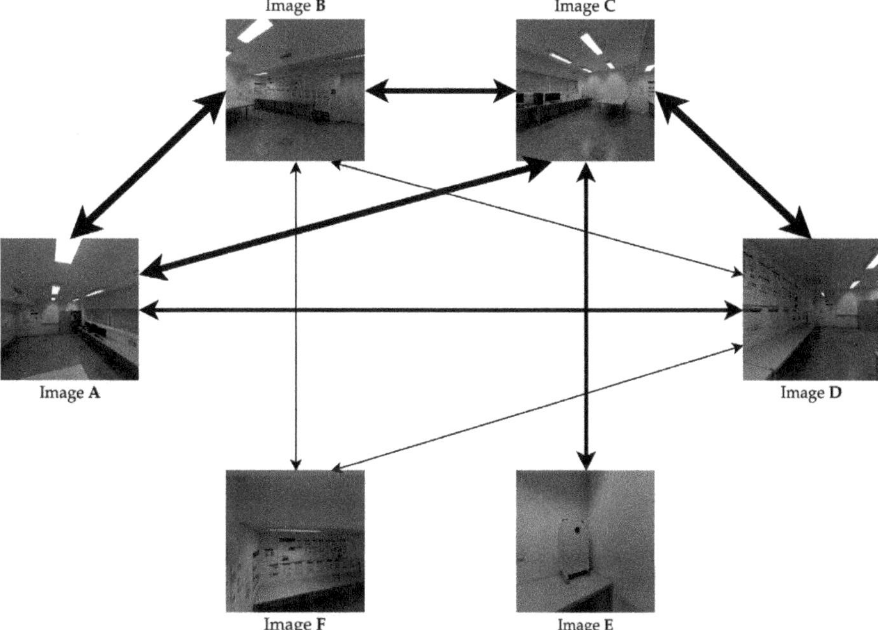

Figure 4. Example of a color-correction weighted graph. The nodes are represented by the individual images, the edges are represented by the arrows, and the thickness of the edges represents their weight.

3.4. Computation of Color-Correction Paths

We propose to select one image from the set of images as the reference image. This reference image is used as the color palette model to guide the color-correction procedure. The selection is arbitrary and is left to the user. Section 3.7 will provide details on how this selection is carried out and the advantages of this approach. With the color-correction weighted graph created in the previous section, the information about all the pairs of images available to carry out the color-correction procedure is structured in a graph. In the graph, there are images that have a direct connection (edge in the graph) with the selected reference image. In these cases, the algorithm can perform a direct color correction. On the other hand, in most cases, there are images in the graph that do not have a direct connection to the reference image. In these cases, it is not possible to perform a direct color correction. To overcome this, we propose to compute color-correction paths from each image in the set to the reference image. These paths are automatically computed from the color-correction weighted graph, described in Section 3.3. We chose the Dijkstra's shortest path algorithm [62] to find the path with the minimum cost. For this reason, the computed color-correction path for each image will have the highest sum of pairwise mappings, thus increasing the accuracy of the pairwise color correction.

We use the color-correction paths to color correct all images with reference to this reference image, resulting in images more similar in color to the reference image and also with higher color consistency between each other. Section 3.7 will provide details on how the color correction of each image of the dataset is performed through the computed color correction paths. It is noteworthy that every image from the dataset should have at least one edge in the graph, otherwise the path selection algorithm will not be able to compute. Figure 5 shows the path computed by the algorithm from image **E** (in green) to the reference image **F** (in red). Image **E** has no connection (edge) with the reference image **F**. However, the path selection algorithm selects an adequate path that goes through image pairs with high overlap.

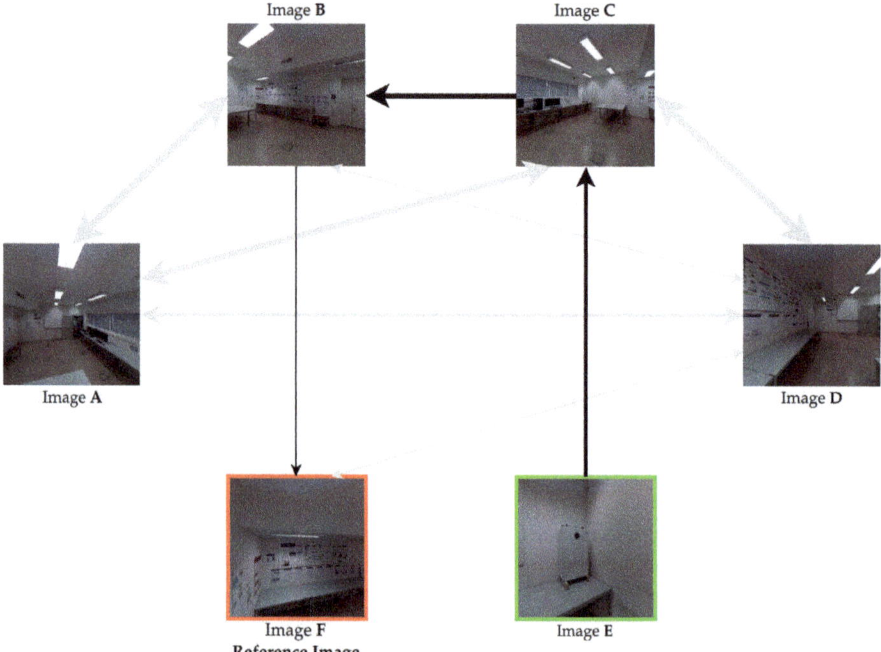

Figure 5. Example of the shortest path from image **E** to the reference image **F**, represented by the unidirectional black arrows. The nodes are represented by the individual images, the edges are represented by the arrows, and the thickness of the edges represents their cost. The bidirectional gray arrows are the edges of the graph that were not selected for this path.

3.5. Computation of Joint Image Histograms

A joint image histogram (JIH) is a 2D histogram created from the color observations of the pairwise mappings between a target image **T**, and a source image **S**. An entry in this histogram is a particular combination of color intensities for a given pairwise mapping. In this section, a JIH is computed for each channel of each pair of images in the computed color correction paths, using the pairwise mappings. A JIH is represented by $\mathbf{JIH}(x, y)$, where x and y represent all possible values of colors in **T** and **S**, being defined in the discrete interval $[0, 2^n - 1]$, where n is the bit depth of the images.

Figure 6 shows an example of a JIH of the red channel from a given pair of images. The red dots represent the observations according to the pairwise mappings, and the observation count is represented by the color intensity of each point, which means that the higher the color intensity, the higher the number of observations in that cell. To visualize the amount of noise coming from the pairwise mappings, gray dots are drawn to represent discarded pairwise mappings, but obviously are not used to estimate the CMFs. The JIHs are used to estimate the CMFs, which will then be used to transform the colors of each image from the dataset, thus producing the color-corrected images.

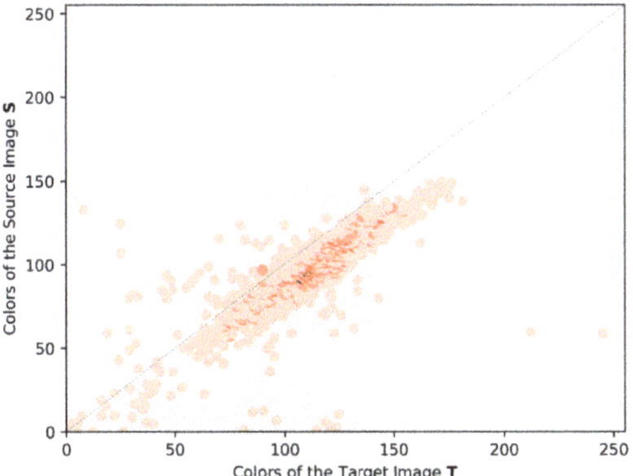

Figure 6. Example of a JIH of the red channel from a pair of images. The red dots represent the observations of the pairwise mappings in the red channel, and the color intensity represents the histogram count. The gray dots represent the discarded pairwise mappings.

3.6. Estimation of Color Mapping Functions

A color mapping function (CMF) is a function that maps the colors of a source image **S** to a target image **T**. Using this function, the colors of the target image **T** are transformed, resulting in a color corrected image $\hat{\mathbf{T}}$, which is more similar in color to the source image **S**. Each JIH created before is used as the input data to estimate a CMF. Since there are three JIHs for every pair of images combined from the color-correction paths, one for each channel of the RGB color space, there are also three CMFs.

The set of three CMFs, one for each color channel, is used to color correct all pixels of the input image. The RGB channels of the input image are color corrected independently by each one of the CMFs. In each step of the pairwise color correction, the CMF receives as input the color value x of each pixel of the target image **T** and then returns the color-corrected value $\hat{x}$. The original color value x of the pixel is transformed to the color-corrected value $\hat{x}$. After performing the color correction for all pixels of the target image and for each channel independently, the channels are merged to generate the corrected image $\hat{\mathbf{T}}$. In this context, the CMF can be formulated as

$$\hat{x} = \hat{f}(x), \quad \forall\, x \in [0, 2^n - 1],\qquad(3)$$

where $\hat{f}$ is the estimated CMF for each JIH given as input data, and $\hat{x}$ is the resultant color of the color-corrected image $\hat{\mathbf{T}}$ for a given color x of the target image **T**.

Since the CMF is a function, it cannot map a single element of its domain to multiple elements of its codomain. For the entire range of values, there is one and only one resulting corrected value $\hat{x}$ for each color of the target image x. However, for each of the values in x for a typical JIH, there are several observations of y. For this reason, it must be assumed that there is a considerable amount of noisy observations in a JIH.

This work proposes to estimate the CMF using a regression analysis to fit the color observations from the JIH. We created a composed model of the support vector regressor (SVR) [63] called composed support vector regressor (CSVR), which combines the linear kernel and the radius basis function kernel. This composed model was created because the radius basis function kernel is not able to extrapolate in the columns of the JIH where there are only a few or no observations, thus becoming very unpredictable in those columns and, most of the time, ruining the color-correction procedure. We estimate one SVR with the linear kernel and another SVR with the radius basis function kernel. Then, we discover the first intersection x_0 and the last intersection x_n between the two functions. Subsequently, we apply the linear function for the intervals $x < x_0$ and $x > x_n$, and the radius basis function for the interval $x_0 \le x \le x_n$. Figure 7 presents an example of the CMF estimation using the CSVR regression model for the given JIHs of a pair of images. The red, green, and blue curves represent the CMFs estimated for the red, green, and blue channels, respectively. For each channel, the CMF appears to fit the data of the JIH quite well, following the greater peaks of observations. The amount of noisy mappings (gray dots) eliminated by the filtering procedure is significant (see Figures 6 and 7). Without the filtering procedure, the JIHs would be more dispersed, which would cause a negative impact on the effectiveness of the regression analysis.

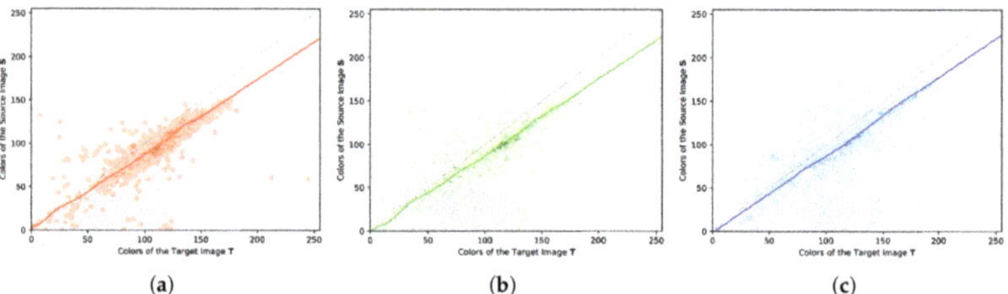

Figure 7. Example of the estimation of three CSVR Functions. (**a**) CMF estimation for the red channel represented by the red curve, (**b**) CMF estimation for the green channel represented by the green curve and (**c**) CMF estimation for the blue channel represented by the blue curve. The gray dots represent the discarded pairwise mappings.

We compute only the JIHs and the CMFs that will be used in the color-correction procedure, according to the computed color correction paths. As the number of images in a dataset increases, the combinations of pairs of images increase exponentially. For this reason, the computation burden is limited by not computing the JIHs and the CMFs for every pair of images in the dataset, as many of them are not used in the color-correction procedure.

3.7. Sequential Pairwise Color Correction

The objective of the color correction in an image pair, also called pairwise color correction, is to transform the colors of the target image, effectively making them more similar to the colors of the source image. In Section 3.4, we computed the color-correction paths between each image in the set and the reference image. We propose a sequential pairwise color correction. We use the CMFs estimated between the pairs of images formed sequentially along each color-correction path. Each path contains a list of images that must be crossed in order. The first image in the list is the image to be color corrected (target image), and the last image in the list is the reference image. We travel across the list, carrying out a color-correction step in the target image, using the CMFs of each consecutive pair of images. For each color-correction path, this approach performs n steps of pairwise color correction in a path with a depth of n. For example, let $[E, C, B, F]$ be the path shown in Figure 5 from the target image E to reference image F, passing through images C and B. For this path, (E, C), (C, B) and (B, F) are the pairs of images that were sequentially formed along it. In this context, let $\hat{f}_{C \rightarrow E}$ be the CMF from source image C to target image E, let $\hat{f}_{B \rightarrow C}$ be the CMF from source image B to target image C and finally, let $\hat{f}_{F \rightarrow B}$ be the CMF from source image F to target image B. The source and target image symbolization of each CMF is only to demonstrate the color-correction order between each pair of images. Since the path of this example has a depth of 3, the color-correction procedure will perform 3 pairwise steps using the CMFs presented above. In the first step, we color correct image E using the CMF $\hat{f}_{C \rightarrow E}(E)$, generating the first corrected image E', which is similar in color to image C. In the second step, we color correct image E' using the CMF $\hat{f}_{B \rightarrow C}(E')$, resulting in the second-corrected image E''. Since the first corrected image E' already has the color similar to image C, in this step, the second corrected image E'' became similar in color to image B. For the third and final step for this example, we apply the CMF $\hat{f}_{F \rightarrow B}(E'')$ to color correct the image E'', producing the third and final corrected image E'''. After the three color-correction steps performed in image **E**, finally the final corrected image **E'''** is similar in color to the reference image **F**. To summarize the sequential pairwise color-correction process in this example, the sequential use of the three CMFs to produce the final corrected image **E'''** can be expressed as

$$\mathbf{E'''} = \hat{f}_{\mathbf{F} \to \mathbf{B}}\left(\hat{f}_{\mathbf{B} \to \mathbf{C}}\left(\hat{f}_{\mathbf{C} \to \mathbf{E}}(\mathbf{E}) \right) \right). \tag{4}$$

Figure 8 depicts the three steps performed in this example by the sequential pairwise color correction to produce the final corrected image **E'''**. On the first row are the four images that composed the color correction path from target image **E** (in green) to reference image **F** (in red), from left to right. On the second row are the color-corrected images **E'**, **E''** and **E'''**, after each step of the sequential pairwise color correction.

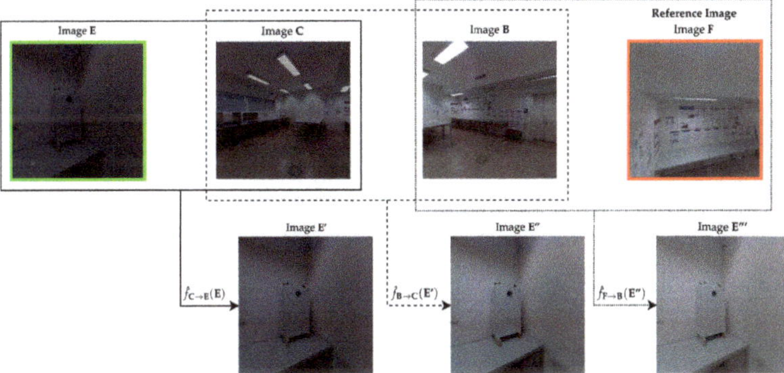

Figure 8. Example of the sequential pairwise color correction for a color correction path from a target image (in green) to a reference image (in red). On the first row are the four images that compose the color-correction path. On the second row are the color-corrected images, after each step of the sequential pairwise color correction. Each box (solid, dashed and dotted) represents a step of the sequential pairwise color correction, showing the CMF used and its input image. To explain our point, we drastically changed the brightness of the images on purpose, because some changes in the color of the images along the steps of the color correction procedure are not always easy to notice.

In short, for each one of the computed paths, we perform the sequential pairwise color correction described above, producing the final corrected images, which are more similar in color to the reference image. If the target image has no color-correction path to the reference image, no color correction is performed and the image remains the same. Note that if the graph is fully connected and all the computed color-correction paths have a depth of 1, the problem is reduced to a simple pairwise color correction, meaning that each image is color corrected directly with the selected reference image. This special case occurs when all the images share a sufficient amount of photometrical information with each other, meaning that all pairs of images can be reliably used in the color-correction procedure.

The assumption behind the selection of the reference image is to have a model image that is selected using any arbitrary criterion to transform the colors of all images to be more similar to the reference one. With this arbitrary selection, the user can have control over the overall appearance of the color corrected mesh, and for that reason, we understand this approach as an advantage rather than a shortcoming of the proposed approach.

3.8. Image Selection for Texture Mapping

The main goal of this paper is not only to produce color-corrected images, but also to improve the visual quality of a textured mesh. To produce the textured meshes using the color-corrected images, we implemented two image selection methods: random selection and largest projection area selection. Both methods select, based on a specific criterion, one image among all available to colorize each face. This process is based on cropping, from the selected image, the projection of the face. Regarding the criterion used for each method, the random selection picks a random image among all available to colorize that face. As for the largest projection area selection, its criterion consists of selecting the image where the area of the face projection is the largest from the available ones. Furthermore, the usage of the information associated with the filtering of pairwise mappings (Section 3.2) is also important to produce high-quality textured meshes, because it assists the image-selection algorithm to avoid coloring a face with an image in which that face is occluded or with an excessively oblique viewpoint. In this context, this information is incorporated in the image-selection algorithms.

Figure 9 shows the textured meshes produced by both image-selection methods, using the original images and the color-corrected images. Figure 9a,d illustrate the image-selection methods by coloring each face with a representative color for each selected image, meaning that faces with the same color have the same selected image. The random selection method, shown in Figure 9a, uses several images to colorize the faces that represent the same surface, resulting in several texture seams artifacts. This method clearly is not the smartest approach in order to produce high-quality textured meshes; however, it facilitates the analysis of the impact of the color-corrected images in the textured mesh because it amplifies the perception of the color dissimilarities between the images. Figure 9b,c present the textured mesh produced by the random selection algorithm, using the original images and the color-corrected images, respectively. Comparing both images, the proposed approach increased considerably the visual quality of the textured mesh, eliminating most of the texture seams. As for the largest projection area selection, shown in Figure 9d, this method uses a clever criterion, producing higher-quality textured meshes with fewer image transitions throughout a surface. Figure 9e,f show the textured mesh produced by the largest projection area algorithm in the same viewpoint as for the previous algorithm, using the original images and the color-corrected images, respectively. The color-correction images combined with a better image selection algorithm clearly produced a high-quality textured mesh. Note that since the 3D mesh is a closed room, we used the back face culling method [64] to hide the faces of the mesh which are facing away from the point of view. It is possible to observe that our proposed color-correction approach produces higher-quality textured meshes, reducing significantly the number of unappealing texture seam artifacts.

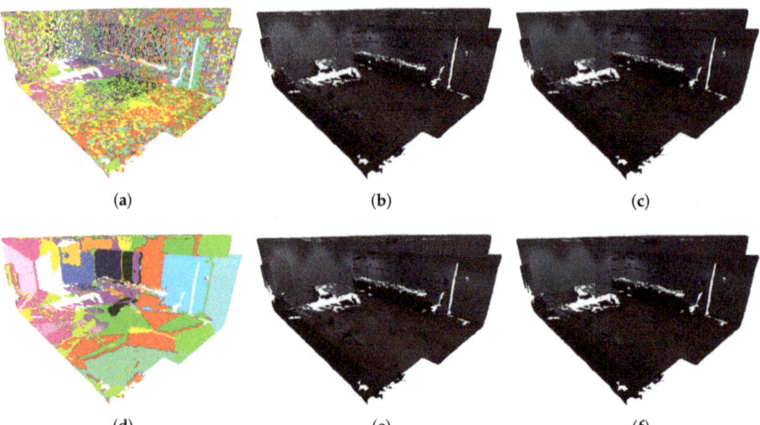

Figure 9. Textured meshes produced by each image selection method. (**a**,**d**) The result of the random selection and the largest projection area selection methods, respectively. (**b**,**c**) The resultant textured meshes produced by the random selection criterion using the original images and the color corrected images, respectively. (**e**,**f**) The resultant textured meshes produced by the largest projection area criterion using the original images and the color corrected images, respectively.

4. Results

In this section, we analyze the influence of the proposed approach on the visual quality of the textured 3D meshes. We also compare the proposed approach with state-of-the-art approaches in quantitative and qualitative evaluations. Section 4.1 presents an image-based qualitative evaluation. Section 4.2 presents an image-based quantitative evaluation. Finally, Section 4.3 presents a mesh-based qualitative evaluation.

There are no metrics that evaluate the overall visual quality of textured meshes. As such, we base our quantitative evaluation on the assessment of how similar in color the images that are used to produce the textured meshes are. To do this, we use two image-similarity metrics to evaluate the color similarity: peak signal-to-noise ratio ($PSNR$) and $CIEDE2000$. We compute the color similarity, not between the images, but instead between pairs of associated pixels of both images. This is done using the pairwise mappings discussed in Section 3.1. To carry out a robust evaluation, the metrics use the filtered pairwise mappings, to avoid incorrect associations (see Section 3.2). The $PSNR$ metric [28] measures color similarity, meaning that the higher the score values, the more similar the images. The $PSNR$ metric between image **A** and image **B** can be formulated as

$$PSNR(\mathbf{A}, \mathbf{B}) = 20 * \log_{10}(L/RMS), \tag{5}$$

where L is the largest possible value in the dynamic range of an image, and RMS is the root mean square difference between the two images. The $CIEDE2000$ metric [65] was adopted as the most recent color difference metric from the International Commission on Illumination (CIE). This metric measures color dissimilarity, meaning that the lower the score values, the more similar the images. The $CIEDE2000$ metric between image **A** and image **B** can be formulated as

$$CIEDE2000(\mathbf{A}, \mathbf{B}) = \sqrt{\left(\frac{\Delta L'}{k_L S_L}\right)^2 + \left(\frac{\Delta C'}{k_C S_C}\right)^2 + \left(\frac{\Delta H'}{k_H S_H}\right)^2 + R_T \left(\frac{\Delta C'}{k_C S_C}\right)\left(\frac{\Delta H'}{k_H S_H}\right)}, \tag{6}$$

where $\Delta L'$, $\Delta C'$ and $\Delta H'$ are the lightness, chroma and hue differences, respectively. S_L, S_C and S_H are compensations for the lightness, chroma and hue channels, respectively. k_L, k_C and k_H are constants for the lightness, chroma and hue channels, respectively. Finally, R_T is a hue rotation term. Since that both $PSNR$ and $CIEDE2000$ are pairwise image metrics, we carry out the evaluation over all image pairs in the dataset. The displayed score values are the simple mean and the standard deviation for all image pairs. Additionally, we present a weighted mean based on the number of pairwise mappings of each pair of images. In this weighted mean, image pairs with higher number of mappings are more important to the final score value.

The qualitative evaluation of each color-correction algorithm is performed on both the images and the textured meshes. For the image-based assessment, we compare the images with respect to each other. For the mesh-based evaluation, the visual quality of the textured meshes produced by the approaches is analyzed, from different viewpoints, using both image-selection criteria discussed in Section 3.8.

Table 1 lists all the evaluated algorithms in this paper. Algorithm #1 is the baseline approach: the original images without any color correction. Algorithms #2 through #4 are pairwise color correction approaches, designed to color correct single pairs of images. As discussed in Section 2, pairwise approaches are not able to tackle the problem of the color correction of image sets. However, we presented a graph-based approach that is able to convert a multi-image color correction problem into a set of simple pairwise color correction problems. In this way, we can use classic pairwise color-correction approaches for comparison purposes. Algorithm #2 uses simple Gaussians to model the global color distribution from a source image onto a target image, in the $l\alpha\beta$ color space [32]. Algorithm #3 estimates the CMFs using a root-polynomial regression [41]. Algorithm #4 uses the Vandermonde matrix to compute the coefficients of a polynomial that is applied as a CMF [49]. Algorithms #5 and #6 are color-correction approaches for image sets (three or more images) and are based on global optimization methods. These are both often used as baseline methods for evaluating color correction approaches for image sets [30,31,54,58,66,67]. Algorithm #5 employs an optimization method to determine a global gain compensation to minimize the color difference in the overlapped regions of the images [39]. Algorithm #6 estimates three global transformations per image, one for each color channel, by minimizing the color difference between overlapped regions through histogram matching [51]. Finally, Algorithm #7 is the proposed approach—sequential pairwise color correction approach. As described in Section 3, the threshold using the number of pairwise mappings t_{npm} is the only tunable parameter for the experiments conducted in this section. We used $t_{npm} = 400$, a value obtained by experimenting several possibilities and evaluating the result.

Table 1. State-of-the-art algorithms compared with the proposed approach.

Algorithm #	Reference	Description
#1	-	Baseline (original images)
#2	Reinhard et al. [32]	Global Color Transfer—Pairwise Approach
#3	Finlayson et al. [41]	Root-Polynomial Regression—Pairwise Approach
#4	De Marchi et al. [49]	Vandermonde Matrices—Pairwise Approach
#5	Brown and Lowe [39]	Gain Compensation—Optimization Approach
#6	Moulon et al. [51]	Histogram Matching—Optimization Approach
#7	This paper	Sequential Pairwise Color Correction

4.1. Image-Based Qualitative Evaluation

A qualitative evaluation of the approaches (Table 1) is presented to analyze the color similarity between the images, once the color-correction procedure is completed. After the color-correction procedure performed by each approach, it is expected that the color similarity between all images of the set has increased. Figure 10 shows the corrected images produced by each approach. Image A was selected as the reference image. We selected three other images (B, C, and D) of the 23 images available, for visualization purposes. Algorithm #1 is the baseline approach, and thus the target images B, C, and D are the original ones without any color correction. Algorithm #2 increased the contrast of the images, but did not increase the color similarity with respect to the reference image. Algorithms #3 and #4 did not produce satisfactory results since the images have degenerate colors. We believe that this is due to occlusions in the scene, which result in incorrect image correspondences. Note that these two approaches have internal image correspondence procedures, and therefore, they are not taking advantage of the filtering procedure detailed in Section 3.2. In algorithms #5 and #6, which are multi-image global optimization approaches, changes are hardly noticeable in the color corrected images, compared with the original images. The probable cause is the different sizes of the overlapped regions and the occlusions that usually occur in indoor scenarios, which produce incorrect correspondences and do not allow the optimizer to significantly reduce the color differences between all images. Another probable cause is because these algorithms use linear models, which in most of the cases are not flexible enough to deal with high color discrepancies. The proposed approach #7 produced the images with higher color similarity with respect to the reference image. Note that some images have no overlapping region with the reference image (image D), or only a small overlapping region (images B and C).

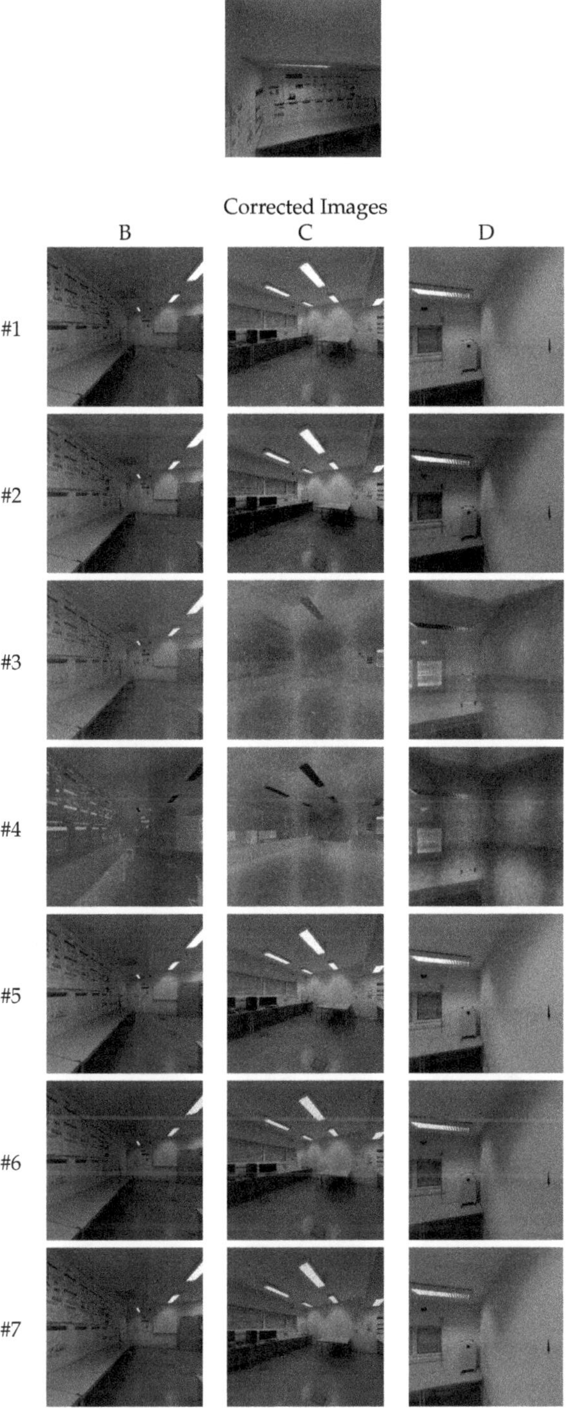

Figure 10. Image-based qualitative evaluation: corrected images **B–D** (columns) produced by each algorithm (rows), using image **A** as reference image.

The complexity of this dataset shows that the sequential pairwise color correction, alongside a reliable pairwise mappings computation and a robust CMF estimation, played a major role to increase the color similarity between the images and the reference image. Note that this complexity is due to several images from the same scene with no overlapping regions, and several occlusions in the scene.

4.2. Image-Based Quantitative Evaluation

Table 2 presents the results of the image-based quantitative evaluation, for each approach. Algorithm #1 is the baseline approach, i.e., shows the level of similarity between the original images. Pairwise approaches (algorithms #2 through #4) worsened the color similarity between the images in both metrics. Those algorithms are designed to color correct only single pairs of images and, therefore, did not produce accurate results across the image set. Multi-image optimization-based algorithms #5 and #6 achieved good results and increased the color similarity between the images. However, the linear models used in these two algorithms are not flexible, and this is a probable reason why they did not achieve better results. The proposed approach #7 outperformed all other approaches. In the $PSNR$ metric, we observe an improvement in the similarity between the images of around 16% (simple mean) and 13% (weighted mean) relative to the original images. In the $CIEDE2000$ metric, the color similarity is improved by around 29% in the simple mean and by around 26% in the weighted mean. These results prove that the proposed pairwise-based method is able to color correct several images from a same scene, and to achieve better results than state-of-the-art multi-image global optimization approaches.

Table 2. Image-based quantitative evaluation: simple mean, weighted mean, and standard deviations of the $PSNR$ and $CIEDE2000$ scores of all images for each algorithm (rows). The best results are highlighted in bold.

Alg.	$PSNR$			$CIEDE2000$		
	μ	μ_w	σ	μ	μ_w	σ
#1	24.72	24.28	3.22	6.02	5.83	1.48
#2	21.28	21.51	4.69	9.09	8.16	4.64
#3	17.33	16.74	4.46	16.28	17.30	7.18
#4	17.14	16.94	3.71	17.04	17.22	6.28
#5	26.22	25.45	3.10	5.23	5.14	1.09
#6	26.91	26.40	3.21	5.27	5.05	1.18
#7	**28.69**	**27.45**	3.51	**4.29**	**4.33**	0.83

4.3. Mesh-Based Qualitative Evaluation

In this section, we evaluate and compare the visual quality of the textured meshes produced by each algorithm (Table 1). As described in Section 3.8, we use two different image selection criteria to carry out the texture mapping process: random image selection and largest projection area image selection. Figure 11 points out with the red arrows the parts of the 3D mesh where the viewpoints were taken. The first viewpoint (Figure 11a) observes a corner of the room, with a high range of colors due to the posters. The second viewpoint (Figure 11b) sees a part of the room with a variety of objects (and color palettes), such as a cabinet, a whiteboard, and a few posters. Finally, the last viewpoint (Figure 11c) shows a wall that has a high luminosity from the lights of the room. The white regions of each viewpoint are holes (no triangles) in the geometry of the 3D meshes. These viewpoints show the highest difficulties of this dataset in terms of color differences. This dataset contains 24 images, and thus, the amount of transitions between images is very high, especially using the random image selection criterion. For that reason, the textured meshes using the original images contain a high number of texture seams. Figure 9a,d show where the transitions occur in the mesh, using representative colors. Figures 12 and 13 show three different viewpoints using the random and largest projection area image selection criteria, respectively.

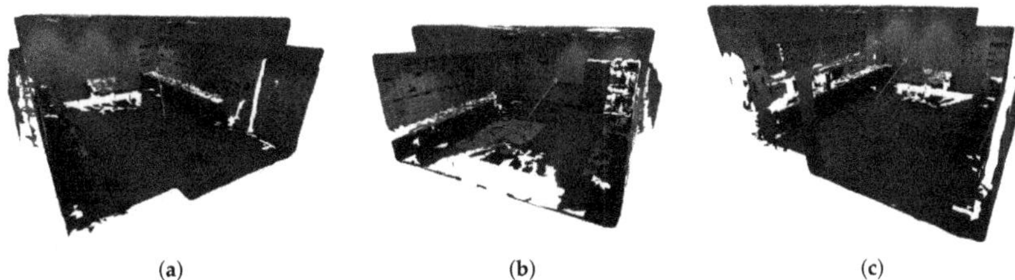

(a)   (b)   (c)

Figure 11. The three viewpoints analyzed in the mesh-based qualitative evaluation. The viewpoints are exemplified in the textured mesh produced by the baseline algorithm (#1) using the random image selection criterion. (**a–c**) Red arrows point out the parts of the 3D mesh where the viewpoints 1, 2, and 3 were taken, respectively.

Analyzing the original images—algorithm #1 (see Figure 12, first row, and Figure 13, first row), we can see a high number of texture seams due to the color inconsistency between the images. Even with the largest projection area image selection criterion, the texture presents many visual artifacts, especially in viewpoint 3. Algorithm #2 increased the color difference between the images, resulting in even more noticeable texture seams, especially in views 1 and 3. Furthermore, as was expected, the textured meshes produced by algorithms #3 and #4 are very degenerated due to the unsuccessful color correction of the images. Algorithms #5 and #6 were also ineffective and reduced the texture seams in some specific regions, while aggravating others. For example, in Figure 12, at the top-left part of the front wall in viewpoint 3, algorithm #6 reduced the texture seams. However, on the right side, the same algorithm aggravated even more the texture seams. The proposed approach #7 outperformed all other algorithms and produced the highest-quality textured meshes with almost no texture seams, even using the random image selection criterion. These results are consistent with the image-based evaluation (see Sections 4.1 and 4.2), that showed the proposed approach obtaining images with more similar color. These results also prove the robustness of the proposed approach, and show it is able to produce high-quality textured meshes, from datasets with many challenges, such as high amount of occlusions, high color range complexity, overlapped regions with varied sizes, and lack of overlapped regions between several pairs of images, among others.

Random Image Selection

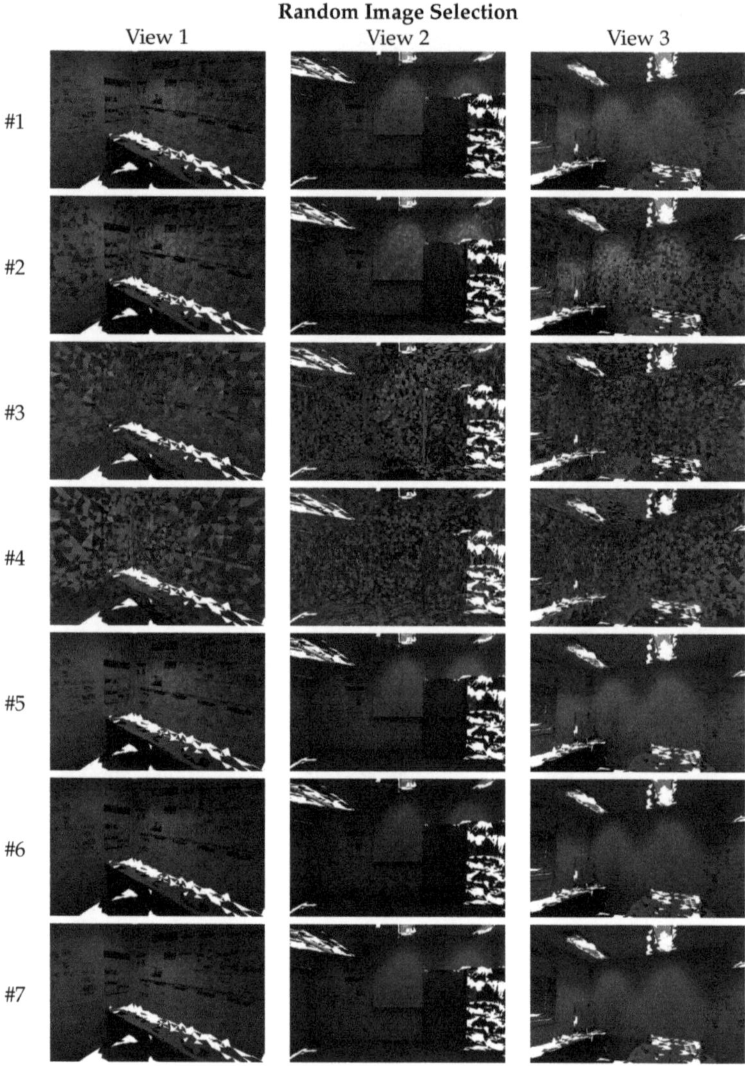

Figure 12. Mesh-based qualitative evaluation: textured meshes produced by each approach (rows) from three different viewpoints, and using the **random image selection technique**.

Largest Projection Area Image Selection

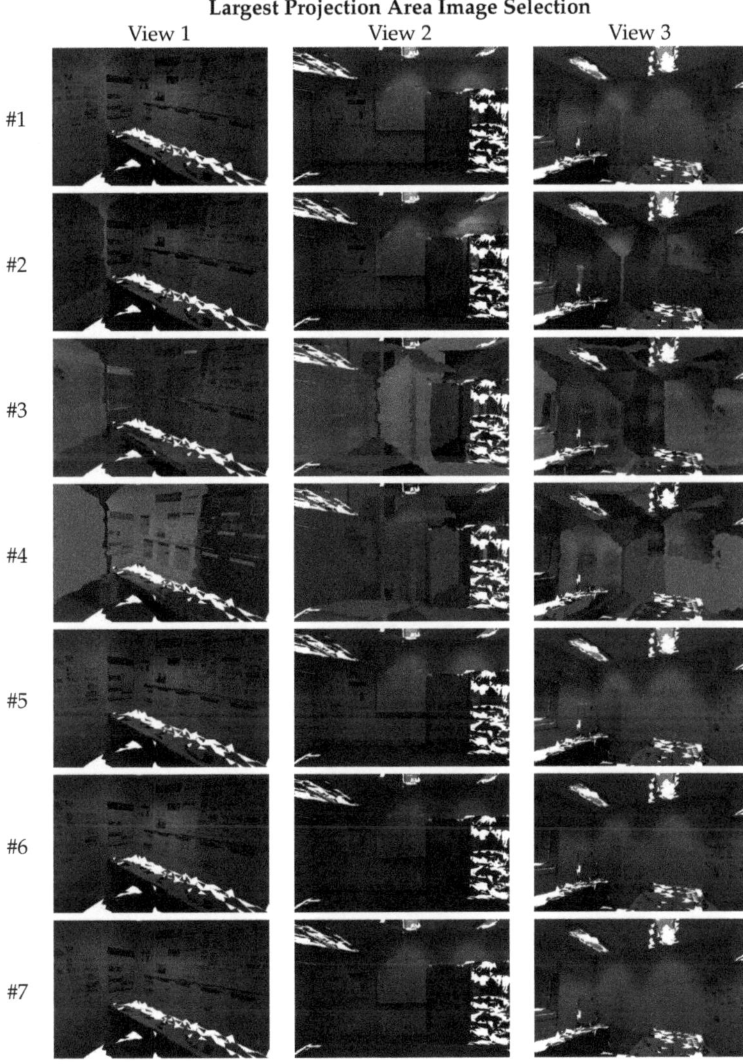

Figure 13. Mesh-based qualitative evaluation: textured meshes produced by each approach (rows) from three different viewpoints, and using the **largest projection area image selection technique**.

5. Conclusions

The key contribution of this paper is a novel sequential pairwise color-correction approach capable of color correcting multiple images from the same scene. The sequential pairwise color-correction approach consists of selecting one image as the reference color palette model and computing a color correction path, from each image to the reference image, through a weighted graph. The color-correction weighted graph is based on the number of pairwise mappings between image pairs. Each image is color corrected by sequentially applying the CMFs of each consecutive pair of images along a color-correction path. CMFs are computed using a regression analysis called composed support vector regressor (CSVR). This procedure increases not only the color similarities of all images with respect to the reference image, but also the color similarities across all images in the set.

Results demonstrate that the proposed approach outperforms other methods in an indoor dataset using both qualitative and quantitative evaluations. More importantly, results show that the color-corrected images improved the visual quality of the textured meshes. The approach is able to correct the color differences between all images of the dataset; this is even more noticeable when using a random image selection criterion for texture mapping, a method that creates a high number of transitions between adjacent triangles. Even in this condition, the color correction results in high-quality textured meshes with little and hardly noticeable texture.

For future work, we plan to explore different metrics for the color-correction weighted graph, such as the standard deviation of a 2D Gaussian that fits the JIH data. We also intend to explore different path-selection algorithms for the color-correction paths. Furthermore, we plan to acquire larger and more complex datasets, for example, several rooms of a building, to further investigate the impact of the sequential pairwise color-correction approach, and better assess its robustness.

Author Contributions: Conceptualization, P.D. and M.O.; Funding acquisition, P.D. and M.O.; Methodology, L.D., D.C., P.D. and M.O.; Software, L.D., D.C., T.M. and M.O.; Writing—original draft, L.D.; Writing—review and editing, D.C., T.M., P.D. and M.O. All authors have read and agreed to the published version of the manuscript.

Funding: This work has been partially supported by Portugal 2020, project I&DT n° 45382 in collaboration with VR360, by National Funds through the FCT - Foundation for Science and Technology, in the context of the project UIDB/00127/2020, by FCT - Portuguese Foundation for Science and Technology, in the context of Ph.D. scholarship 2020.07345.BD, and by FCT - Portuguese Foundation for Science and Technology, in the context of Ph.D. scholarship 2022.10977.BD.

Institutional Review Board Statement: Not applicable.

Informed Consent Statement: Not applicable.

Data Availability Statement: Not applicable.

Acknowledgments: We thank everyone involved for their time and expertise.

Conflicts of Interest: The authors declare no conflict of interest.

Abbreviations

The following abbreviations are used in this manuscript:

BTF	Brightness Transfer Function
CMF	Color Mapping Function
CSVR	Composed Support Vector Regressor
EM	Expectation Maximization
GMMs	Gaussian Mixture Models
CIE	International Commission on Illumination
JIH	Joint Image Histogram
LiDAR	Light Detection and Ranging
PSNR	Peak Signal-to-Noise Ratio
S-CIELAB	Spatial CIELAB
SfM	Structure from Motion
SIFT	Scale-Invariant Feature Transform
SVR	Support Vector Regressor

References

1. Pérez, L.; Rodríguez, Í.; Rodríguez, N.; Usamentiaga, R.; García, D.F. Robot Guidance Using Machine Vision Techniques in Industrial Environments: A Comparative Review. *Sensors* **2016**, *16*, 335. [CrossRef] [PubMed]
2. Zhang, Y.; Chen, H.; Waslander, S.L.; Yang, T.; Zhang, S.; Xiong, G.; Liu, K. Toward a More Complete, Flexible, and Safer Speed Planning for Autonomous Driving via Convex Optimization. *Sensors* **2018**, *18*, 2185. [CrossRef] [PubMed]

3. Ma, X.; Wang, Z.; Li, H.; Zhang, P.; Ouyang, W.; Fan, X. Accurate Monocular 3D Object Detection via Color-Embedded 3D Reconstruction for Autonomous Driving. In Proceedings of the 2019 IEEE/CVF International Conference on Computer Vision (ICCV), IEEE, Seoul, Republic of Korea, 27 October–2 November 2019; pp. 6850–6859.

4. Di Angelo, L.; Di Stefano, P.; Guardiani, E.; Morabito, A.E.; Pane, C. 3D Virtual Reconstruction of the Ancient Roman Incile of the Fucino Lake. *Sensors* **2019**, *19*, 3505. [CrossRef] [PubMed]

5. Vázquez-Arellano, M.; Griepentrog, H.W.; Reiser, D.; Paraforos, D.S. 3-D Imaging Systems for Agricultural Applications—A Review. *Sensors* **2016**, *16*, 618. [CrossRef]

6. Shen, Z.; Ding, F.; Jolfaei, A.; Yadav, K.; Vashisht, S.; Yu, K. DeformableGAN: Generating Medical Images With Improved Integrity for Healthcare Cyber Physical Systems. *IEEE Trans. Netw. Sci. Eng.* **2022**, 1–13. [CrossRef]

7. Afanasyev, I.; Sagitov, A.; Magid, E. ROS-based SLAM for a gazebo-simulated mobile robot in image-based 3D model of Indoor environment. *Lect. Notes Comput. Sci.* **2015**, *9386*, 273–283. [CrossRef]

8. Newcombe, R.A.; Izadi, S.; Hilliges, O.; Molyneaux, D.; Kim, D.; Davison, A.J.; Kohli, P.; Shotton, J.; Hodges, S.; Fitzgibbon, A. KinectFusion: Real-time dense surface mapping and tracking. In Proceedings of the 2011 10th IEEE International Symposium on Mixed and Augmented Reality, ISMAR 2011, Basel, Switzerland, 26–29 October 2011; pp. 127–136. [CrossRef]

9. Han, J.; Shao, L.; Xu, D.; Shotton, J. Enhanced computer vision with Microsoft Kinect sensor: A review. *IEEE Trans. Cybern.* **2013**, *43*, 1318–1334. [CrossRef]

10. Fan, H.; Yao, W.; Fu, Q. Segmentation of sloped roofs from airborne LiDAR point clouds using ridge-based hierarchical decomposition. *Remote Sens.* **2014**, *6*, 3284–3301. [CrossRef]

11. Henn, A.; Gröger, G.; Stroh, V.; Plümer, L. Model driven reconstruction of roofs from sparse LIDAR point clouds. *ISPRS J. Photogramm. Remote Sens.* **2013**, *76*, 17–29. [CrossRef]

12. Xu, H.; Chen, C.Y. 3D Facade Reconstruction Using the Fusion of Images and LiDAR: A Review. In *Proceedings of the New Trends in Computer Technologies and Applications*; Chang, C.Y., Lin, C.C., Lin, H.H., Eds.; Springer: Singapore, 2019; pp. 178–185.

13. Becker, S.; Haala, N. Refinement of building fassades by integrated processing of LIDAR and image data. In Proceedings of the PIA 2007—Photogrammetric Image Analysis, Proceedings, Munich, Germany, 19–21 September 2007; pp. 7–12.

14. Li, B.; Zhao, Y.; Wang, X. A 2D face image texture synthesis and 3D model reconstruction based on the Unity platform. In Proceedings of the 2020 IEEE International Conference on Artificial Intelligence and Computer Applications (ICAICA), IEEE, Dalian, China, 27–29 June 2020; pp. 870–873. [CrossRef]

15. Bhattad, A.; Dundar, A.; Liu, G.; Tao, A.; Catanzaro, B. View Generalization for Single Image Textured 3D Models. In Proceedings of the 2021 IEEE/CVF Conference on Computer Vision and Pattern Recognition (CVPR), IEEE, Virtual, 20–25 June 2021; pp. 6077–6086.

16. Fu, Y.; Yan, Q.; Liao, J.; Zhou, H.; Tang, J.; Xiao, C. Seamless Texture Optimization for RGB-D Reconstruction. *IEEE Trans. Vis. Comput. Graph.* **2021**, *14*, 1. [CrossRef]

17. He, H.; Yu, J.; Cheng, P.; Wang, Y.; Zhu, Y.; Lin, T.; Dai, G. Automatic, Multiview, Coplanar Extraction for CityGML Building Model Texture Mapping. *Remote Sens.* **2021**, *14*, 50. [CrossRef]

18. Marroquim, R.; Pfeiffer, G.; Carvalho, F.; Oliveira, A.A.F. Texturing 3D models from sequential photos. *Vis. Comput.* **2012**, *28*, 983–993. [CrossRef]

19. Liu, L.; Nguyen, T.Q. A Framework for Depth Video Reconstruction From a Subset of Samples and Its Applications. *IEEE Trans. Image Process.* **2016**, *25*, 4873–4887. [CrossRef]

20. Neugebauer, P.J.; Klein, K. Texturing 3D Models of Real World Objects from Multiple Unregistered Photographic Views. *Comput. Graph. Forum* **2001**, *18*, 245–256. [CrossRef]

21. Guo, Y.; Wan, J.; Zhang, J.; Xu, K.; Lu, M. Efficient registration of multiple range images for fully automatic 3D modeling. In Proceedings of the 2014 International Conference on Computer Graphics Theory and Applications (GRAPP), Lisbon, Portugal, 5–8 January 2014; pp. 1–8.

22. Lempitsky, V.S.; Ivanov, D.V. Seamless Mosaicing of Image-Based Texture Maps. In Proceedings of the 2007 IEEE Conference on Computer Vision and Pattern Recognition, Minneapolis, MN, USA, 17–22 June 2007; pp. 1–6.

23. Allène, C.; Pons, J.P.; Keriven, R. Seamless image-based texture atlases using multi-band blending. In Proceedings of the 2008 19th International Conference on Pattern Recognition, Tampa, FL, USA, 8–11 December 2008; pp. 1–4.

24. Gal, R.; Wexler, Y.; Ofek, E.; Hoppe, H.; Cohen-Or, D. Seamless Montage for Texturing Models. *Comput. Graph. Forum* **2010**, *29*, 479–486. [CrossRef]

25. Callieri, M.; Cignoni, P.; Corsini, M.; Scopigno, R. Masked Photo Blending: Mapping dense photographic dataset on high-resolution 3D models. *Comput. Graph.* **2008**, *32*, 464–473. [CrossRef]

26. Kehl, W.; Navab, N.; Ilic, S. Coloured signed distance fields for full 3D object reconstruction. In Proceedings of the British Machine Vision Conference 2014, Nottingham, UK, 1–5 September 2014.

27. Oliveira, M.; Lim, G.H.; Madeira, T.; Dias, P.; Santos, V. Robust texture mapping using rgb-d cameras. *Sensors* **2021**, *21*, 3248. [CrossRef] [PubMed]

28. Xu, W.; Mulligan, J. Performance evaluation of color correction approaches for automatic multi-view image and video stitching. In Proceedings of the IEEE Computer Society Conference on Computer Vision and Pattern Recognition, San Francisco, CA, USA, 13–18 June 2010; pp. 263–270. [CrossRef]

29. Park, J.; Tai, Y.W.; Sinha, S.N.; Kweon, I.S. Efficient and robust color consistency for community photo collections. In Proceedings of the IEEE Computer Society Conference on Computer Vision and Pattern Recognition, Las Vegas, NV, USA, 27–30 June 2016; pp. 430–438. [CrossRef]

30. Xia, M.; Yao, J.; Gao, Z. A closed-form solution for multi-view color correction with gradient preservation. *ISPRS J. Photogramm. Remote Sens.* **2019**, *157*, 188–200. [CrossRef]

31. Yang, J.; Liu, L.; Xu, J.; Wang, Y.; Deng, F. Efficient global color correction for large-scale multiple-view images in three-dimensional reconstruction. *ISPRS J. Photogramm. Remote Sens.* **2021**, *173*, 209–220. [CrossRef]

32. Reinhard, E.; Ashikhmin, M.; Gooch, B.; Shirley, P. Color transfer between images. *IEEE Comput. Graph. Appl.* **2001**, *21*, 34–41. [CrossRef]

33. Zhang, M.; Georganas, N.D. Fast color correction using principal regions mapping in different color spaces. *Real-Time Imaging* **2004**, *10*, 23–30. [CrossRef]

34. Xiao, X.; Ma, L. Color transfer in correlated color space. In Proceedings of the VRCIA 2006: ACM International Conference on Virtual Reality Continuum and Its Applications, Hong Kong, China, 14–17 June 2006; pp. 305–309. [CrossRef]

35. Oliveira, M.; Sappa, A.D.; Santos, V. Color correction for onboard multi-camera systems using 3D Gaussian mixture models. In Proceedings of the 2012 IEEE Intelligent Vehicles Symposium, Madrid, Spain, 3–7 June 2012; pp. 299–303. [CrossRef]

36. Tai, Y.W.; Jia, J.; Tang, C.K. Local color transfer via probabilistic segmentation by expectation-maximization. In Proceedings of the 2005 IEEE Computer Society Conference on Computer Vision and Pattern Recognition, CVPR 2005, San Diego, CA, USA, 20–25 June 2005; Volume I, pp. 747–754. [CrossRef]

37. Oliveira, M.; Sappa, A.D.; Santos, V. A probabilistic approach for color correction in image mosaicking applications. *IEEE Trans. Image Process.* **2015**, *24*, 508–523. [CrossRef]

38. Yin, J.; Cooperstock, J. Color correction methods with applications to digital projection environments. In Proceedings of the International Conference in Central Europe on Computer Graphics, Visualization and Computer Vision, WSCG 2004, Plzen, Czech Republic, 2–6 February 2004; pp. 499–506.

39. Brown, M.; Lowe, D.G. Automatic Panoramic Image Stitching Automatic 2D Stitching. *Int. J. Comput. Vis.* **2007**, *74*, 59–73. [CrossRef]

40. Li, Y.; Wang, Y.; Huang, W.; Zhang, Z. Automatic image stitching using SIFT. In Proceedings of the 2008 International Conference on Audio, Language and Image Processing, Shanghai, China, 7–9 July 2008; pp. 568–571. [CrossRef]

41. Finlayson, G.D.; Mackiewicz, M.; Hurlbert, A. Color Correction Using Root-Polynomial Regression. *IEEE Trans. Image Process.* **2015**, *24*, 1460–1470. [CrossRef] [PubMed]

42. Zheng, Y.; Lin, S.; Kambhamettu, C.; Yu, J.; Kang, S.B. Single-image vignetting correction. *IEEE Trans. Pattern Anal. Mach. Intell.* **2009**, *31*, 2243–2256. [CrossRef] [PubMed]

43. Kim, S.J.; Pollefeys, M. Robust radiometric calibration and vignetting correction. *IEEE Trans. Pattern Anal. Mach. Intell.* **2008**, *30*, 562–576. [CrossRef] [PubMed]

44. Jia, J.; Tang, C.K. Image registration with global and local luminance alignment. *Proc. IEEE Int. Conf. Comput. Vis.* **2003**, *1*, 156–163. [CrossRef]

45. Jia, J.; Tang, C.K. Tensor voting for image correction by global and local intensity alignment. *IEEE Trans. Pattern Anal. Mach. Intell.* **2005**, *27*, 36–50. [CrossRef] [PubMed]

46. Jia, Y.; Sun, J.; Tang, C.K.; Shum, H.Y. Bayesian correction of image intensity with spatial consideration. *Lect. Notes Comput. Sci.* **2004**, *3023*, 342–354. [CrossRef]

47. Yamamoto, K.; Oi, R. Color correction for multi-view video using energy minimization of view networks. *Int. J. Autom. Comput.* **2008**, *5*, 234–245. [CrossRef]

48. Sajadi, B.; Lazarov, M.; Majumder, A. ADICT: Accurate direct and inverse color transformation. *Lect. Notes Comput. Sci.* **2010**, *6314 LNCS*, 72–86. [CrossRef]

49. De Marchi, S. Polynomials arising in factoring generalized Vandermonde determinants: An algorithm for computing their coefficients. *Math. Comput. Model.* **2001**, *34*, 271–281. [CrossRef]

50. Hwang, Y.; Lee, J.Y.; Kweon, I.S.; Kim, S.J. Probabilistic moving least squares with spatial constraints for nonlinear color transfer between images. *Comput. Vis. Image Underst.* **2019**, *180*, 1–12. [CrossRef]

51. Moulon, P.; Duisit, B.; Monasse, P. Global Multiple-View Color Consistency. In Proceedings of CVMP 2013, London, UK, 5–7 November 2013.

52. Lowe, D.G. Distinctive Image Features from Scale-Invariant Keypoints. *Int. J. Comput. Vis.* **2004**, *60*, 91–110. [CrossRef]

53. Moulon, P.; Monasse, P.; Perrot, R.; Marlet, R. OpenMVG: Open Multiple View Geometry. In *Proceedings of the Reproducible Research in Pattern Recognition*; Kerautret, B., Colom, M., Monasse, P., Eds.; Springer International Publishing: Cham, Switzerland, 2017; pp. 60–74.

54. Xiong, Y.; Pulli, K. Color and luminance compensation for mobile panorama construction. In Proceedings of the Proceedings of the 18th ACM international conference on Multimedia—MM '10, Firenze, Italy, 25–29 October 2010; p. 1547. [CrossRef]

55. Cabral, R.; De la Torre, F.; Costeira, J.P.; Bernardino, A. Unifying Nuclear Norm and Bilinear Factorization Approaches for Low-Rank Matrix Decomposition. In Proceedings of the 2013 IEEE International Conference on Computer Vision, Sydney, Australia, 1–8 December 2013; pp. 2488–2495. [CrossRef]

56. HaCohen, Y.; Shechtman, E.; Goldman, D.B.; Lischinski, D. Optimizing color consistency in photo collections. *ACM Trans. Graph.* **2013**, *32*, 1–10. [CrossRef]
57. Ullman, S. The interpretation of structure from motion. *Proc. R. Soc. Lond. Ser. B Biol. Sci.* **1979**, *203*, 405–426.
58. Shen, T.; Wang, J.; Fang, T.; Zhu, S.; Quan, L. Color correction for image-based modeling in the large. In Proceedings of the Computer Vision—ACCV 2016, Lecture Notes in Computer Science (Including Subseries Lecture Notes in Artificial Intelligence and Lecture Notes in Bioinformatics), Taipei, Taiwan, 20–24 November 2016; Volume 10114 LNCS, pp. 392–407. [CrossRef]
59. Coelho, D.; Dal'Col, L.; Madeira, T.; Dias, P.; Oliveira, M. A Robust 3D-Based Color Correction Approach for Texture Mapping Applications. *Sensors* **2022**, *22*, 1730. [CrossRef]
60. Sturm, P. Pinhole Camera Model. In *Computer Vision: A Reference Guide*; Ikeuchi, K., Ed.; Springer: Boston, MA, USA, 2014; pp. 610–613. [CrossRef]
61. Bertalmio, M.; Bertozzi, A.; Sapiro, G. Navier-stokes, fluid dynamics, and image and video inpainting. In Proceedings of the 2001 IEEE Computer Society Conference on Computer Vision and Pattern Recognition. CVPR 2001, Kauai, HI, USA, 8–14 December 2001; Volume 1, p. I. [CrossRef]
62. Dijkstra, E.W. A Note on Two Problems in Connexion with Graphs. *Numer. Math.* **1959**, *1*, 269–271. [CrossRef]
63. Platt, J. Probabilistic outputs for support vector machines and comparisons to regularized likelihood methods. *Adv. Large Margin Classif.* **1999**, *10*, 61–74.
64. Rusinkiewicz, S.; Levoy, M. QSplat: A Multiresolution Point Rendering System for Large Meshes. In Proceedings of the 27th Annual Conference on Computer Graphics and Interactive Techniques SIGGRAPH '00, New Orleans, LA, USA, 23–28 July 2000; pp. 343–352. [CrossRef]
65. Sharma, G.; Wu, W.; Dalal, E.N. The CIEDE2000 color-difference formula: Implementation notes, supplementary test data, and mathematical observations. *Color Res. Appl.* **2005**, *30*, 21–30. [CrossRef]
66. Li, L.; Xia, M.; Liu, C.; Li, L.; Wang, H.; Yao, J. Jointly optimizing global and local color consistency for multiple image mosaicking. *ISPRS J. Photogramm. Remote Sens.* **2020**, *170*, 45–56. [CrossRef]
67. Li, Y.; Yin, H.; Yao, J.; Wang, H.; Li, L. A unified probabilistic framework of robust and efficient color consistency correction for multiple images. *ISPRS J. Photogramm. Remote Sens.* **2022**, *190*, 1–24. [CrossRef]

Article

Real-Time LiDAR Point-Cloud Moving Object Segmentation for Autonomous Driving

Xing Xie [1], Haowen Wei [2] and Yongjie Yang [1,*]

[1] School of Information Science and Technology, Nantong University, Nantong 226019, China
[2] Department of Computer Science, Columbia University, New York, NY 10027, USA
* Correspondence: yang.yj@ntu.edu.cn

Abstract: The key to autonomous navigation in unmanned systems is the ability to recognize static and moving objects in the environment and to support the task of predicting the future state of the environment, avoiding collisions, and planning. However, because the existing 3D LiDAR point-cloud moving object segmentation (MOS) convolutional neural network (CNN) models are very complex and have large computation burden, it is difficult to perform real-time processing on embedded platforms. In this paper, we propose a lightweight MOS network structure based on LiDAR point-cloud sequence range images with only 2.3 M parameters, which is 66% less than the state-of-the-art network. When running on RTX 3090 GPU, the processing time is 35.82 ms per frame and it achieves an intersection-over-union(IoU) score of 51.3% on the SemanticKITTI dataset. In addition, the proposed CNN successfully runs the FPGA platform using an NVDLA-like hardware architecture, and the system achieves efficient and accurate moving-object segmentation of LiDAR point clouds at a speed of 32 fps, meeting the real-time requirements of autonomous vehicles.

Keywords: moving object segmentation; LiDAR; CNN; FPGA

1. Introduction

Presently, the key to autonomous navigation in autonomous driving systems is the ability to recognize static and moving objects in the environment, and to support predicting the future state of the environment, avoiding collisions, and planning tasks. Moving object segmentation (MOS) algorithms improve environment perception [1], localization [2], and future state prediction [3] by distinguishing between moving and static objects in 3D LiDAR point cloud data. However, the MOS task is computationally intensive and the network model is complex [4,5], so it is very important to meet the real-time processing requirements of autonomous driving applications.

Most of the existing LiDAR-based point-cloud semantic segmentation networks predict the semantic labels of point clouds, such as vehicles, buildings or roads from a single frame. However, comparing to images, LiDAR provides better object location information via consecutive frames, so that we can also use LiDAR to distinguish moving objects. Recently proposed point-based [6] or voxel-based [1] segmentation networks, although superior in performance, are structurally complicated and computationally expensive. In the recent work [4], a segmentation network based on range images was adopted, the 3D LiDAR point cloud was projected onto a 2D plane, the range images of consecutive frames were used as the intermediate representation, and a 2D convolutional neural network (CNN) was used. This network performs the moving segmentation task. Furthermore, almost all state-of-the-art point-cloud moving object segmentation networks target GPUs that may not suitable for edge computing. From a computation point of view, edge deep learning accelerators (such as NVIDIA Deep Learning Accelerator (NVDLA) [7] and Xilinx DPU [8]) do not accelerate all common operations. Therefore, designing neural networks compatible with edge deep learning accelerators is critical for real-time embedded applications.

Citation: Xie, X.; Wei, H.; Yang, Y. Real-Time LiDAR Point-Cloud Moving Object Segmentation for Autonomous Driving. *Sensors* **2023**, *23*, 547. https://doi.org/10.3390/s23010547

Academic Editors: Miaohui Wang, Guanghui Yue, Jian Xiong and Sukun Tian

Received: 1 December 2022
Revised: 29 December 2022
Accepted: 29 December 2022
Published: 3 January 2023

In this paper, we propose a lightweight multi-branch network structure to solve the problem of 3D LiDAR point-cloud moving object segmentation, which can run in real time on GPU. Figure 1 shows an example scene of our segmentation, red boxes are moving cars, the yellow box is a parked car, and moving objects are represented by red masks, which also verifies the feasibility of our method. Furthermore, the MOS computing system is built for autonomous vehicles, which can perform point-cloud pre-processing and neural network segmentation. Since only post-processing steps are left to the automotive electronic Control Unit ECU, this solution significantly alleviates the computation burden of ECU, thereby reducing the decision making and vehicle reaction latency. Our moving object segmentation network achieved 32 frames per second (fps) on FPGA. The contributions of this paper are summarized below:

(1) To our knowledge, this is one of the first end-to-end FPGA implementations for a real time LiDAR point-cloud moving-object segmentation deep learning platform, a LiDAR is directly connected to the processing system (PS) side. After pre-processing, the point cloud is stored in the DDR memory, which is accessible by the hardware accelerator on the programmable logic (PL) side.

(2) A light-weight and real-time moving-object segmentation network is proposed, targeting to NVDLA. Hardware-friendly layers are used (i.e., by replacement of deconvolution with bi-linear interpolation) to greatly reduce the complexity of computation. Its IoU score on the SemanticKITTI test set is 51.3%. The inference time on NVIDIA RTX 3090TI is about 35.82 ms.

(3) An efficient moving-object segmentation network architecture is implemented on the ZCU104 MPSoC FPGA platform, which enables real-time processing at 32 frames per second (fps).

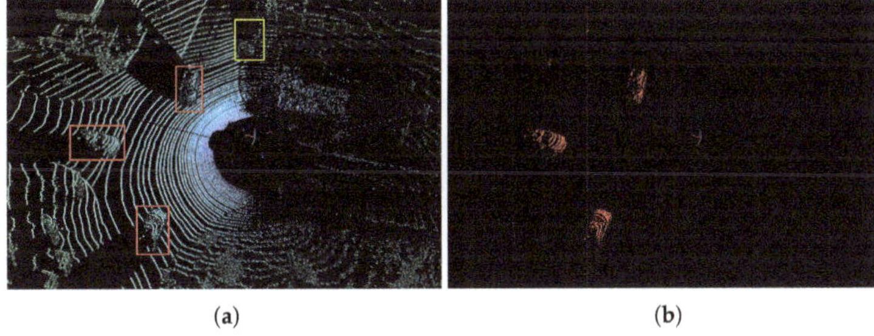

(a)   (b)

Figure 1. Moving object segmentation using our approach. (**a**) Raw Point Cloud ; (**b**) Segmented Point Cloud.

The rest of this paper is organized as follows: Section 2 summarizes the existing research results of moving-object segmentation in LiDAR point clouds and the FPGA implementation of the segmentation network. In Section 3, the proposed moving-object segmentation network model of LiDAR point cloud and its training details are described. The FPGA implementation and its results are discussed in Sections 4 and 5, respectively. Finally, Section 6 summarizes the whole paper.

2. Related Work

2.1. LiDAR Point-Cloud Moving Object Segmentation

Existing LiDAR point cloud moving object segmentation networks can be categorized into two groups: computer-vision-based [9–13] and LiDAR-sensor-based [14–16]. However, the processing of LiDAR data remains challenging due to the uneven distribution and sparsity of LiDAR point clouds. Here, we mainly study the MOS problem of 3D LiDAR point cloud data.

In recent years, great progress has been made in semantic segmentation based on LiDAR sensor point cloud data [17–24], such as the point-cloud compression methods in [23,24] and so on. Semantic segmentation is a key step in the segmentation of moving objects in LiDAR point clouds. However, most of the existing semantic segmentation convolutional neural networks can only predict the semantic labels of point clouds, such as vehicles, buildings, and people, but cannot distinguish between actual moving objects and static objects, such as moving cars and parked ones.

The state-of-the-art scene flow method, FlowNet3D [25], is designed based on PointNet [6] and PointNet++ [26], which directly processes the original irregular 3D points without any pre-processing, and estimates each LiDAR point for two consecutive frames. The translational flow vectors include moving vehicles and pedestrians. Although these methods perform well on small point clouds, processing power becomes inefficient on larger point cloud datasets, requiring longer runtime. In addition, there are various 3D point-cloud-based semantic segmentation methods, such as SpSequencenet [27], KPConv [28], and SPVConv [29], which are also able to achieve state-of-the-art performance in semantic segmentation tasks. Among them, SpSequencenet [27] uses changes in sequence point clouds to predict moving objects. However, one problem with all networks based on operating directly on the point cloud is the dramatic increase in processing power and memory requirements, causing the point cloud to become larger. Therefore, training is difficult and cannot meet the real-time requirements of the automatic driving system.

Chen et al. [4] developed LMNet, which utilizes the residual between the current frame and the previous frame to be used as an additional input to the semantic segmentation network to achieve class-independent moving object segmentation, as well as in RangeNet++ [17] and SalsaNext [18] for performance evaluation. These networks are capable of real-time moving object segmentation running faster than the frame rate of the LiDAR sensor used. Mohapatra et al. [30] introduced a computationally efficient moving object segmentation framework based on LiDAR bird's eye view (BEV) space. The work in [31] utilizes a dual-branch structure to fuse the spatio-temporal information of LiDAR scans to improve the performance of MOS. In contrast, Kim et al. [32] proposed a network architecture that fuses motion features and semantic features, achieving improvements in computational speed and performance metrics. In the recent work of [33], the autoregressive system identification (AR-SI) theory was used to significantly improve the segmentation effect of the traditional encoder–decoder structure, and the model was deployed in embedded devices for actual measurement.

2.2. FPGA Implementations of Segmentation Networks

Advanced driver-assistance systems (ADAS) are rapidly being integrated into almost all new vehicles. The LiDAR point cloud segmentation algorithm must meet the real-time requirements, which may not be well met by standard CPUs or GPUs. FPGA has the advantages of high energy efficiency ratio and flexible reconfiguration, which can realize the high energy efficiency deployment of semantic segmentation networks in ADAS. Current researchers [5,34–39] mainly study and analyze the lightweight semantic segmentation network algorithm and the accelerated computation combined with the resource characteristics of customized hardware platform. Xilinx will support Continental's new advanced LiDAR sensor ARS 540 through the Zynq UltraScale+ MPSoC platform, partnering to create the automotive industry's first mass-produced 4D imaging sensor, paving the way for L5-level autonomous driving systems. In Ref. [16], a LiDAR sensor is directly connected to FPGA through an Ethernet interface, realizing a deep learning platform of end-to-end 3D point cloud semantic segmentation based on FPGA, which can process point-cloud segmentation in real time.

3. Proposed Network

The design method of the moving object segmentation network is mainly inspired by [4]. The residual image is used as an additional input to the designed semantic segmentation network to achieve moving object segmentation. In the following, we will describe our method in detail.

3.1. Spherical Projection of LiDAR Point Cloud

Following previous work [16,17], we use a 2D neural network convolution to extract features from the range view (RV) of LiDAR. Specifically, we project the LiDAR point (x, y, z) onto a sphere and finally convert it to image coordinates (u, v), defined as:

$$
\begin{pmatrix} u \\ v \end{pmatrix} = \begin{pmatrix} \frac{1}{2}\left[1 - \arctan(y, x)\pi^{-1}\right]w \\ \left[1 - \left(\arcsin(zr^{-1}) + f_{up}\right)f^{-1}\right]h \end{pmatrix}
\tag{1}
$$

where (u, v) are the image coordinates, (h, w) are the desired range image according to the height and width, r represents the range of each point as $r = \sqrt{x^2 + y^2 + z^2}$, and $f = |f_{\text{down}}| + |f_{up}|$ for the sensor's vertical field of view.

We use Equation (1) to extract the range index r, 3D point coordinates (x, y, z) and intensity value i for each point projected to (u, v), and take them as features to be superimposed along the channel dimension. Therefore, we can directly input these features into the network, and then transform point-cloud moving segmentation into image moving segmentation.

3.2. Residual Images

As in ref. [4], the residual image and range view based on LiDAR point cloud are used as the input of the segmentation network, and the temporal information in the residual image is used to distinguish the static object and the pixels on the moving object, so the actual moving object and the static object can be distinguished.

Assuming that there are N time series of LiDAR scans in the SLAM history, $S_j = \{p_i \in \mathbb{R}^4\}$ and M points are represented as homogeneous coordinates, i.e., $p_i = (x, y, z, 1)$. $T_N^{N-1}, \ldots, T_1^0$ is denoted as the transformation matrix between $N + 1$ scan poses, i.e., $T_k^l \in \mathbb{R}^{4 \times 4}$. Equation (2) represents the coordinate system in which the kth scan transformed into the lth scan

$$
S^{k \to l} = \left\{ \prod_{j=k}^{l+1} T_j^{j-1} p_i \mid p_i \in S_k \right\}
\tag{2}
$$

In Ref. [4], in order to generate the residual image and fuse it into the current range image, transformation and re-projection are required. First, the transformation estimate defined in the ego-motion is compensated according to Equation (2) by transforming the previous scan to the current given local coordinate system, and next, the $S^{k \to l}$ of the past scans are re-projected to the current range image view using Equation (1). In order to calculate the residual $d_{k,i}^l$ for each pixel i, we use the normalized absolute difference between the range of the current frame and the transformed frame to calculate, as defined by

$$
d_{k,i}^l = \frac{\left| r_i - r_i^{k \to l} \right|}{r_i}
\tag{3}
$$

where r_i is the range value from p_i to the current frame at the image coordinates (u_i, v_i), and $r_i^{k \to l}$ is the range value from the transformed scan to the pixel in the same image. In the scene of moving objects, the displacement of the moving car is relatively large compared to the static background, and the residual image is obvious, while the residual image of the slowly moving object is blurred and the residual pattern is not obvious. Therefore, direct use of residual images for moving object segmentation cannot achieve good performance. Finally, we concatenate the residual image and the range view as the input of the segmentation network, and each pixel fuses the spatial and temporal information.

3.3. Network Architecture

In this paper, our proposed network is mainly divided into two branches: context path and spatial path, which respectively extract feature information and then fuse these feature information. The architecture of our proposed CNN for point-cloud moving object segmentation is shown in Figure 2. The backbone module utilized in the context path branch is ResNet-18 [40] for eight times fast down-sampling. Subsequently, the extracted features are fed into the Atrous Spatial Pyramid Pooling (ASPP) [41] module in order to connect features from different perceptual domains. ASPP builds convolution kernels with different receptive fields through different dilated rates to increase the receptive fields of the network and enhance the ability of the network to obtain multi-scale context, so as to obtain good performance. However, from the perspective of hardware (GPU or FPGA), dilated convolution has low efficiency and slow inference speed, and the larger the dilated rate, the longer the convolution processing time. Therefore, our network keeps the dilated rate as 2, and this method can simulate the function of ASPP without reducing the computational efficiency of GPU and NVDLA. Next, a global context module (GCM) is introduced to extract contextual information and guide feature learning of the current path. GCM consists of a global average pooling layer and a 1×1 convolution layer that extracts global context features.

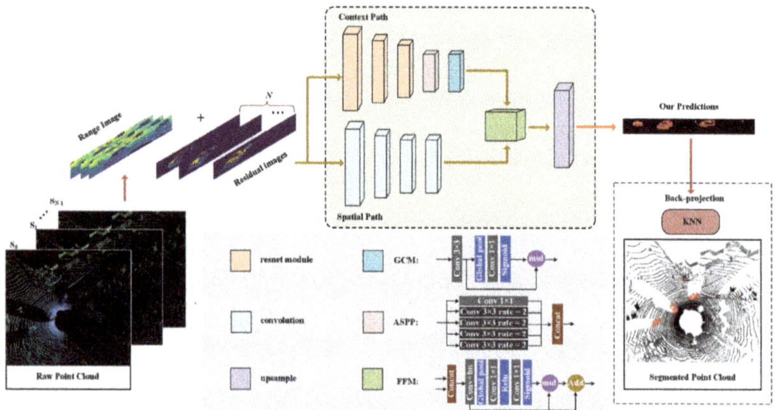

Figure 2. Architecture of the proposed CNN for point-cloud moving object segmentation.

The spatial path mainly retains rich spatial information to generate high-resolution feature maps, which only contains four convolutional layers. The first three convolutional layers are stride = 2, and 1/8 feature maps are extracted. The feature fusion module (FFM) [42] is used to fuse the features of context branch and spatial branch at different scales. Therefore, the features of the two channels cannot be simply weighted, but superimposed by concatenation method. FFM combines the attention mechanism for feature fusion, mainly including global pooling layer, 1×1 convolution layer, ReLU activation layer and Sigmoid layer. At the end of the network, in order to output the moving object segmentation results of the original image size, the mainstream high-performance segmentation networks, U-NET [43] and FCN [44], use layer skip connection for up-sampling. This requires a GPU or FPGA for more computation and data movement. Therefore, we up-sample the FFM output eight times using a bi-linear interpolation algorithm.

3.4. Training Details

We implemented a 3D LiDAR point-cloud moving-object segmentation network using PyTorch and trained on a single NVIDIA RTX 3090TI GPU. We use the method of [4] to train the network, process all point clouds according to Equations (1)–(3), and generate 64×2048 range views and residual images respectively. The residual images are then concatenated with the current range image and used as input to a 2D convolutional neural network. Trained with the new binary masks, the proposed method can separate moving and static objects label maps. During training, the network is trained with an initial learning rate of 0.01 and a weight decay of 1×10^{-4}.

3.5. Dataset and Evaluation

SemanticKITTI [45] is a large-scale dataset for semantic scene understanding of 3D LiDAR point cloud sequences, including semantic segmentation and semantic scene completion. The dataset contains 28 annotated categories such as pedestrians, vehicles, parking lots, roads, buildings, etc., which further distinguishes static objects from moving objects. The raw odometry data consists of 22 sequences of point cloud data. We follow the same protocol in [17], where the sequences 00–10 are used for training and the sequence 08 is used for validation. The remaining sequences 11–21 are used as the test set. All classes are reorganized into two types: moving and non-moving/static objects according to [4]. The former one contains actually moving vehicles and pedestrians, all other classes are non-moving/static objects.

In order to evaluate the MOS performance, we follow the official guidance, using the Jaccard index or IoU [46], which is:

$$\text{IoU} = \frac{TP}{TP + FP + FN} \tag{4}$$

where TP, FP, and FN correspond to the number of true-positive, false-positive, and false-negative predictions for the moving classes.

Referring to the evaluation method proposed in [4], we use IoU to evaluate the accuracy of moving object segmentation in LiDAR point clouds. Table 1 shows the MOS performance compared to the state-of-the-art on the SemanticKITTI test set. Table 2 shows the results of the validation set. Since all operations of our network are supported by NVDLA, the network complexity is reduced and no semantic information is added. Thus, when our proposed model uses N = 8 residual images, the best performance IoU score (51.3%) is obtained, but slightly lower than the baseline LMNet [4] on the test benchmark.

For qualitative evaluation, Figure 3 shows the qualitative results of different methods on the SemanticKITTI test set. Meanwhile, the qualitative results of the point clouds are shown in Figure 4. In the intersection in the figure below, there are a large number of moving objects and non-moving/stationary objects such as moving vehicles and walking people; our method can distinguish between actual moving vehicles and pedestrians, while other methods cannot detect slow-moving objects.

Table 1. MOS performance compared to the state-of-the-art on the SemanticKITTI test set.

Methods	IoU (%)
SceneFlow [25]	28.7
SpSequenceNet [27]	43.2
SalsaNext [18]	46.6
LMNet [4]	62.5
BiSeNet [42]	45.1
Ours	51.3

Table 2. MOS performance compared to the state-of-the-art on the SemanticKITTI validation set.

Methods	IoU (%)
SalsaNext [18]	48.6
LMNet [4]	65.3
BiSeNet [42]	46.1
Ours	52.4

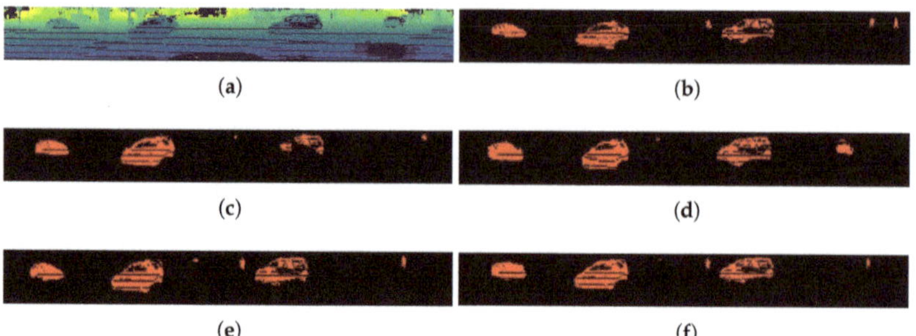

Figure 3. Qualitative results of different methods on the SemanticKITTI test set, where red pixels correspond to moving objects (range view images). (**a**) Range Image; (**b**) Ground Truth Labels; (**c**) BiSeNet (retrained); (**d**) SalsaNext (retrained); (**e**) LMNet; (**f**) ours.

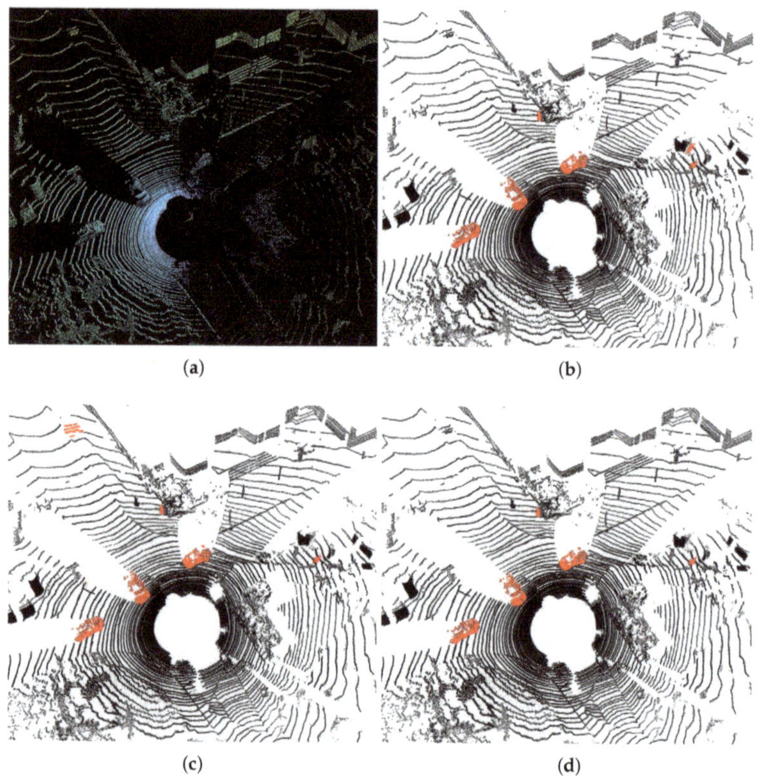

Figure 4. Qualitative results shown as point clouds. (**a**) Raw Point Cloud; (**b**) Ground Truth Labels, and (**c**,**d**) prediction results, where red points correspond to the class moving (**c**) LMNet; (**d**) Ours.

3.6. Ablation Studies

In this section, some ablation experiments on the validation set (sequence 08) of the SemanticKITTI dataset are conducted to analyze the effect of each component's performance shown in Table 3. As shown in Table 3, it can be observed that the IoU of the dual-branch architecture can be increased by 9.6% compared to that of the single-branch architecture.

On this basis, FFM, GCM, ASPP, and their combination are added. It is worth noting that the IoU of our proposed final setup can achieve 52.4%.

Table 3. Ablation study of components on the validation set. CP: Context Path; SP: Spatial Path; GCM: Global Context Module; FFM: Feature Fusion Module.

Methods	IoU (%)
CP	33.6
CP + SP + Sum	43.2
CP + SP + FFM	45.1
CP + SP + FFM + GCM	47.4
CP + SP + FFM + GCM+ASPP	52.4

In our design, the K-Nearest Neighbor (KNN) post-processing is used to back-project the 2D prediction result to the 3D point cloud. In order to verify the attractive performance of the KNN post-processing in our proposed design, the comparison with regard to the back-projection between the Conditional Random Field (CRF) post-processing and the KNN post-processing is provided in Table 4. As shown in Table 4, compared to CRF post-processing, KNN post-processing results in better IoU performance. Figure 5 shows the qualitative results of different post-processing methods for MOS on the validation set. The qualitative results prove that the KNN can handle the blurred boundary of moving objects in a better way. This also obeys our expectation. Due to the fact that, during dimension reduction, different 3D points belonging to different categories might project into the same pixel in the 2D range image. Considering the principle of CRF, it contributes little to solve this issue if it is applied to a 2D range image. While KNN counts nearest points in 3D space rather than 2D.

Table 4. Comparison of the post-processing between the KNN and the CRF on the validation set.

	Methods	IoU (%)
(1)	KNN	52.4
(2)	CRF	49.1

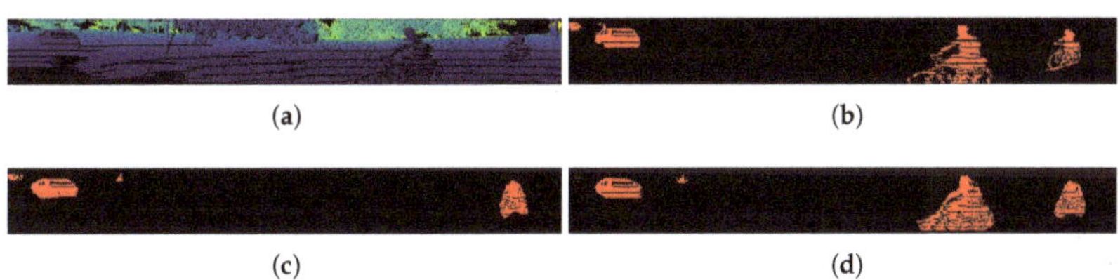

(a)   (b)

(c)   (d)

Figure 5. Qualitative results of different post-processing methods for MOS on the validation set, where red pixels correspond to moving objects (range view images). (**a**) Range Image; (**b**) Ground Truth Labels; (**c**) CRF; (**d**) KNN.

3.7. Run-Time Evaluation on GPU

In autonomous driving systems, the processing speed of the moving object segmentation network must meet real-time requirements. To get a fair performance evaluation, all measurements are evaluated on the SemanticKITTI dataset 08 sequence using a single NVIDIA RTX 3090TI-24GB card and the network performances are shown in Table 5. Compared to the state-of-the-art network LMNet [4], our model clearly shows better performance, the running time is 35.82 ms, and the amount of network parameters is about 2.3 M, which is reduced to 1/3 of that in LMNet [4].

Table 5. Run-time performance on the SemanticKITTI validation set.

	Processing Time	Speed	Parameters
BiSeNet [42]	56.3 ms	18 fps	13.76 M
LMNet [4]	42.21 ms	23 fps	6.71 M
MotionSeg3D [31]	116.71 ms	8 fps	6.73 M
Ours (GPU)	35.82 ms	28 fps	2.3 M
Ours (FPGA)	31.69 ms	32 fps	2.3 M

Figure 6 shows the inference speed vs. IoU on the validation set. For practical use in embedded systems on autonomous vehicles, the IoU of our design in GPU and FPGA is sacrificed to achieving the higher inference speed compared to the other methods. Note that our implementation runs significantly faster than the 10 Hz sampling rate of mainstream LiDAR sensors [47].

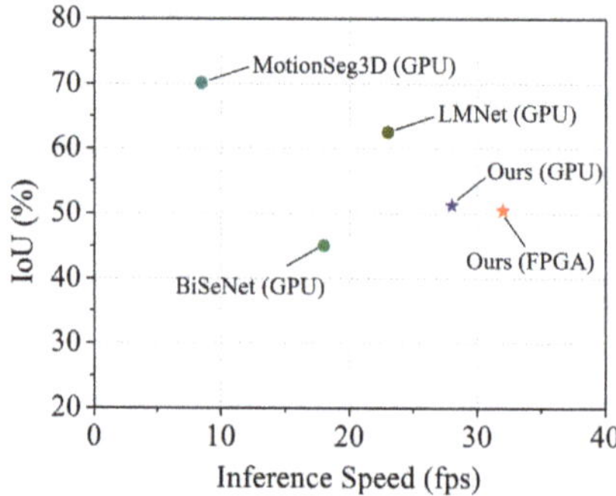

Figure 6. Inference speed vs. IoU on the SemanticKITTI validation set. Red star indicates our method in FPGA, the blue star indicates our method in GPU, and colored dots represent other methods.

4. Hardware Architecture

The hardware architecture of the point cloud moving object segmentation network is shown in Figure 7. It consists of processing system (ARM core) and programmable logic (FPGA) parts. ARM core is used to complete pre-processing and post-processing, such as point cloud reading, image resizing [48], result showing, etc. On the FPGA side, an NVIDIA Deep Learning Accelerator (NVDLA) like system is implemented. We tailor it and adopt it into FPGA.

The core of the convolutional engine is the MAC array. In this work, the size of the MAC array is chosen to be 32×32. To improve the processing speed and alleviate the bandwidth requirement between FPGA and DDR memory, NVDLA adopts a Ping-Pong buffer. Different from a micro-controller in NVDLA, we implement a finite state machine (FSM) to control the running order of CNN operations. In this study, we quantize the neural network to INT8 for high computation efficiency.

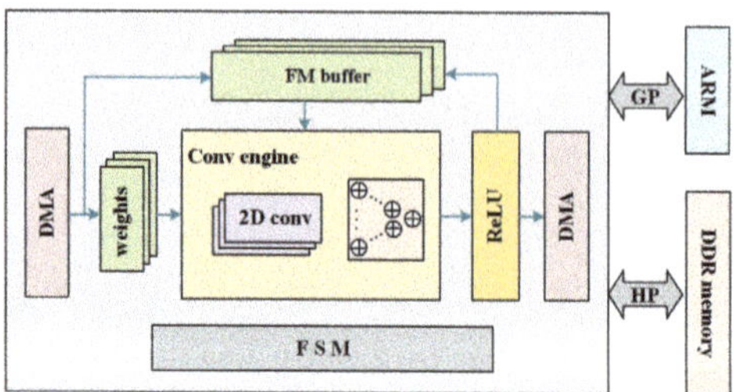

Figure 7. Hardware architecture of the CNN accelerator on the FPGA.

5. Results and Discussion

The target hardware platform is the Zynq UltraScale+ MPSoC ZCU104 development board. Figure 8 exhibits the overall system setup with LiDAR connected to the FPGA board directly, which demonstrates our experiment setup. The LiDAR driver is implanted in the ARM processor on ZCU104 board and connected to LiDAR via UDP protocol on the Ethernet port. The ARM processor receives each set of point cloud data from LiDAR and stores it into DDR memory for NVDLA fetching.

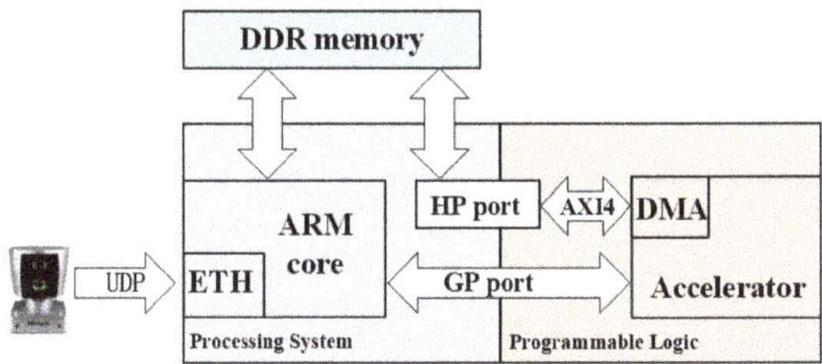

Figure 8. Overall system setup with LiDAR connected to the FPGA board directly.

The hardware resource usage of our proposed neural network is shown in Table 6. This design has used 91.84% of the DSP resources, if the parallelism is increased, a larger FPGA needs to be used. Table 7 shows the run-time performance of the proposed approach on SemanticKITTI Dataset. It can be observed that when running at 250 MHz, this accelerator's processing speed is 32 fps. The estimated power consumption of the FPGA implementation is 12.8 W. The only real-time solution currently available, SalsaNext [18], runs on Nvidia Quadro P6000 GPUs and requires 600–650 W PC power support. Therefore, our solution provides a balanced and practical approach for running LiDAR point cloud moving object segmentation tasks on embedded devices. Since there are few 3D point-cloud moving object segmentation implementations on FPGA, the performance and hardware resource utilization comparison with similar works is not yet available.

Table 6. FPGA Resource utilization for the CNN accelerator.

FPGA Resource	Used	Available	Utilization
LUT	102,707	230,400	44.58%
FF	114,221	460,800	24.79%
DSP	1587	1728	91.84%
BRAM	162	312	51.92%

Table 7. Run-time Performance of the Proposed Approach on SemanticKITTI Dataset.

Device	Precision	Processing time	Speed
GPU	FP32	35.82 ms	28 fps
FPGA	INT8	31.69 ms	32 fps

6. Conclusions

In this study, we proposed a lightweight CNN architecture for LiDAR point-cloud moving object segmentation. Edge deep learning accelerators are designed with their limitations and computational efficiency in mind, and our proposed network structure fully supports all operations of NVDLA. On SemanticKITTI dataset, the processing time of the network on RTX 3090TI GPU is 35.82 ms and the IoU score of 51.3% is achieved. When compared to the state-of-the-art network, the network achieves a similar error performance, but using only 34% of the parameters. In addition, the proposed network successfully targets the MPSoC FPGA platform using NVDLA hardware architecture. The system successfully achieves efficient and accurate moving-object segmentation of LiDAR point clouds at 32 fps, which meets the real-time requirements of autonomous vehicles.

However, some potential problems need to be solved in the future. First of all, in order to reduce the computational complexity of our proposed network, we plan to simplify the original structure and remove the time-consuming cross-layer connections while ensuring its high performance. Secondly, due to acceleration of the model inference, the low-level details mostly sacrificed, which leads to a considerable decrease in accuracy. In view of this, the spatial path would be improved by capturing the low-level details with wide channels and shallow layers in our future works.

Author Contributions: Conceptualization, X.X. and H.W.; methodology, X.X., H.W.; software, X.X.; formal analysis, X.X., H.W.; data curation, H.W. and X.X.; writing—original draft preparation, X.X; writing—review and editing, Y.Y.; supervision, X.X.; project administration, Y.Y. All authors have read and agreed to the published version of the manuscript.

Funding: This research received no external funding.

Institutional Review Board Statement: Not applicable.

Informed Consent Statement: Not applicable.

Data Availability Statement: Not applicable.

Conflicts of Interest: The authors declare no conflict of interest.

References

1. Yin, Z.; Tuzel, O. VoxelNet: End-to-End Learning for Point Cloud Based 3D Object Detection. In Proceedings of the 2018 IEEE/CVF Conference on Computer Vision and Pattern Recognition (CVPR), Salt Lake City, UT, USA, 18–23 June 2018; pp. 4490–4499.
2. Zhang, C.; Luo, W.; Urtasun, R. Efficient Convolutions for Real-Time Semantic Segmentation of 3D Point Clouds. In Proceedings of the 2018 International Conference on 3D Vision (3DV), Verona, Italy, 5–8 September 2018; pp. 399–408.
3. Yin, H.; Wang, Y.; Ding, X.; Tang, L.; Xiong, R. 3D LiDAR-Based Global Localization Using Siamese Neural Network. *IEEE Trans. Intell. Transp. Syst.* **2019**, *21*, 1380–1392. [CrossRef]
4. Chen, X.; Li, S.; Mersch, B.; Wiesmann, L.; Stachniss, C. Moving Object Segmentation in 3D LiDAR Data: A Learning-based Approach Exploiting Sequential Data. *IEEE Robot. Autom. Lett.* **2021**, *6*, 6529–6536. [CrossRef]
5. Lyu, Y.; Bai, L.; Huang, X. Real-Time Road Segmentation Using LiDAR Data Processing on an FPGA. In Proceedings of the 2018 IEEE International Symposium on Circuits and Systems (ISCAS), Florence, Italy, 27–30 May 2018.
6. Qi, C.R.; Su, H.; Mo, K.; Guibas, L.J. PointNet: Deep Learning on Point Sets for 3D Classification and Segmentation. In Proceedings of the 2017 IEEE Conference on Computer Vision and Pattern Recognition (CVPR), Honolulu, HI, USA, 21–26 July 2017; pp. 77–85.
7. Nvidia. Nvidia Deep Learning Accelerator. 2018. Available online: https://developer.nvidia.com/deep-learning-accelerator (accessed on 2 July 2022).
8. Xilinx. Dpuczdx8g for Zynq Ultrascale+ Mpsocs Product Guide. 2021. Available online: https://docs.xilinx.com/r/en-US/pg338-dpu?tocId=Bd4R4bhnWgMYE6wUISXDLw (accessed on 3 August 2022).
9. Barnes, D.; Maddern, W.; Pascoe, G.; Posner, I. Driven to distraction: Self-supervised distractor learning for robust monocular visual odometry in urban environments. In Proceedings of the 2018 IEEE International Conference on Robotics and Automation (ICRA), Brisbane, Australia, 21–25 May 2018; pp. 1894–1900.
10. Patil, P.W.; Biradar, K.M.; Dudhane, A.; Murala, S. An end-to-end edge aggregation network for moving object segmentation. In Proceedings of the IEEE/CVF Conference on Computer Vision and Pattern Recognition (CVPR), Seattle, WA, USA, 13–19 June 2020; pp. 8146–8155.
11. Cai, Y.; Luan, T.; Gao, H.; Wang, H.; Li, Z. YOLOv4-5D: An Effective and Efficient Object Detector for Autonomous Driving. *IEEE Trans. Instrum. Meas.* **2021**, *70*, 1–13. [CrossRef]
12. Bell, A.; Mantecón, T.; Díaz, C.; del-Blanco, C.R.; Jaureguizar, F.; García, N. A Novel System for Nighttime Vehicle Detection Based on Foveal Classifiers with Real-Time Performance. *IEEE Trans. Intell. Transp. Syst.* **2021**, *23*, 5421–5433. [CrossRef]
13. Wang, H.; Chen, Y.; Cai, Y.; Chen, L.; Li, Y.; Sotelo, M.A.; Li, Z. SFNet-N: An Improved SFNet Algorithm for Semantic Segmentation of Low-light Autonomous Driving Road Scenes. *IEEE Trans. Intell. Transp. Syst.* **2022**, *23*, 21405–21417. [CrossRef]
14. Yan, J.; Chen, D.; Myeong, H.; Shiratori, T.; Ma, Y. Automatic Extraction of Moving Objects from Image and LIDAR Sequences. In Proceedings of the 2014 2nd International Conference on 3D Vision, Tokyo, Japan, 8–11 December 2014.
15. Postica, G.; Romanoni, A.; Matteucci, M.; Shiratori, T.; Ma, Y. Robust moving objects detection in lidar data exploiting visual cues. In Proceedings of the 2016 IEEE/RSJ International Conference on Intelligent Robots and Systems (IROS), Daejeon, Republic of Korea, 9–14 October 2016.
16. Xie, X.; Bai, L.; Huang, X. Real-Time LiDAR Point Cloud Semantic Segmentation for Autonomous Driving. *Electronics* **2022**, *11*, 11. [CrossRef]

17. Milioto, A.; Vizzo, I.; Behley, J.; Stachniss, C. RangeNet ++: Fast and Accurate LiDAR Semantic Segmentation. In Proceedings of the 2019 IEEE/RSJ International Conference on Intelligent Robots and Systems (IROS), Macau, China, 3–8 November 2019; pp. 4213–4220.

18. Cortinhal, T.; Tzelepis, G.; Aksoy, E.E. SalsaNext: Fast, Uncertainty-aware Semantic Segmentation of LiDAR Point Clouds for Autonomous Driving. *arXiv* **2020**, arXiv:2003.03653.

19. Li, S.; Chen, X.; Liu, Y.; Dai, D.; Stachniss, C.; Gall, J. Multiscale interaction for real-time lidar data segmentation on an embedded platform. *arXiv* **2020**, arXiv:2008.09162.

20. Wu, B.; Zhou, X.; Zhao, S.; Yue, X.; Keutzer, K. SqueezeSegV2: Improved Model Structure and Unsupervised Domain Adaptation for Road-Object Segmentation from a LiDAR Point Cloud. In Proceedings of the 2019 International Conference on Robotics and Automation (ICRA), Montreal, QC, Canada, 20–24 May 2019; pp. 4376–4382.

21. Zhang, Y.; Zhou, Z.; David, P.; Yue, X.; Foroosh, H. PolarNet: An Improved Grid Representation for Online LiDAR Point Clouds Semantic Segmentation. In Proceedings of the 2020 IEEE/CVF Conference on Computer Vision and Pattern Recognition (CVPR), Seattle, WA, USA, 13–19 June 2020; pp. 9598–9607.

22. Xu, C.; Wu, B.; Wang, Z.; Zhan, W.; Vajda, P.; Keutzer, K.; Tomizuka, M. *SqueezeSegV3: Spatially-Adaptive Convolution for Efficient Point-Cloud Segmentation*; Springer: Berlin/Heidelberg, Germany, 2020.

23. Sun, X.; Wang, M.; Du, J.; Sun, Y.; Cheng, S.S.; Xie, W. A Task-Driven Scene-Aware LiDAR Point Cloud Coding Framework for Autonomous Vehicles. *IEEE Trans. Ind. Inform.* **2022**. [CrossRef]

24. Sun, X.; Wang, S.; Wang, M.; Cheng, S.S.; Liu, M. An Advanced LiDAR Point Cloud Sequence Coding Scheme for Autonomous Driving. In Proceedings of the 28th ACM International Conference on Multimedia (MM '20), Seattle, WA, USA, 12–16 October 2020; pp. 2793–2801.

25. Liu, X.; Qi, C.R.; Guibas, L.J. Flownet3d: Learning scene flow in 3d point clouds. In Proceedings of the 2019 IEEE/CVF Conference on Computer Vision and Pattern Recognition (CVPR), Long Beach, CA, USA, 15–20 June 2019; pp. 529–537.

26. Qi, C.R.; Yi, L.; Su, H.; Guibas, L.J. PointNet++: Deep Hierarchical Feature Learning on Point Sets in a Metric Space. *arXiv* **2017**, arXiv:1706.02413.

27. Shi, H.; Lin, G.; Wang, H.; Hung, T.Y.; Wang, Z. Spsequencenet: Semantic segmentation network on 4d point clouds. In Proceedings of the IEEE/CVF Conference on Computer Vision and Pattern Recognition (CVPR), Seattle, WA, USA, 13–19 June 2020; pp. 4573–4582.

28. Thomas, H.; Qi, C.; Deschaud, J.; Marcotegui, B.; Goulette, F.; Guibas, L. KPConv: Flexible and Deformable Convolution for Point Clouds. In Proceedings of the 2019 IEEE/CVF International Conference on Computer Vision (ICCV), Seoul, Republic of Korea, 27–28 October 2019.

29. Tang, H.; Liu, Z.; Zhao, S.; Lin, Y.; Lin, J.; Wang, H.; Han, S. Searching Efficient 3D Architectures with Sparse Point-Voxel Convolution. In Proceedings of the 16th European Conference on Computer Vision (ECCV), Edinburgh, UK, 23–28 August 2020.

30. Mohapatra, S.; Hodaei, M.; Yogamani, S.; Milz, S.; Gotzig, H.; Simon, M.; Rashed, H.; Maeder, P. LiMoSeg: Real-time Bird's Eye View based LiDAR Motion Segmentation. *arXiv* **2021**, arXiv:2111.04875.

31. Sun, J.; Dai, Y.; Zhang, X.; Xu, J.; Rui, A.; Gu, W.; Chen, X. Efficient Spatial-Temporal Information Fusion for LiDAR-Based 3D Moving Object Segmentation. *arXiv* **2022**, arXiv:2207.02201.

32. Kim, J.; Woo, J.; Im, S. RVMOS: Range-View Moving Object Segmentation Leveraged by Semantic and Motion Features. *IEEE Robot. Autom. Lett.* **2022**, *7*, 8044–8051. [CrossRef]

33. He, Z.; Fan, X.; Peng, Y.; Shen, Z.; Jiao, J.; Liu, M. EmPointMovSeg: Sparse Tensor Based Moving Object Segmentation in 3D LiDAR Point Clouds for Autonomous Driving Embedded System. *IEEE Trans. Comput. Aided Des. Integr. Circuits Syst.* **2022**, *42*, 41–53. [CrossRef]

34. Bai, L.; Lyu, Y.; Huang, X. RoadNet-RT: High Throughput CNN Architecture and SoC Design for Real-Time Road Segmentation. *IEEE Trans. Circ. Syst. Regul. Pap.* **2020**, *68*, 704–714. [CrossRef]

35. Bai, L.; Zhao, Y.; Elhousni, M.; Huang, X. DepthNet: Real-Time LiDAR Point Cloud Depth Completion for Autonomous Vehicles. *IEEE Access* **2020**, *8*, 7825–7833. [CrossRef]

36. Shen, J.; You, H.; Qiao, Y.; Mei, W.; Zhang, C. Towards a Multi-array Architecture for Accelerating Large-scale Matrix Multiplication on FPGAs. In Proceedings of the 2018 IEEE International Symposium on Circuits and Systems (ISCAS), Florence, Italy, 27–30 May 2018; pp. 1–5.

37. Wu, G.; Yong, D.; Miao, W. High performance and memory efficient implementation of matrix multiplication on FPGAs. In Proceedings of the International Conference on Field-Programmable Technology, FPT 2010, Beijing, China, 8–10 December 2010.

38. Chang, J.W.; Kang, S.J. Optimizing FPGA-based convolutional neural networks accelerator for image super-resolution. In Proceedings of the 2018 23rd Asia and South Pacific Design Automation Conference (ASP-DAC), Jeju, Republic of Korea, 22–25 January 2018; pp. 343–348.

39. Lyu, Y.; Bai, L.; Huang, X. ChipNet: Real-Time LiDAR Processing for Drivable Region Segmentation on an FPGA. *IEEE Trans. Circ. Syst. Regul. Pap.* **2019**, *66*, 1769–1779. [CrossRef]

40. He, K.; Zhang, X.; Ren, S.; Sun, J. Deep Residual Learning for Image Recognition. In Proceedings of the 2016 IEEE Conference on Computer Vision and Pattern Recognition (CVPR), Las Vegas, NV, USA, 27–30 June 2016; pp. 770–778.

41. Chen, L.-C.; Papandreou, G.; Schroff, F.; Adam, H. Rethinking atrous convolution for semantic image segmentation. *arXiv* **2017**, arXiv:1706.05587.
42. Yu, C.; Wang, J.; Peng, C.; Gao, C.; Yu, G.; Sang, N. BiSeNet: Bilateral Segmentation Network for Real-time Semantic Segmentation. In Proceedings of the 15th European Conference on Computer Vision (ECCV), Munich, Germany, 8–14 September 2018.
43. Ronneberger, O.; Fischer, P.; Brox, T. U-net: Convolutional networks for biomedical image segmentation. In Proceedings of the International Conference on Medical Image Computing and Computer-Assisted Intervention (MICCAI), Munich, Germany, 5–9 October 2015; pp. 234–241.
44. Long, J.; Shelhamer, E.; Darrell, T. Fully convolutional networks for semantic segmentation. *IEEE Trans. Pattern Anal. Mach. Intell.* **2015**, *39*, 640–651.
45. Behley, J.; Garbade, M.; Milioto, A.; Quenzel, J.; Behnke, S.; Stachniss, C.; Gall, J. SemanticKITTI: A Dataset for Semantic Scene Understanding of LiDAR Sequences. In Proceedings of the 2019 IEEE/CVF International Conference on Computer Vision (ICCV), Seoul, Republic of Korea, 27–28 October 2019.
46. Everingham, M.; Gool, L.V.; Williams, C.; Winn, J.; Zisserman, A. The Pascal Visual Object Classes (VOC) Challenge. *Int. J. Comput. Vis.* **2010**, *88*, 303–338. [CrossRef]
47. Geiger, A.; Lenz, P.; Urtasun, R. Are we ready for autonomous driving? The KITTI vision benchmark suite. In Proceedings of the 2012 IEEE Conference on Computer Vision and Pattern Recognition, Providence, RI, USA, 16–21 June 2012; pp. 3354–3361.
48. Bradski, G. The OpenCV Library. Available online: https://github.com/opencv/opencv/wiki/CiteOpenCV (accessed on 20 August 2022).

Article

3D Reality-Based Survey and Retopology for Structural Analysis of Cultural Heritage

Sara Gonizzi Barsanti [1,*], Mario Guagliano [2] and Adriana Rossi [1]

[1] Department of Engineering, Università deli Studi della Campania Luigi Vanvitelli, 81031 Aversa, Italy
[2] Department of Mechanical Engineering, Politecnico di Milano, 20156 Milan, Italy
* Correspondence: sara.gonizzibarsanti@unicampania.it

Abstract: Cultural heritage's structural changes and damages can influence the mechanical behaviour of artefacts and buildings. The use of finite element methods (FEM) for mechanical analysis is largely used in modelling stress behaviour. The workflow involves the use of CAD 3D models and the use of non-uniform rational B-spline (NURBS) surfaces. For cultural heritage objects, altered by the time elapsed since their creation, the representation created with the CAD model may introduce an extreme level of approximation, leading to wrong simulation results. The focus of this work is to present an alternative method intending to generate the most accurate 3D representation of a real artefact from highly accurate 3D reality-based models, simplifying the original models to make them suitable for finite element analysis (FEA) software. The approach proposed, and tested on three different case studies, was based on the intelligent use of retopology procedures to create a simplified model to be converted to a mathematical one made by NURBS surfaces, which is also suitable for being processed by volumetric meshes typically embedded in standard FEM packages. This allowed us to obtain FEA results that were closer to the actual mechanical behaviour of the analysed heritage asset.

Keywords: 3D modelling; 3D survey; retopology; NURBS; FEA; convergence analysis

Citation: Gonizzi Barsanti, S.; Guagliano, M.; Rossi, A. 3D Reality-Based Survey and Retopology for Structural Analysis of Cultural Heritage. *Sensors* **2022**, *22*, 9593. https://doi.org/10.3390/s22249593

Academic Editors: Miaohui Wang, Guanghui Yue, Jian Xiong and Sukun Tian

Received: 8 November 2022
Accepted: 29 November 2022
Published: 7 December 2022

Publisher's Note: MDPI stays neutral with regard to jurisdictional claims in published maps and institutional affiliations.

1. Introduction

Accurate 3D reality-based documentation is a must have for proper preservation and conservation in the cultural heritage field, as it is a prerequisite for more analyses. This documentation and these analyses are of more importance in contemporary times because of atmospheric agents, the growing of the cities and of the density of constructions, carelessness over the centuries, and the present political instability in certain areas that have all affected and strongly influenced the solidity of our heritage. It is hence fundamental to complete diagnostic studies aimed at valuing the level of decay of cultural heritage for selecting the appropriate preservation methods. However, it is challenging to calculate how a historical artefact suffers for environmental agents (e.g., earthquakes, pollution, wind, and rain) or human factors (e.g., construction in the environments, vehicular traffic, dense tourism). Hence, it is mandatory to find the best pipeline to obtain results as close as possible to reality. Finite element analysis (FEA) is a recognised technique used in engineering for various purposes (e.g., for modelling stress behaviour under mechanical and thermal loads), starting from a CAD 3D model made by non-uniform rational B-spline (NURBS) surfaces. Once imported in an FEA software, these 3D closed models can be meshed using modules capable of transforming a surface model to a volumetric one, which has nodes allocated in both the exterior and the interior volume, joined by simple volumes such as tetrahedron, pyramids, prisms, or hexahedral. In the mechanical area, this workflow is applicable because the 3D digital object to be simulated is close to the reality, within strict tolerances. Contrarywise, when applied to 3D models of cultural heritage (CH) objects or structures, the representation with a CAD model introduces a disproportionate level of approximation that can lead to incorrect simulation outcomes. Today, the 3D documentation of CH has been extensively matured through active sensors or passive approaches such as photogrammetry. The model obtained through these techniques is a surface formed by millions of triangles and is not suitable for direct use in FEA because the software is not able to manage so many polygons, and the computational complexity of FEA grows exponentially with the number of nodes representing the simulated object. Hence, a simplification is needed, and then a transformation of the superficial 3D meshes in volumetric models is needed to be meshed in the FEA software accordingly. Preliminary experiments were carried out on real CH artefacts surveyed with active or passive methods [1,2] for simulating stress behaviour and predicting critical damages. Analysing the results, few issues were made evident: (a) the creation of a volumetric model to be used in the FEA software from the raw 3D data is not yet clearly defined and may greatly affect the result, (b) the balance between geometric resolution and the accuracy and precision level of the simulated results is often not compatible with the shape of a 3D reality-based model.

The approaches used to generate the volumetric model from the acquired 3D point cloud are different: (a) using CAD software for the drawing of a new surface model following the superficial mesh originated by the acquired 3D cloud [3]; (b) using the triangular mesh generated by the 3D survey [4]; (c) generating a volumetric model from the 3D point cloud without preliminary surface meshing; (d) using the 3D reality-based model as the basis for a BIM/HBIM for FEA [5–7]; (e) creating new tools (Figure 1).

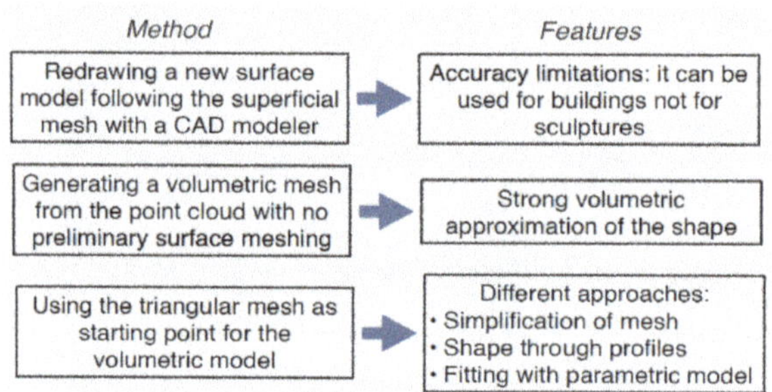

Figure 1. Pros and cons of the different processes used for the generation of volumetric models of cultural heritage for FEA.

The first methodology was used, for example, to simulate the structural behaviour of the Trajan's Markets [3] and in many other applications [8–11]. In some cases, the mathematical model was improved with the insertion of patches of reality-based meshes [12,13] that also used a strong discretization of the model using both a 3D CAD model and a 3D composite beam one [14,15]. Some projects used a joined system to obtain the 3D model, starting from the use of reality-based techniques, GPR, and radar [15]. Extracting or drawing cross-sections and profiles from the 3D reality-based model are ways to produce the CAD model extrapolated from a reality-based one, as in [16–18]. This procedure has its limitation related to the re-draw of the model in a CAD environment, but it can be applied to CH buildings for which the structure and the geometric details can be replicated through a CAD drawing using profiles. On the contrary, it cannot be used for statues, whose geometry is more complex and cannot be reduced through elements such as beam, truss, or shell, which are used for modelling in FEA.

The second approach has a variety of different methods: (i) plain simplification of the triangular mesh before converting it to a volumetric one, which may have important differences between the real shape and the simulated one [19]; (ii) simplified depiction of the shape as discretized profiles offering a low-resolution representation of the interior and the exterior of the structure, from which generate a volumetric model can be created [5,7]; the fitting of the acquired 3D model with a parametric one suitable to be transformed to volume [20].

The use of the triangular mesh for the structural analysis to assess the stability of a marble statue is highlighted in [21], analysing the mechanical stresses generated on the statue and the pedestal materials. The procedure begins with the subtraction of the mesh from a prismatic block shape and is called FEA in situ (the algorithm performs the analysis directly on the mesh without passing through the volumetric model). Some tests on mechanical objects were presented to corroborate this method.

The third strategy does not even take into consideration the mesh because it creates a volumetric approximation of the shape of the original 3D model from the raw 3D cloud of points [22], later compared by the same authors to other approaches [6,23]. The fourth methodology uses a 3D reality-based model to produce HBIM models to be used in FEA, implying two levels of approximation: (i) the first one related to the drawing of a BIM model from 3D reality-based mesh; (ii) creating the volumes for FEA starting from the BIM models. In [24], the authors investigated the use of BIM models to assist sites with monitoring and management, extrapolating thematic information for structural analysis, even if the authors did not provide a finite element analysis on the structure. Another example is the creation of the HBIM from both archival data and a laser scan survey. The model created was then segmented and utilised for structural analysis [25]. The Masegra Castle, located near the city of Sondrio in Italy, served as a test object for an original procedure called Cloud-to-BIM-to-FEM [26]. In this case, the basis for the HBIM was an accurate survey combining geometrical features, diagnostic analysis based on destructive and non-destructive inspections, material data, element interconnections, and architectural and structural considerations. This model was then converted into a finite element model with a geometric rationalization, considering irregularities and anomalies (e.g., verticality deviation and variable thickness).

Directly using the 3D reality-based models in FEA has its advantages because it overcomes all the approximations seen in the previous works. It is necessary to simplify the 3D meshes originated by this type of survey to make them suitable for FEA software. The best procedure for this, maintaining the accuracy of the 3D reality-based model, is using retopology, which implies a strong simplification of the mesh connected to the topological rearrangement of it, hence the creation of a new topology for the 3D model [27,28]. The retopologised mesh is normally based on quadrangular element (quads). Its main advantage is the reorganization of the polygonal superficial elements of the meshes for their better distribution on the surface. This procedure allows for the application of a huge simplification of the meshes without losing the initial accuracy of the models. The more organised structure of the elements on the mesh helps also in the conversion of it in NURBS, while maintaining a better coherence with the digitized artefact. This is valuable when working with reality-based 3D models of cultural heritage, usually accurate and precise but with a complex geometry.

The method proposed is based on the simplification of 3D meshes and their export into mathematical ones, close as much as possible to the real shape of the object surveyed but suitable to be converted into rationally complex volumetric 3D models. This permits us to obtain volumetric models of cultural heritage artefacts, increasing the closeness of the resultant NURBS model with the acquired one and, in the meantime, reducing the number of NURBS patches necessary for describing it. This methodology can be useful for experts such as archaeologists, architects, restores, and structural engineers, given that the lack of funding usually affects the restoration's interventions. An accurate pipeline can help in locating probable causes for future problems, allowing a more effective conservation of the artefact.

Three different case studies are discussed, showing the accuracy of the methods and the application on real cultural heritage objects.

2. Materials and Methods

The most complex part of executing structural analysis on cultural heritage artefacts is related to the geometrical complexity of the object analysed and the fact that they, especially buildings, are built with different construction techniques and different materials. This circumstance leads to the consideration of different points when dealing with structural analysis in the field of cultural heritage. The intrinsic characteristics of the different elements involved in the pipeline and the non-linear and non-symmetric geometry of the structures influence the choice of the procedure. Therefore, the methodology proposed took into consideration the use of retopology for the decimation of the reality-based 3D models.

3. 3D Modelling, Post-Processing, and Orientation of the Model

The test objects were surveyed with photogrammetry and laser scanning, considering the final aim of the modelling. The GSD (ground sample distance) and the accuracy of the scanner used were taken into consideration to obtain precise and accurate models and to obtain a value to compare the following stages of the pipeline proposed. For the validation of the methodology, a lab steel specimen for mechanical testing and a violin were used. Then, the methodology was applied to a statue of the Uffizi Museum of Florence (Figure 2a–c). The differences in these test objects were that, for the lab specimen, the analytical results of the mechanical tests were known, and for the violin, the results of FEA were compared to direct tests in the laboratory. Once the procedure was tested and validated, it was applied to a statue for which no analytical comparison was available.

The lab specimen has a cylindrical shape with a groove and is 148.28 mm long, with the larger diameter of 20 mm and the smaller of 7.4 mm. The choice of this object was made considering (i) its shape that originated from the revolution of a profile defined by an analytical function. This allowed us to calculate the stresses in critical points relatively easily, thus permitting the assessment of the results of an FE run with respect to the analytical solution. The results obtained can be used as theoretical reference; (ii) the physical object can be used for laboratory tests in different stress conditions, experimentally measuring its mechanical behaviour, which can be used as an experimental reference; (iii) the 3D model of the object can be generated with 3D digitization, and the FEA can be applied to different instances of the 3D model created with different 3D simplification methods, allowing us to prove the value of the proposed method. The specimen was surveyed with a structured light Solutionix Rexcan CS device (Table 1).

(a)

(b)

Figure 2. *Cont.*

(c)

Figure 2. The three tested objects: (**a**) the lab specimen, (**b**) the contemporary violin, and (**c**) the statue of the Gladiator in the Uffizi Gallery.

Table 1. Specification of the active 3D device used in the experiments.

Element	Description
Camera resolution	2.0 Mega/5.0 Mega pixel
Distance among points	0.0035~0.2 mm
Lenses	12, 25 and 50 mm
Working distance	570 mm
Principle of scan	Optical Triangulation
Dimensions	$400 \times 110 \times 210$ mm
Weight	40 N
Light	Blue LED
Unit	mm

The blue-light sensor for the pattern projection is suitable for digitizing small and medium not-totally Lambertian objects and is considered the most precise type of sensor for 3D digitization in mechanical engineering. The survey was carried out by locating the optical head on a base connected to a rotating plate (TA-300) composed of two axes, one for rotation around the vertical direction and one for the oscillation. The rotation allows for movement of $\pm\,180°$, and the axis of oscillation allows the inclination of the plane, where the specimen is located, up to 45° with respect to the vertical direction. With this scanning range, it is possible to limit to the minimum blind spots by reducing the scanning time up to 40%. The range device can mount three different lenses, for which the specifications are summarized in Table 2. Given the size of the test object, wide-angle lenses (12 mm) that were calibrated using the appropriate calibration table fixed on the turntable were chosen.

Table 2. Identification of the different lenses available.

Lens	FOV/Diagonal	Distance among Points	Estimated Uncertainty
50 mm	85 (S) mm	0.044 mm	0.010 mm
25 mm	185 (S) mm	0.097 mm	0.020 mm
12 mm	370 (S) mm	0.200 mm	0.030 mm

The 3D device works with proprietary software for the acquisition and alignment phases. For the survey, the Multiscan setting was used with an oscillation of $\pm 30°$ and $\pm 150°$ rotations for a total of 36 scans for each position of the object on the turntable (12 scans, one for each rotation for the three different oscillations, $-30°$, $0°$, and $30°$). For the survey of the steel specimen, two groups of 36 scans were performed, automatically aligned by the Solutionix software, with a final RMS error of 18 µm.

The contemporary violin was surveyed with a six-axis arm laser scanner with a tolerance of 0.03 mm by the company that provided the mesh model. The violin was used because it was possible to conduct some acoustic test in the laboratory on the two separate sides of the artefact, which were then compared to the FEA on the retopologised models.

The statue was surveyed through photogrammetry using a Mirrorless APS-C SONY ILCE 6000 camera coupled with a 16 mm lens, which acquired the images even when the distance between the object and the camera was short. The distortion parameters of the lenses were corrected through the automatic calibration of the camera and the lens during the alignment phase in Agisoft Metashape, the software used to create the 3D model. The setting of the camera was ISO equal to 1000 and the focal length at 5.6. The GSD obtained was 1 mm (The survey was performed by Dott. Umair Shafquat Malik of Politecnico di Milano during a joint project with the Indiana University (coordinator Prof. Bernie Frisher). In 2019, the Uffizi Gallery in Florence with an agreement with Indiana University (IU-USA), started 3D digitization of its complete Roman and Greek sculptural collection in which the Reverse Engineering and Computer Vision group of Politecnico di Milano was involved as a technical partner, under the coordination of Prof. Gabriele Guidi [Virtual World Heritage Lab 2019].).

After the survey, the models were post-processed with the correction of the topological errors and of their orientation on a suitable reference system, with the XY plane corresponding to the base of the model and the z-axis passing from the centre of gravity of the artefact. The post-processed meshes were then simplified with retopology, converted in closed NURBS and then into volumetric models. The use of retopology is also valuable when converting the superficial meshes in NURBS because the process tends to generate a higher number of patches when the original mesh is topologically unorganized. Hence, the rearrangement of the initial topology of the mesh can be seen as a preliminary condition for minimizing the number of NURBS patches of the converted model. In this way, the mesh embedded in FEA software works better. After all these passages, the 3D models were finally prepared for the finite element analysis.

3.1. Simplification of the Models: Triangular and Retopology Method

3D models can be simplified with different strategies. The first approach is based on triangular simplification, and the second one involves the transformation of the original triangular mesh into a quadrangular one, its retopology and projection of the nodes on the original triangular mesh (Figure 3), according to the method described in [29]. The geometrical complexity of the models was exemplified in terms of nodes of the mesh (or vertexes) rather than in terms of polygons because the counting of polygons on a mesh is different if the shape is triangular or quadrangular. For the latter, the number of nodes and polygons is approximately the same, while for triangular meshes the number of polygons is approximately double that of the nodes (This is obvious since a squared element, once divided in two parts on one of its diagonals, produce two triangles.). It must be borne in mind that the number of vertexes of a mesh is what defines the surface sampling;

therefore, it was used as an indicator for the level of detail of each mesh, independently of its triangular or quadrangular arrangement.

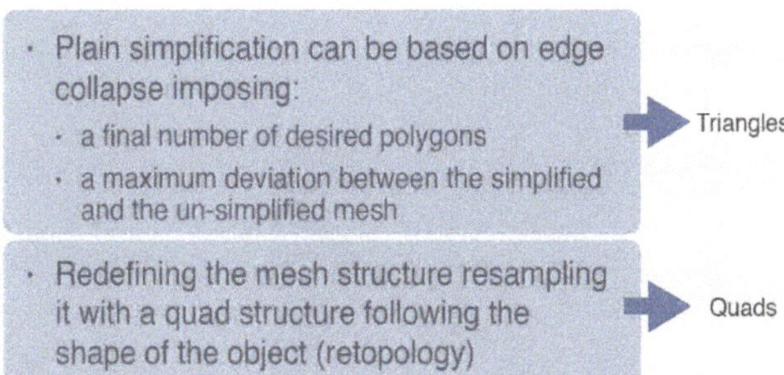

Figure 3. Comparison between the triangular and the retopology simplification.

3.1.1. Triangular Simplification

The triangular mesh is a set of vertexes V = (v1, v2, ... , vk) and faces F = (f1, f2, ... , fn). The simplification process obtains a surface M' as similar as possible to the initial high resolution mesh M by lessening the number of the element on the surface. The simplification process is usually controlled by a set of user-defined quality criteria that can preserve specific properties of the original mesh as much as possible (e.g., geometric distance, visual appearance, etc.).

There are different approaches, the majority of which involves the degradation of the mesh to reduce the number of polygons [30,31]:

- Vertex decimation: it iteratively removes vertices and the adjacent faces. It preserves the mesh topology. The sequential optimization process manages the removal of points from the triangulation, leading to a gradual increase of its overall approximation error [32].
- Energy function optimization: the algorithm assigns an energy function to the number of nodes, the approximation error, and the length of the edges that regulate the regularity of the mesh. It produces higher results, minimizing the energy and solving the mesh optimization problems but increasing the computational cost.
- The agglomeration of vertices or vertex clustering: it partitions the mesh vertices into clusters and merges all the vertices in a cluster into one single vertex.
- Region merging–face clustering: it works on coplanarity. As a planarity threshold is set, the neighbourhood of each triangle is evaluated, and all the triangles that are inside the threshold are merged.

In this work, the first algorithm, implemented in the Polyworks software package (IMCompress), was used.

3.1.2. Retopology

For the retopology process, this feature is available both in open-source packages, such as InstantMeshes [33] or Blender, or in commercial software packages, such as ZBrush by Pixologic, used in this work because this software is built specifically for rebuilding the topology of the models. Blender showed some problems in managing big files, while InstantMeshes, even if powerful, gave sometimes inadequate meshes as results, with holes and missing parts, especially when strongly decreasing the number of nodes. Moreover, ZBrush has the option of projecting the retopologised model onto the high resolution one, increasing the adherence of the two models.

The tool used for retopology was the ZRemesher, which is optimized to work on all kinds of structures and shapes but will by default produce better results with organic shapes, since Zbrush is specifically for the creation of video game characters. In the ZRemesher palette, there is the possibility to select the number of polygons desired for the retopology and the choice of increasing the coherence of the two models. This can be performed by selecting the "adapt" button and increasing the value of the "adapt size" slider. This is an important parameter for ZRemesher because it defines the polygon distribution on the model, and it can drastically increase the quality of the topology by giving more flexibility to the algorithm (Figure 4a,b).

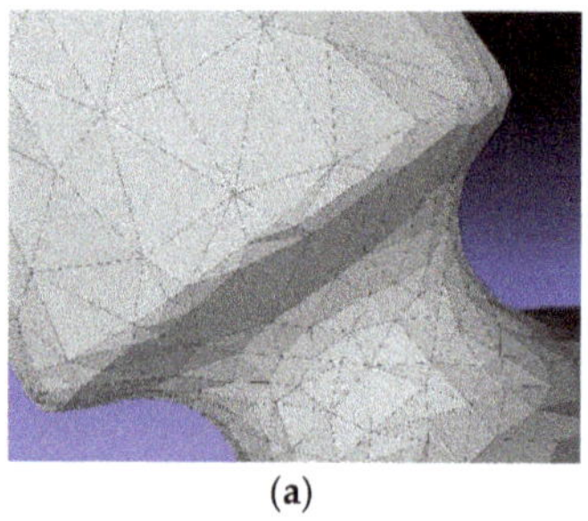
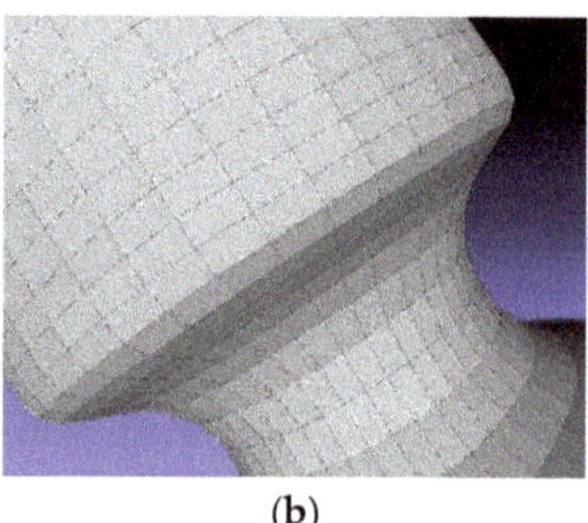

(a)   (b)

Figure 4. Difference between the triangular simplification (**a**) 9A and retopology simplification (**b**) on the lab specimen.

This function defines a vertex ratio based on the curvature of the mesh. A low setting provides polygons that are as square as possible and almost the same size, a number of final polygons closer to the number set in the selection tool, but it can introduce topology irregularities where the geometry is more complex. A high adaptive size means obtaining polygons that are rectangular in shape to best fit the mesh's curvature and for which density can vary along the mesh surface even if the program creates smaller polygons where the geometry requires. With a higher value of this parameter, there is less control on the final number of desired polygons after retopology. The adaptive size quantity goes from 0 to 100; it is not a unit, but it is a number that is only referred to for the different quantities and different settings of the quadrangular elements on the retopologised mesh.

Thus, by increasing the value of the adaptive size, the quality of the retopology increases but the program is more elastic regarding the target number of final polygons. This happens because, when the desired number of polygons is set, the software distributes them equally on the surface and then analyses the curvature, deforming the shape of the polygons or changing their density to be more adherent to the initial mesh.

3.2. NURBS Conversion

A mesh represents 3D surfaces with a series of discreet faces, similar to how pixels form an image. NURBS, on the contrary, are mathematical surfaces that can represent complex shapes with no granularity that is in mesh. The conversion from mesh to NURBS is implemented in CAD software or similar software (e.g., 3DMax, Blender, Rhinoceros, Maya, Grasshopper, etc.), and it transforms a mesh composed by polygons or faces to a faceted NURBS surface. In detail, it creates one NURBS surface for each face of the mesh and then merges everything into a single polysurface.

Depending on the mesh, the conversion works in different ways:

- If the starting point is a triangular mesh, and while, by definition, triangles are plane, the conversion creates trimmed or untrimmed planar patches. The degree of the patches is a 1×1 surface trimmed in the middle to form a triangle.
- If the starting point is a quadrangular mesh, the conversion creates four-sided untrimmed degree 1 NURBS patches, meaning that the edges of the mesh are the same as the outer boundaries of the patches.

- Considering the theory, a quadrangular mesh is more suitable to be converted into NURBS (Figure 5a,b). For this work, the MeshToNurb tool implemented in Rhinoceros was used.

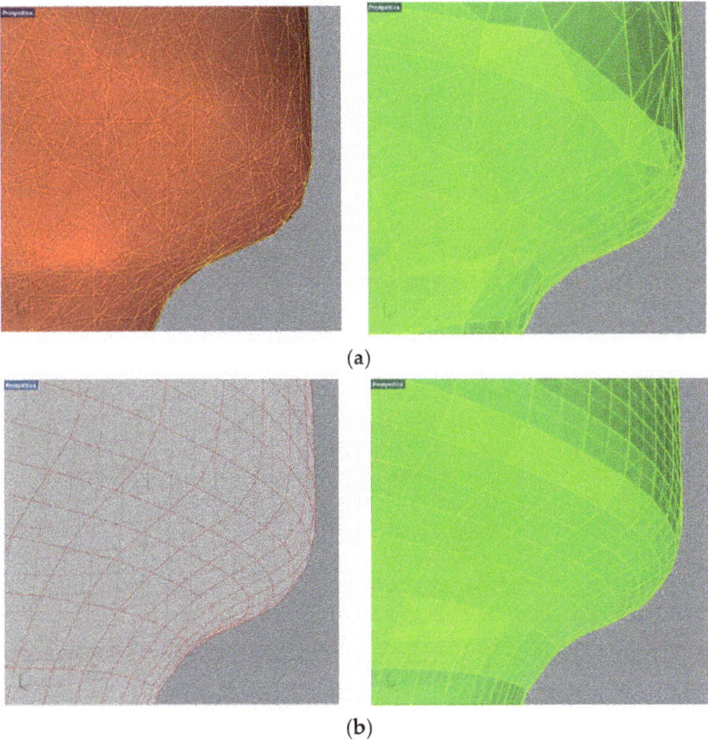

Figure 5. The comparison between a simplified triangular mesh and its corresponding NURBS (**a**) and the retopologised mesh with its corresponding NURBS (**b**). The meshes are represented in red (left side of the images) and the NURBS in green (right side of the images).

3.3. Finite Element Analysis

The analysis was carried out on the models meshed using given elements, and they were different if a 2D or a 3D problem was evaluated. The 2D elements are the triangular and the quadrangular, while the 3D elements are the tetrahedral and the hexahedral; the hexahedral ones are more accurate (e.g., deform in a lower strain energy state) but it is more difficult to mesh a 3D volume with this kind of element if it is not segmented [34]. The 3D elements can be linear or quadratic; the difference is that the quadratic ones have nodes also on the mid side, varying in number from four nodes (linear tetrahedron) to 20 nodes (quadratic hexahedron), and the shape functions vary from linear to quadratic, allowing a more accurate description of the geometry and the displacement of the nodes. In the present study, the linear elements were chosen to simplify FE modelling from the geometrical model and to consider the convergence of the results by increasing the mesh density to assess the accuracy of the results.

4. Results

To validate the proposed method, some tests were conducted both on a case with a known analytical solution, using this one as a term of comparison, and on a case with experimental measurements conducted, also in this case to use the latter as a term of comparison to check the goodness of the developed finite element model and to check if the mesh size is adequate to obtain reliable results.

In fact, the best way to assess if a finite element model has the proper mesh size is to perform convergence studies by increasing the element count in the model and assuring that the result of interest graphically converges to a stable value. Mesh convergence in finite element analysis is related to the smallest dimension of the element of the mesh (how many elements are required in a model) to ensure that the results of the analysis are not influenced by the changing size of the mesh. At least three element dimensions are required to compute a convergence test, which happens when an asymptotic behaviour of the solution shows up, meaning that the difference among the results becomes smaller or equal. To determine mesh convergence, a curve of a stress parameter is required, plotted against different sizing in mesh density. The laboratory steel specimen was chosen as the test object because of its simple geometry and because it is possible to calculate the analytical result for the stress analysis; a simple traction analysis was carried out. Three different simplifications were used, triangular, retopology with the adaptive size parameter of 30 (low value), and retopology with the adaptive size parameter of 100 (highest value). It was decided to opt for these values to compare the results of a retopology with a distribution of the elements that was more rigid. Hence, less adaptation to the geometry of the object surveyed was performed, and a retopology with a final number of elements was more erratic (the control on the target number set in the software was less accurate) but with more adaptation to the geometry of the object. The high-resolution model was composed by 345,026 polygons and was simplified starting from the lower number of polygons accepted by the software, 500 polygons. Then, the number was doubled until the highest possible number in the retopology software, 95 K polygons, with a final number of nine models for the triangular and the adaptive size of 30 for retopology simplification and eight for the adaptive size of 100 for retopology, because, with this parameter, a simplification of 500 polygons did not give a proper result. All the models obtained were then compared with the high-resolution model (Tables 3 and 4; Figures 6 and 7).

Table 3. The mean in mm of the simplified models for each simplification method compared to the high-resolution one.

	500	1000	2000	4000	8000	16,000	32,000	64,000	95,000
adapt_30	0.1914	0.1063	0.0568	0.0246	0.0138	0.0078	0.0043	0.0024	0.0019
adapt_100		0.2707	0.1337	0.0457	0.0164	0.0078	0.0038	0.0022	0.0017
triangular	0.1678	0.0631	0.0269	0.0131	0.0091	0.0037	0.0011	0.0005	0.0003

Table 4. The standard deviation in mm of the simplified models for each simplification method compared to the high-resolution one.

	500	1000	2000	4000	8000	16,000	32,000	64,000	95,000
adapt_30	0.145	0.087	0.056	0.033	0.003	0.024	0.017	0.011	0.009
adapt_100		0.162	0.084	0.036	0.023	0.021	0.015	0.01	0.008
triangular	0.15	0.07	0.038	0.03	0.024	0.014	0.009	0.005	0.002

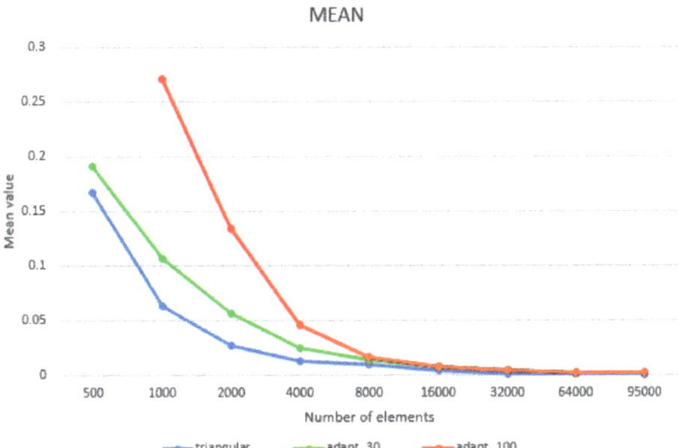

Figure 6. The mean in mm of the simplified models for each simplification method compared to the high-resolution one. Expressed in a graph.

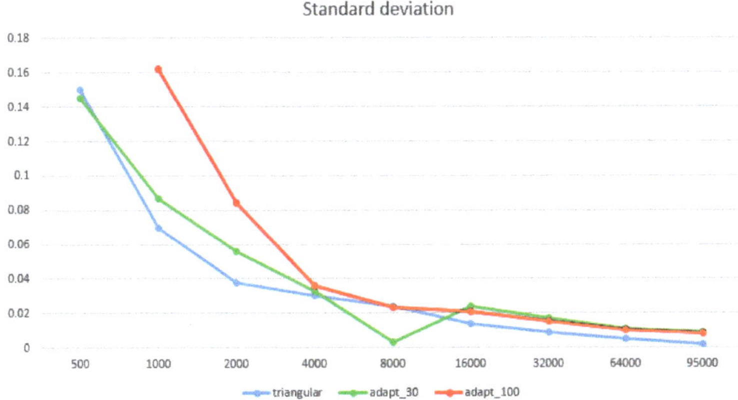

Figure 7. The standard deviation in mm of the simplified models for each simplification method compared to the high-resolution one. Expressed in a graph.

The metric comparison reported in the tables shows that both the mean value and the standard deviation of the point distance between the vertexes of the reference model and the mesh of each simplified model were, as expected, higher for the retopologised model than for the triangular mesh because, with retopology, a smoothness is added to the models (The mean of the distribution gives the average position of the cloud along the normal direction, i1 and i2, and the standard deviation gives a local estimate of the point cloud roughness $\sigma1(d)$ and $\sigma2(d)$ along the normal direction. If outliers are expected in the data, i1 and i2 can be defined as the median of the distance distribution and the roughness is measured by the inter-quartile range. The local distance between the two clouds LM3C2(i) is then given by the distance between i1 and i2. Hence, the mean or median is the estimate of the local average position of each cloud [35], p. 8).

Given the fact that the σ of the range device is equal to 0.015 mm, the simplification aimed at creating a situation of strong difference in the final number between the different models; however, it maintained a deviation between the simplified model and high-resolution mesh not exceeding 0.2 mm, which was equal to the graphic error on paper.

Another important parameter that was considered was the number of final patches created during the automatic process of conversion from meshes to NURBS. As shown in Table 3, the number of patches in the triangular models was higher than in the other models. This can be easily explained by the fact that the patches in the NURBS had a quadrangular shape and converting a triangular mesh to NURBS involves the subdivision of each patch in two. Comparing the models derived from the two different retopology processes, it is interesting to note that the adaptive size 100 implies a number of patches slightly higher than the other process for the models with 1 K and 2 K nodes, while, by increasing the number of nodes, the result is reversed (except for the 95 K model, singularity explained with the fact that with these settings this model has a number of nodes higher than 95 K, in detail, more than 3 K nodes more than the adapt 30 model). This can be explained in a better coherence of these models on the high resolution one. The reprojection of the retopologised models on the target one permits a better geometry and a better arrangement of the elements than the patches on the surface.

The NURBS were then exported in *. step extension to create a volumetric model.

The analysis carried on was a traction analysis. The first step was to calculate the analytical result for traction on this specific object:

$$\sigma = K_t * \sigma_n$$

where

$$K_t = 1.66 - theoretical\ stress\ concentration\ factor\ under\ tensile\ axial\ load\ N.$$

$$\sigma_n = N/A = 7MPa = nominal\ stress\ (= load/cross\ sec\ tion\ area)$$

with

$$A = area\ of\ the\ cross\ section\ (smaller\ diameter\ 7.4mm)$$

$$N = 300N - applied\ tensile\ force$$

Thus, the analytical result is

$$\sigma = K_t * \sigma_n = 1.66 * 7 = 11.62\ MPa$$

The tensile test analysis was performed on each simplified model, imposing the following parameters:

- Young's Modulus for steel 200,000 MPa;
- Poisson Ratio 0.3;
- Tensile load on Z axis 300 N;
- Displacement as boundary conditions on the other plane face, components X = 0, Y = 0 and Z = 0;
- Meshing element: 10-nodes tetrahedrons (quadratic shape functions);
- Element size: from 0.1 to 2.5 mm;
- Size function adaptive;
- Fast transition in filling the volume.

The results are summarized in Table 5.

To better understand and read the results and to easier analyse the convergence for each model in the three simplifications processes, it was decided to convert them in percentage, giving the maximum rate of ±3% of the analytical result, as shown in Tables 6–8 and Figures 8–10.

Table 5. The results expressed in MPa of the FEA analysis on traction on the different models of the laboratory specimen. The different colours highlight the values that were recurrent in the analysis even with different element dimensions.

0.1 mm size mesh Ansys									
	500	1000	2000	4000	8000	16,000	32,000	64,000	95,000
triangular	36,784	26,421	18,862	17,555	17,927	16,192	14,874	/	/
adapt_30	32,464	29,033	24,009	19,641	17,293	16,056	15,317	14,328	13,974
adapt_100		15,895	15,252	14.5	15,715	15,424	14,583	13,945	13,785
0.5 mm size mesh Ansys									
	500	1000	2000	4000	8000	16,000	32,000	64,000	95,000
triangular	17,835	14,506	14,269	13,312	13,521	13,303	12,711	12,201	/
adapt_30	16,171	16,803	14,395	14,243	14,079	12,584	12.66	12,067	12,304
adapt_100	/	13,704	13,718	12,912	12,686	12,525	12,527	11,983	12,305
1 mm size mesh Ansys									
	500	1000	2000	4000	8000	16,000	32,000	64,000	95,000
triangular	15,478	13,401	12.9	12.05	12,459	12,707	11,943	12.04	/
adapt_30	13,522	12.27	13,282	12,034	12,916	11,584	11,855	12,084	12,304
adapt_100	/	12,415	12,187	11,809	15,575	11,714	11,792	12,026	12,305
1.5 mm size mesh Ansys									
	500	1000	2000	4000	8000	16,000	32,000	64,000	95,000
triangular	13,985	13,722	11,935	12,102	11,986	11,499	11,634	12,037	/
adapt_30	12,954	11,493	10,486	11,502	11,299	11,616	11,855	12,084	12,304
adapt_100	/	12,503	12,253	11.63	11,546	11,695	11,792	12,026	12,305
2 mm size mesh Ansys									
	500	1000	2000	4000	8000	16,000	32,000	64,000	95,000
triangular	12,792	11,119	11,338	11,388	12,375	11,714	12,358	12.12	/
adapt_30	12,286	9,4354	10.75	10,885	11,299	11,616	11,855	12,084	12,304
adapt_100	/	12,467	11,385	11,826	11,567	11,695	11,792	12,026	21,305
2.5 mm size mesh Ansys									
	500	1000	2000	4000	8000	16,000	32,000	64,000	95,000
triangular	12,838	10,759	11.08	11.18	11,734	11,505	12,383	12.12	/
adapt_30	94,614	10,324	95,181	10,629	11,299	11,616	11,855	12,084	12,304
adapt_100	/	12,418	11,347	11,489	11,479	11,695	11,792	12,026	12,305

Table 6. Convergence analysis for the triangular simplified models. In yellow, the results gone to convergence are expressed in percentage in relation to the result given by the analysis and the analytical result calculated for the specimen.

	500	1000	2000	4000	8000	16,000	32,000	64,000
0.1	2.17	1.27	0.62	0.51	0.54	0.39	0.28	−1.00
0	0.53	0.25	0.23	0.15	0.16	0.14	0.09	0.05
1	0.33	0.15	0.11	0.04	0.07	0.09	0.03	0.04
1.5	0.20	0.18	0.03	0.04	0.03	−0.01	0.00	0.04
2	0.10	−0.04	−0.02	−0.02	0.06	0.01	0.06	0.04
2.5	0.10	−0.07	−0.05	−0.04	0.01	−0.01	0.07	0.04

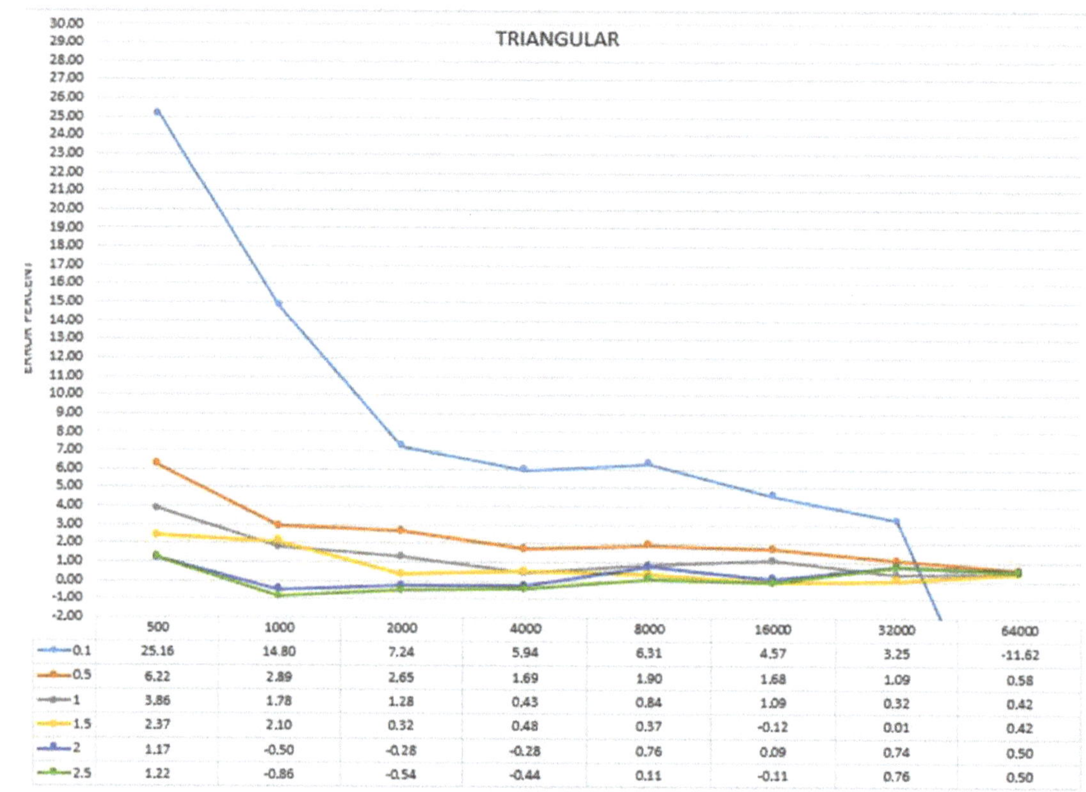

	500	1000	2000	4000	8000	16000	32000	64000
0.1	25.16	14.80	7.24	5.94	6.31	4.57	3.25	-11.62
0.5	6.22	2.89	2.65	1.69	1.90	1.68	1.09	0.58
1	3.86	1.78	1.28	0.43	0.84	1.09	0.32	0.42
1.5	2.37	2.10	0.32	0.48	0.37	-0.12	0.01	0.42
2	1.17	-0.50	-0.28	-0.28	0.76	0.09	0.74	0.50
2.5	1.22	-0.86	-0.54	-0.44	0.11	-0.11	0.76	0.50

Figure 8. Convergence analysis for the triangular simplified models. In yellow, the results gone to convergence are expressed in percentage in relation to the result given by the analysis and the analytical result calculated for the specimen. Expressed in a graph. (Left is ERROR PERCENT).

Table 7. Convergence analysis for the models simplified using retopology with the adaptive size fixed at 30. Highlighted in yellow are the results on convergence expressed in percentage in relation to the result given by the analysis and the analytical result calculated for the specimen.

	500	1000	2000	4000	8000	16,000	32,000	64,000	95,000
0.1	1.79	1.50	1.07	0.69	0.49	0.38	0.32	0.23	0.20
0.5	0.39	0.45	0.24	0.23	0.21	0.08	0.09	0.04	0.06
1	0.16	0.06	0.14	0.04	0.11	0.00	0.02	0.04	0.06
1.5	0.11	−0.01	−0.10	−0.01	−0.03	0.00	0.02	0.04	0.06
2	0.06	−0.19	−0.07	−0.06	−0.03	0.00	0.02	0.04	0.06
2.5	−0.19	−0.11	−0.18	−0.09	−0.03	0.00	0.02	0.04	0.06

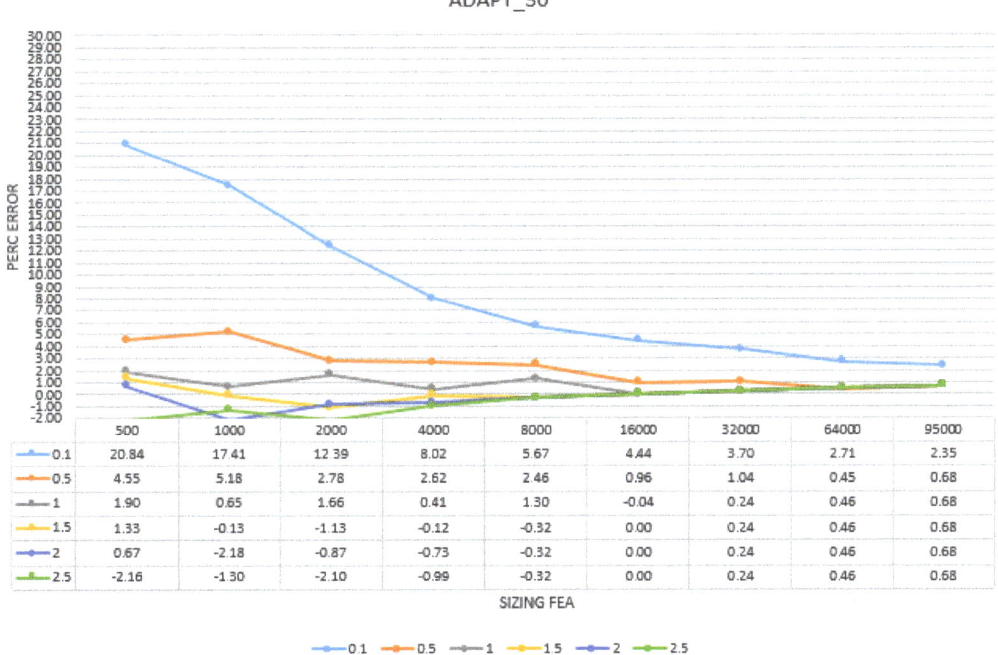

Figure 9. Convergence analysis for the models simplified using retopology with the adaptive size fixed at 30. Highlighted in yellow are the results on convergence expressed in percentage in relation to the result given by the analysis and the analytical result calculated for the specimen. Expressed in a graph.

Table 8. Convergence analysis for the models simplified with retopology and the adaptive size parameter fixed at 100. In yellow, the results gone to convergence are expressed in percentage in relation to the result given by the analysis and the analytical result calculated for the specimen.

	1000	2000	4000	8000	16,000	32,000	64,000	95,000
0.1	0.37	0.31	0.25	0.35	0.33	0.23	0.20	0.19
0.5	0.18	0.18	0.11	0.09	0.08	0.08	0.03	0.06
1	0.07	0.05	0.02	0.00	0.01	0.01	0.03	0.06
1.5	0.08	0.05	0.00	−0.01	0.01	0.01	0.03	0.06
2	0.07	−0.02	0.02	0.00	0.01	0.01	0.03	0.06
2.5	0.07	−0.02	−0.01	−0.01	0.01	0.01	0.03	0.06

The results gave important information regarding the best solution to be adopted when using a reality-based mesh as a starting point of FEA analysis. The triangular simplified models provided the worst results regarding both the analysis itself and the convergence; the results were not homogenous, and only the 16 K model showed a real convergence, starting from the 1.5 mm volumetric mesh. The retopology method showed much better results. The models produced using the low adaptive size parameter converged from the 8 K model meshed with 1.5 mm element size to the 32 K models, meshed from 1 to 2.5 mm element size. These results are much better than the ones obtained with the triangular superficial mesh.

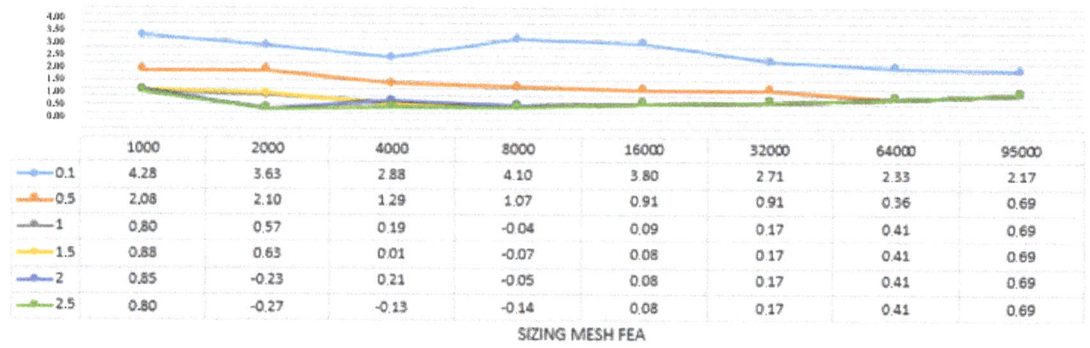

	1000	2000	4000	8000	16000	32000	64000	95000
0.1	4.28	3.63	2.88	4.10	3.80	2.71	2.33	2.17
0.5	2.08	2.10	1.29	1.07	0.91	0.91	0.36	0.69
1	0.80	0.57	0.19	-0.04	0.09	0.17	0.41	0.69
1.5	0.88	0.63	0.01	-0.07	0.08	0.17	0.41	0.69
2	0.85	-0.23	0.21	-0.05	0.08	0.17	0.41	0.69
2.5	0.80	-0.27	-0.13	-0.14	0.08	0.17	0.41	0.69

Figure 10. Convergence analysis for the models simplified with retopology and the adaptive size parameter fixed at 100. In yellow, the results gone to convergence are expressed in percentage in relation to the result given by the analysis and the analytical result calculated for the specimen. Expressed in a graph.

Finally, setting the adaptive size parameter at its higher value gave the best results. The convergence test was positive from the 2 K model meshed with 2 and 2.5 mm to the 64 K model, meshed from 0.5 to 2.5 mm element size. Compared to the other outcomes, the ones from this method were the most complete and homogeneous, even considering the results that did not go to convergence.

Another important parameter that came out from this test was that the analysis started to converge when the size of the volumetric model in the FEA software was close to the size of the superficial mesh. This information was fundamental to set the proper methodology for the tests on cultural heritage objects. For the triangular simplified models, given the inhomogeneous arrangement and dimension of the superficial elements, the comparison was conducted considering the mean value of the dimension of the elements. The convergence was reached when the models were meshed with the dimension of the tetrahedrons close to this value (Tables 9–11).

Table 9. The comparison between the dimension of the element of the superficial meshes and the volumetric element in FEA for the triangular simplified models. The column represents the dimension of the superficial mesh element, and the rows represent the dimension of the volumetric elements in the FEA software.

Element Dim	3–10	1.3–6	1–5	0.6–4	0.4–3	0.4–2	0.3–1.2	0.2–1.2
0.1	2.17	1.27	0.62	0.51	0.54	0.39	0.28	−1.00
0.5	0.53	0.25	0.23	0.15	0.16	0.14	0.09	0.05
1	0.33	0.15	0.11	0.04	0.07	0.09	0.03	0.04
1.5	0.20	0.18	0.03	0.04	0.03	−0.01	0.00	0.04
2	0.10	−0.04	−0.02	−0.02	0.06	0.01	0.06	0.04
2.5	0.10	−0.07	−0.05	−0.04	0.01	−0.01	0.07	0.04

For the retopologised models, as said, the models produced with the low adaptive size parameter showed a homogeneous dimension of the elements that are perfectly square. Additionally, in this case, the convergence analysis started when the two different element sizes were almost the same (Table 10).

Table 10. The comparison between the dimension of the superficial and the volumetric element for the retopologised models with the adaptive size parameter set to 30. The columns represent the size of the quadrangular mesh elements, while the rows represent the size of the volumetric ones.

Element Dim	5	3.7	2.7	1.8	1.2	0.9	0.6	0.4	0.3
0.1	1.79	1.50	1.07	0.69	0.49	0.38	0.32	0.23	0.20
0.5	0.39	0.45	0.24	0.23	0.21	0.08	0.09	0.04	0.06
1	0.16	0.06	0.14	0.04	0.11	0.00	0.02	0.04	0.06
1.5	0.11	−0.01	−0.10	−0.01	−0.03	0.00	0.02	0.04	0.06
2	0.06	−0.19	−0.07	−0.06	−0.03	0.00	0.02	0.04	0.06
2.5	−0.19	−0.11	−0.18	−0.09	−0.03	0.00	0.02	0.04	0.06

Finally, with the retopology method and the adaptive size parameter set at its maximum value, given the fact that especially with a lower number of nodes the shape of the elements was rectangular, the comparison was performed with the mean value. Also in this case, the convergence started when the two element sizes were similar (Table 11).

Table 11. The comparison between the two element sizes in the retopologised models with an adaptive size parameter of 100. The columns represent the size of the superficial element, and the rows represent sizes of the volumetric ones.

Element Dim	6 × 1.8	4 × 1.8	2.5 × 1.7	1.4	1 × 0.8	0.7	0.4	0.3
0.1	0.37	0.31	0.25	0.35	0.33	0.23	0.20	0.19
0.5	0.18	0.18	0.11	0.09	0.08	0.08	0.03	0.06
1	0.07	0.05	0.02	0.00	0.01	0.01	0.03	0.06
1.5	0.08	0.05	0.00	−0.01	0.01	0.01	0.03	0.06
2	0.07	−0.02	0.02	0.00	0.01	0.01	0.03	0.06
2.5	0.07	−0.02	−0.01	−0.01	0.01	0.01	0.03	0.06

Summarizing, the models derived from a simplification with triangular superficial elements showed the worst results both in the order of processing time and, more important, structural analysis results. The two retopology methods were equivalent regarding the processing time, but when dealing with the structural results, the method adopted with the high adaptive size parameter gave the best accordance both regarding the convergence analysis and the closeness to the analytical result. This can be explained with the higher adherence of these models to the high-resolution one, even if, with a strong simplification, the models are smoother than the others. The advantage of the rectangular elements and the adaptability of the models with this process make this the best method to be used for the structural analysis of heritage directly using the reality-based meshes. This is clear looking at the highlighted values in the tables above that indicate the percentage of the convergence value of the FE analysis. The highlighted values are the ones that went to convergence; thus, the results expressed in Tables 9–11 related to different types of simplification of the mesh indicate that using a higher adaptive size parameter in the retopology process helps to reach convergence in a more distributed and coherent way.

5. Discussion

Before using this methodology on objects and structures that cannot be tested in the laboratory, to corroborate the results, another validation test was performed on a handmade contemporary violin, the closest to the objects to which the methodology was set for.

The process was applied to evaluate a vibroacoustic numerical model of the violin, based on accurate structural modelling [36]. The violin was surveyed with a six-axis arm laser scanner with a tolerance of 0.03 mm. The post processing started from the correction of the alignment of the different meshes acquired (Figure 11). It was not possible to perfectly correct the misalignment on one side of the model, but the mesh was cleaned and completed where the missing parts or holes were visible.

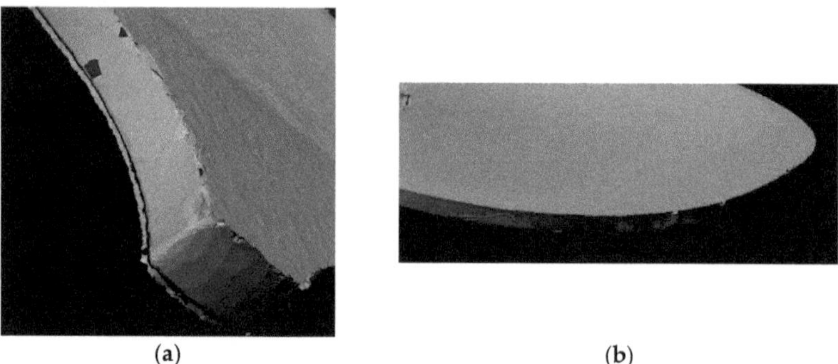

(a)   **(b)**

Figure 11. The initial mesh of the violin: (**a**) the misalignment of the different meshes on one side of the object; (**b**) holes on the surface.

After this step, the superficial mesh was simplified using retopology with the adaptive size parameter set to 100 to obtain a 22 K model for each part of the violin. From this superficial mesh, two NURBS were created and exported in *.iges format (Figure 12a,b).

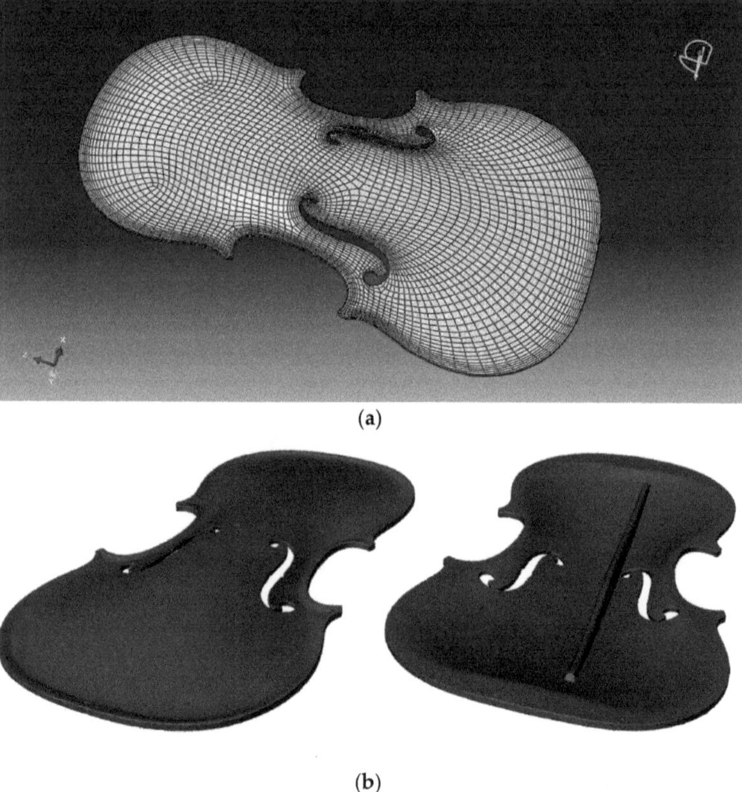

(a)

(b)

Figure 12. The retopologised model of the of the soundboard of the violin (**a**) and the NURBS obtained of both the soundboard and the back of the violin (**b**).

The volumetric models were then meshed in the FEA software, and some modal analyses were provided and then compared to the experimental ones (Figure 13). The results gave good accuracy compared to the laboratory ones. The slight difference between the results must be attributed to the error in the alignment of the initial mesh that caused a reduction of the thickness of one side of the violin of 0.4 mm. This test represents a solid validation of the process presented in this work because it was applied to an object with a complex geometry and for which it was possible to evaluate the results of the FE analysis with a test directly performed on the physical object. Furthermore, the modal analysis results depended on the mass and stiffness distribution along the entire model, and this evidenced the accuracy of the proposed method in respect to the actual mass and geometry of the analysed object.

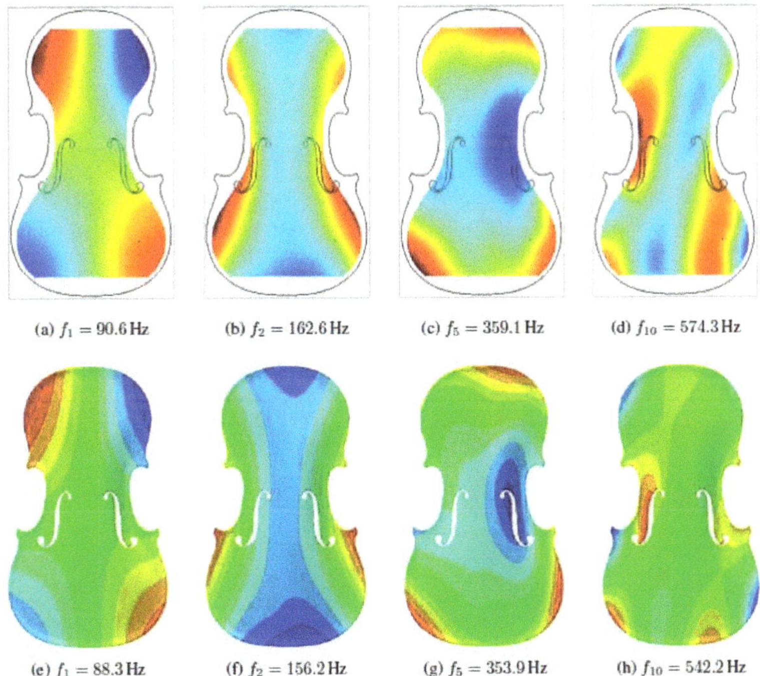

(a) $f_1 = 90.6\,\mathrm{Hz}$ (b) $f_2 = 162.6\,\mathrm{Hz}$ (c) $f_5 = 359.1\,\mathrm{Hz}$ (d) $f_{10} = 574.3\,\mathrm{Hz}$

(e) $f_1 = 88.3\,\mathrm{Hz}$ (f) $f_2 = 156.2\,\mathrm{Hz}$ (g) $f_5 = 353.9\,\mathrm{Hz}$ (h) $f_{10} = 542.2\,\mathrm{Hz}$

Figure 13. First, second, fifth and tenth vibration modes of the free soundboard from the experimental data (upper part of the image, (**a**–**d**)); same mode evaluated through the FEA (lower part of the image, (**e**–**h**)).

After these validation tests, the methodology was applied to the statue of the Gladiator, in the Uffizi Gallery in Florence. The high-resolution model had 1,141,268 faces and was simplified to 30 K nodes. The comparison of the triangular high-resolution model and the retopologised one gave a mean of 0.0001 m and a standard deviation of 0.0004 m (Figure 14).

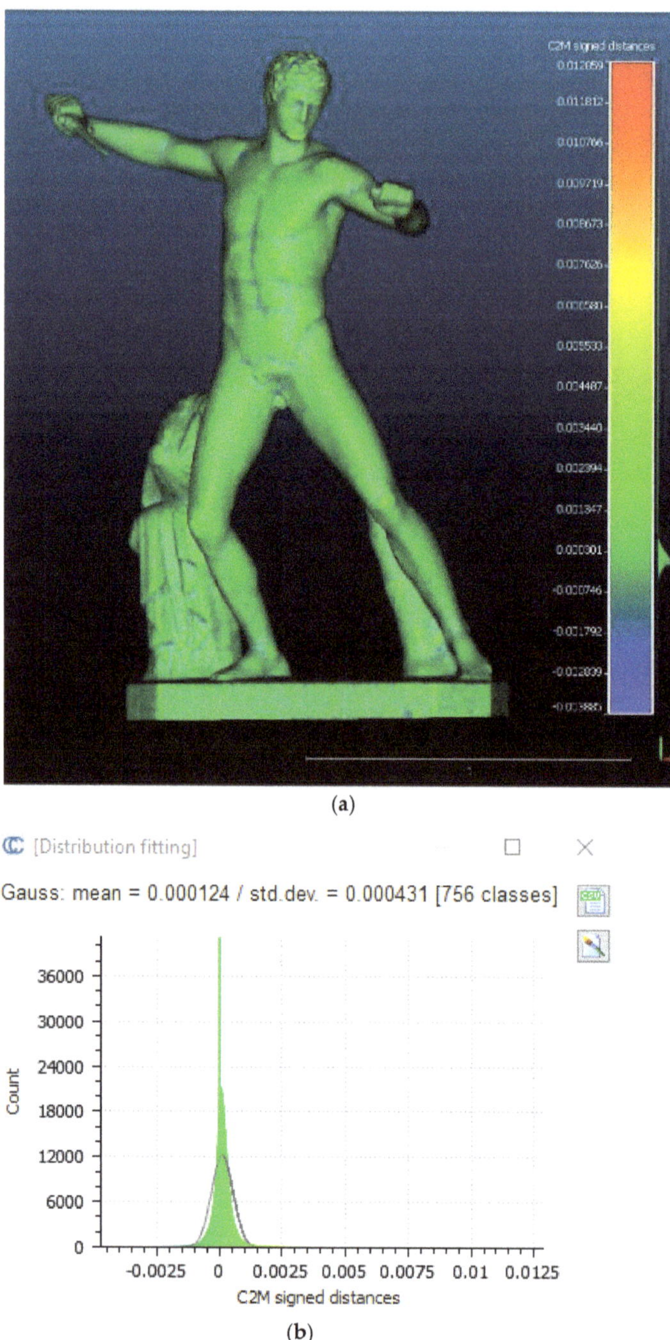

(a)

(b)

Figure 14. The comparison between the high resolution and the simplified retopologised model of the gladiator, where (**a**) is the graphical and visual comparison of the two meshes, and (**b**) is the gaussian distribution of the mean and the standard deviation calculated during the mesh-to-mesh comparison.

The simplified model was converted in NURBS, exported in *.step format and then imported in the FEA software for two different analyses: a static one imposing the gravitational load (self-weight) and a modal one, fixing 10 analyses to determine the natural mode shapes and frequencies of the object during free vibration (Figure 15). The following conditions were imposed in the structural analysis with a dimension of the volumetric element of 16 mm:

- Density: 2500 kg/m^3;
- Young's Modulus for marble 78,000 MPa;
- Poisson Ratio 0.3;
- Gravity on -Z axis;
- Fixed support under the basement of the statue as boundary condition;
- Meshing element: 10-nodes tetrahedrons;
- Element size: 50 mm.

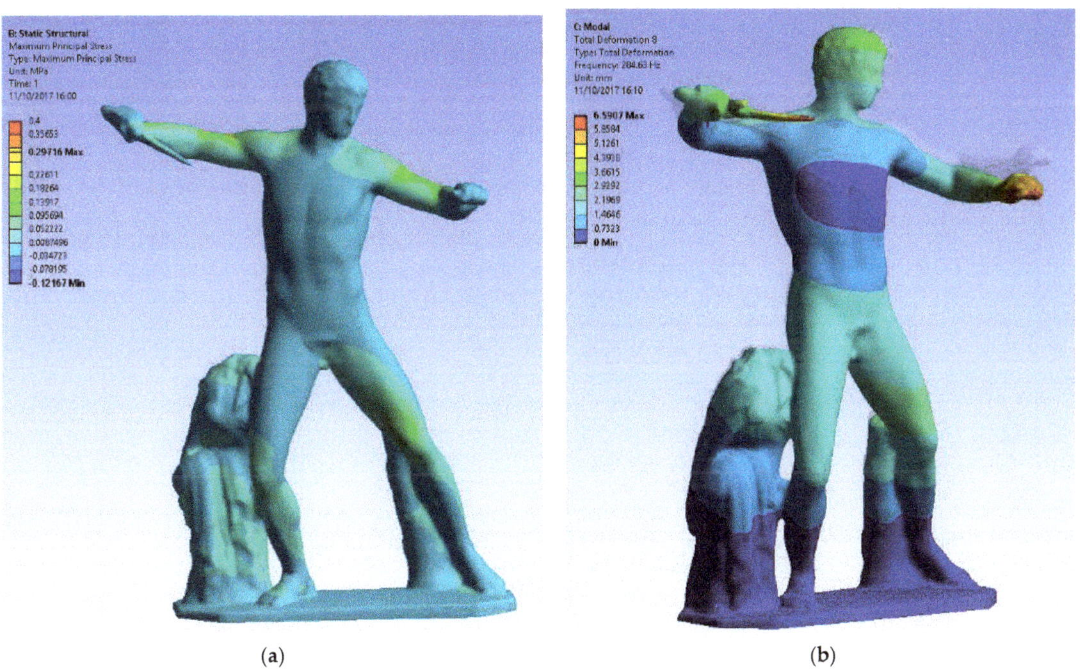

(a)   (b)

Figure 15. The static structural analysis imposing (**a**) the max principal stress under gravity on the –Z axis and (**b**) the modal analysis for the statue of the gladiator.

Having validated the method with the previous analyses, its application to a cultural heritage object indicates one of the most attractive applications of the method, which is to accurately determine the stress state of geometrical details (where the notch effect could case unexpected failures) with the static analysis and the global dynamic behaviour (depending upon the global schematization of the object) by means of an affordable model, both in terms of modelling and regarding the computational time.

The results show that the static analysis of the statue under its self-weight evidenced that there are not local details with severe stress concentration and that the stress is in each single point moderate and not dangerous. Further development could include occasional load due to some movement to check the proper way to move it without causing a dangerous situation.

Regarding the modal analysis, the dynamic behaviour was identified, and this can be an important aid for the evaluation of the behaviour of the statue under seismic loads, to address the proper way to protect the statue from this exceptional event. The results can be used as a term of comparison for simplified models, which respect the global mass and geometric properties of the statue, allowing a faster analysis without losing accuracy. From this point, this model can be considered itself a term of comparison for other models.

6. Conclusions

The use of retopology and the method proposed showed great performances using a model derived from a reality-based survey for finite element analysis. The use of the 3D measurement uncertainty as a simplification criterion allowed a considerable reduction in mesh size, maintaining a high accuracy of the simplified model compared to the high resolution one.

The main purpose of this work was benefit from the high-resolution reality-based models, considering their details, and use them for the structural analysis. The need to retain a high level of formal definition was acquired with the survey and was compatible with FEA software was the most important result.

The laboratory specimen demonstrated that the FEA results always give solutions closer to the analytical reference using volumetric models originated by the proposed method with retopology instead of a model created by a generic simplification of triangular meshes that showed not only a more arbitrary performance in the analysis but also lower accuracy.

Another important result obtained, because of the convergence analysis, was the detection of the best element size to be put in the FEA software to complete the analysis. The statement that using a size of the volumetric element close to the size of the elements of the meshes from which the volumetric model was created provided an important parameter to be set, especially when dealing with models of objects that cannot be tested in the laboratory. More information that can be acquired testing the process on a simple object and comparing the results with the analytical one more the pipeline can be robust. In this way, every single part of the process is verified, and the uncertainty of the process is minimized to the standard approximation of finite element analysis.

The tests on the violin confirmed the process on a more morphologically complex object, with the possibility to compare the results with the one performed in the laboratory.

Thanks to the tests conducted, it was also evident that the transformation of the meshes into NURBS, given the same number of point-nodes, since meshes interpolate points while NURBs approximate internal points constrained at the beginning and the end, worked better in the retopologised models, meaning a smaller number of patches. This is a fundamental point when transforming the superficial meshes into volumetric ones, both in the reduction of the processing time and, most important, in the accuracy of the results.

Applying the methodology to cultural heritage objects confirmed that it is possible to obtain an accurate FE analysis starting with an accurate simplified reality-based model, keeping the level of details and the geometric complexity of the initial high-resolution model.

The research is not closed with this work; some parts must be more deeply analysed. the first is imagining new paths of experimentation, where digital construction acquires the value of a descriptive language of exchange on several levels and is a diriment objective for critical survey and representation. Sharing and making models interoperable therefore directs the subsequent developments of the presented study. Ongoing efforts are focused on the opportunity to consider the properties of the different materials used in constructed artefacts in parallel with a more accurate segmentation analysis of the models. Having the possibility to subdivide the model in its main part (structural or decorative) and giving them the proper parameters (density, Young Modulus, Poisson's ration, element of the mesh) gains a higher accuracy in the analysis, especially when dealing with buildings. This type of test object is more interesting considering FEA because of their structural behaviour and the higher complexity of geometry.

Author Contributions: Conceptualization, S.G.B.; methodology, S.G.B.; software, S.G.B.; validation, S.G.B., M.G. and A.R.; formal analysis, S.G.B. and M.G.; investigation, S.G.B., M.G. and A.R.; resources, S.G.B., M.G. and A.R.; data curation, S.G.B. and M.G.; writing—original draft preparation, S.G.B.; writing—review and editing, S.G.B., M.G. and A.R.; supervision, S.G.B. All authors have read and agreed to the published version of the manuscript.

Funding: This research received no external funding.

Institutional Review Board Statement: Not applicable.

Informed Consent Statement: Not applicable.

Data Availability Statement: Not applicable.

Acknowledgments: The authors want to thank Gabriele Guidi of Politecnico di Milano for the valuable discussion of reality-based modelling and the help during the preparation of the PhD thesis.

Conflicts of Interest: The authors declare no conflict of interest.

References

1. Höllig, K. *Finite Element Methods with B-Splines. Society for Industrial and Applied Mathematics*; Society for Industrial and Applied Mathematics: Philadelphia, PA, USA, 2003. [CrossRef]
2. Alfio, V.S.; Costantino, D.; Pepe, M.; Restuccia Garofalo, A. A Geomatics Approach in Scan to FEM Process Applied to Cultural Heritage Structure: The Case Study of the "Colossus of Barletta". *Remote Sens.* **2022**, *14*, 664. [CrossRef]
3. Brune, P.; Perucchio, R. Roman Concrete Vaulting in the Great Hall of Trajan's Markets: A Structural Evaluation. *J. Archit. Eng.* **2012**, *18*, 332–340. [CrossRef]
4. Castellazzi, G.; Altri, A.M.D.; Bitelli, G.; Selvaggi, I.; Lambertini, A. From Laser Scanning to Finite Element Analysis of Complex Buildings by Using a Semi-Automatic Procedure. *Sensors* **2015**, *15*, 18360–18380. [CrossRef] [PubMed]
5. Shapiro, V.; Tsukanov, I.; Grishin, A. Geometric Issues in Computer Aided Design/Computer Aided Engineering Integration. *J. Comput. Inf. Sci. Eng.* **2011**, *11*, 21005. [CrossRef]
6. Doğan, S.; Güllü, H. Multiple methods for voxel modeling and finite element analysis for man-made caves in soft rock of Gaziantep. *Bull. Eng. Geol Environ.* **2022**, *81*, 23. [CrossRef]
7. Karanikoloudis, G.; Lourenço, P.B.; Alejo, L.E.; Mendes, N. Lessons from Structural Analysis of a Great Gothic Cathedral: Canterbury Cathedral as a Case Study. *Int. J. Archit. Herit.* **2021**, *15*, 1765–1794. [CrossRef]
8. Ferraioli, M.; Abruzzese, D. Seismic Assessment of Four Historical Masonry Towers in Southern Italy. *Cult. Herit. Sci.* **2021**, *2*, 50–60.
9. Erkal, A.; Ozhan, H.O. Value and vulnerability assessment of a historic tomb for conservation. *Sci. World J.* **2014**, *2014*, 357679. [CrossRef]
10. Riveiro, B.; Caamaño, J.C.; Arias, P.; Sanz, E. Photogrammetric 3D modelling and mechanical analysis of masonry arches: An approach based on a discontinuous model of voussoirs. *Autom. Constr.* **2011**, *20*, 380–388. [CrossRef]
11. Milani, G.; Casolo, S.; Naliato, A.; Tralli, A. Seismic assessment of a medieval masonry tower in northern Italy by limit, non-linear static and full dynamic analyses. *Int. J. Archit. Herit.* **2012**, *6*, 489–524. [CrossRef]
12. Zvietcovich, F.; Castaneda, B.; Perucchio, R. 3D solid model updating of complex ancient monumental structures based on local geometrical meshes. *Digit. Appl. Archaeol. Cult. Herit.* **2014**, *2*, 12–27. [CrossRef]
13. Pegon, P.; Pinto, A.V.; Géradin, M. Numerical modelling of stone-block monumental structures. *Comput. Struct.* **2001**, *79*, 2165–2181. [CrossRef]
14. Peña, F.; Meza, M. Seismic Assessment of Bell Towers of Mexican Colonial Churches. *Adv. Mater. Res.* **2010**, *133–134*, 585–590. [CrossRef]
15. Sánchez-Aparicio, L.J.; Rivero, B.; González-Aguilera, D.; Ramos, L.F. The combination of geomatic approaches and operational modal analysis to improve calibration of finite element models: A case of study in Saint Torcato Church (Guimarães, Portugal). *Constr. Build. Mater.* **2014**, *70*, 118–129. [CrossRef]
16. Guarnieri, A.; Pirotti, F.; Pontin, M.; Vettore, A. Combined 3D surveying techniques for structural analysis applications. *Int. Arch. Photogramm. Remote Sens. Spat. Inf. Sci.* **2005**, *36*, 6.
17. Sadholz, A.; Muir, C.; Perucchio, R. A 3D kinematic model for assessing the seismic capacity of the Frigidarium of the Baths of Diocletian. *Digit. Herit.* **2015**, *2*, 89–92. [CrossRef]
18. Chen, S.; Wang, S.; Li, C.; Hu, Q.; Yang, H. A Seismic Capacity Evaluation Approach for Architectural Heritage Using Finite Element Analysis of Three-Dimensional Model: A Case Study of the Limestone Hall in the Ming Dynasty. *Remote Sens.* **2018**, *10*, 963. [CrossRef]
19. Riccardelli, C.; Morris, M.; Wheeler, G.; Soultanian, J.; Becker, L.; Street, R. The Treatment of Tullio Lombardo's Adam: A New Approach to the Conservation of Monumental Marble Sculpture. *Metrop. Mus. J.* **2014**, *49*, 48–116. [CrossRef]

20. Calì, A.; Saisi, A.; Gentile, C. Structural Assessment of Cultural Heritage Buildings Using HBIM and Vibration-Based System Identification. In Proceedings of the 12th International Conference on Structural Analysis of Historical Constructions SAHC 2020, Barcelona, Spain, 16–18 September 2020; Roca, P., Pelà, L., Molins, C., Eds.; Scipedia: Barcelona, Spain, 2021. [CrossRef]
21. Bagnéris, M.; Cherblanc, F.; Bromblet, P.; Gattet, E.; Gügi, L.; Nony, N.; Mercurio, V.; Pamart, A. A complete methodology for the mechanical diagnosis of statue provided by innovative uses of 3D model. Application to the imperial marble statue of Alba-la-Romaine (France). *J. Cult. Herit.* **2017**, *28*, 109–116. [CrossRef]
22. Shapiro, V.; Tsukanov, I. Mesh free simulation of deforming domains. *CAD Comput. Aided Des.* **1999**, *31*, 459–471. [CrossRef]
23. Freytag, M.; Shapiro, V.; Tsukanov, I. Finite element analysis in situ. *Finite Elem. Anal. Des.* **2011**, *47*, 957–972. [CrossRef]
24. Oreni, D.; Brumana, R.; Cuca, B. Towards a methodology for 3D content models: The reconstruction of ancient vaults for maintenance and structural behaviour in the logic of BIM management. In Proceedings of the18th International Conference on Virtual Systems and Multimedia, Milan, Italy, 2–5 September 2012; pp. 475–482. [CrossRef]
25. Dore, C.; Murphy, M.; McCarthy, S.; Brechin, F.; Casidy, C.; Dirix, E. Structural Simulations and Conservation Analysis -Historic Building Information Model (HBIM). *Int. Arch. Photogramm. Remote Sens. Spatial Inf. Sci.* **2015**, *XL-5/W4*, 351–357. [CrossRef]
26. Barazzetti, L.; Banfi, F.; Brumana, R.; Gusmeroli, G.; Previtali, M.; Schiantarelli, G. Cloud-to-BIM-to-FEM: Structural simulation with accurate historic BIM from laser scans. *Simul. Model. Pract. Theory* **2015**, *57*, 71–87. [CrossRef]
27. Gonizzi Barsanti, S.; Guidi, G. A New Methodology for the Structural Analysis of 3D Digitized Cultural Heritage through FEA. *IOP Conf. Ser. Mater. Sci. Eng.* **2018**, *364*, 012005. [CrossRef]
28. Gonizzi Barsanti, S.; Guidi, G. A Geometric Processing Workflow for Transforming Reality-Based 3D Models in Volumetric Meshes Suitable for FEA. *Int. Arch. Photogramm. Remote Sens. Spat. Inf. Sci.* **2017**, *XLII-2/W3*, 331–338. [CrossRef]
29. Guidi, G.; Angheleddu, D. Displacement Mapping as a Metric Tool for Optimizing Mesh Models Originated by 3D Digitization. *J. Comput. Cult. Herit.* **2016**, *9*, 1–23. [CrossRef]
30. Franc, M. *Methods for Polygonal Mesh Simplification*; Technical Report No. DCSE/TR-2002-01; University of Bohemia in Pilsen: Pilsen, Czechoslovakia, 2002.
31. Taime, A.; Saaidi, A.; Satori, K. Comparative Study of Mesh Simplification Algorithms. In *Lecture Notes in Electrical Engineering, Proceedings of the Mediterranean Conference on Information & Communication Technologies 2015, Sadia, Morocco, 7–9 May 2015*; El Oualkadi, A., Choubani, F., El Moussati, A., Eds.; Springer: Cham, Switzerland, 2015; Volume 380.
32. Soucy, M.; Laurendeau, D. A General Surface Approach to the Integration of a Set of Range Views. *IEEE Trans. Pattern Anal. Mach. Intell.* **1995**, *17*, 344–358. [CrossRef]
33. Jakob, W.; Tarini, M. Instant Field-Aligned Meshes. *Siggraph Asia* **2015**, *34*, 15. [CrossRef]
34. Benzley, S.E.; Perry, E.; Merkley, K.; Clarck, B. A Comparison of All Hexagonal and All Tetrahedral Finite Element Meshes for Elastic and Elasto-plastic Analysis. In Proceedings of the 4th International Meshing Roundtable, Pittsburgh, PA, USA, 12–15 October 2008; Sandia National Laboratories: Albuquerque, NM, USA; Springer: Berlin/Heidelberg, Germany, 2008; pp. 179–191.
35. Lague, D.; Brodu, N.; Leroux, J. Accurate 3D comparison of complex topography with terrestrial laser scanner: Application to the Rangitikei canyon (N-Z). *ISPRS J. Photogramm. Remote Sens.* **2013**, *82*, 10–26. [CrossRef]
36. Corradi, R.; Liberatore, A.; Miccoli, S. Experimental Modal Analysis and Finite Element Modelling of A Contemporary Violin. In Proceedings of the ICSV23 23rd International Congress on Sound & Vibration, Athens, Greece, 10–14 July 2016.

Article

PU-MFA: Point Cloud Up-Sampling via Multi-Scale Features Attention

Hyungjun Lee [1] and Sejoon Lim [2,*]

[1] Graduate School of Automotive Engineering, Kookmin University, Seoul 02707, Republic of Korea
[2] Department of Automobile and IT Convergence, Kookmin University, Seoul 02707, Republic of Korea
* Correspondence: lim@kookmin.ac.kr; Tel.: +82-2-910-5469

Abstract: Recently, research using point clouds has been increasing with the development of 3D scanner technology. According to this trend, the demand for high-quality point clouds is increasing, but there is still a problem with the high cost of obtaining high-quality point clouds. Therefore, with the recent remarkable development of deep learning, point cloud up-sampling research, which uses deep learning to generate high-quality point clouds from low-quality point clouds, is one of the fields attracting considerable attention. This paper proposes a new point cloud up-sampling method called Point cloud Up-sampling via Multi-scale Features Attention (PU-MFA). Inspired by prior studies that reported good performance at generating high-quality dense point set using the multi-scale features or attention mechanisms, PU-MFA merges the two through a U-Net structure. In addition, PU-MFA adaptively uses multi-scale features to refine the global features effectively. The PU-MFA was compared with other state-of-the-art methods in various evaluation metrics through various experiments using the PU-GAN dataset, which is a synthetic point cloud dataset, and the KITTI dataset, which is the real-scanned point cloud dataset. In various experimental results, PU-MFA showed superior performance of generating high-quality dense point set in quantitative and qualitative evaluation compared to other state-of-the-art methods, proving the effectiveness of the proposed method. The attention map of PU-MFA was also visualized to show the effect of multi-scale features.

Keywords: 3D vision; deep-learning; point cloud; attention mechanism; point cloud up-sampling

Citation: Lee, H.; Lim, S. PU-MFA: Point Cloud Up-Sampling via Multi-Scale Features Attention. *Sensors* **2022**, *22*, 9308. https://doi.org/10.3390/s22239308

Academic Editors: Miaohui Wang, Guanghui Yue, Jian Xiong and Sukun Tian

Received: 4 October 2022
Accepted: 10 November 2022
Published: 29 November 2022

Publisher's Note: MDPI stays neutral with regard to jurisdictional claims in published maps and institutional affiliations.

1. Introduction

A point cloud is one of the most popular formats for accurately representing 3D geometric information in robotics and autonomous vehicles. Recently, the number of studies using point clouds has been increasing with the development of 3D scanners, such as LiDAR [1,2]. Along with this trend, there is an increasing demand for high-quality point clouds that are low-noise, uniform, and dense. However, the high cost of collecting high-quality point clouds remains problematic. Therefore, point cloud up-sampling, which generates a low-noise, uniform, and dense point set from noisy, non-uniform, and sparse point sets, is an interesting study.

Similar to learning-based image super-resolution studies [3,4], various learning-based point cloud up-sampling studies [5–7] show better performance of generating high-quality dense point set than traditional point cloud up-sampling studies [8,9]. Intuitively, the image super-resolution tasks and the point cloud up-sampling tasks are similar. Unlike the image super-resolution tasks, which process regular format images, the point cloud up-sampling tasks, which process irregular formats, require additional consideration. First, the up-sampled point set should have a uniform distribution and a dense set of points. Next, the up-sampled point set should represent the details of the target 3D mesh surface well [10].

A traditional learning-based point cloud up-sampling study usually consists of a feature extractor and an up-sampler. In addition, most studies use multi-scale features or attention mechanisms. PU-Net [11], 3PU [12], and PU-GCN [7] extract multi-scale features from sparse point sets. These studies have reported that they are excellent for generating dense point sets, but the last feature extracted by the feature extractor has a limitation in that the details of the sparse point set are diluted features because the output of each layer is used as the input of the next layer. Dis-PU [13], PU-EVA [6], and PU-Transformer [5] showed successful performance of generating high-quality dense point set using the self-attention mechanism to learn long-range dependencies between points. However, there is a limit to applying the attention mechanism with limited information because the key, query, and value of the self-attention mechanism are generated from the same input.

Focusing on these limitations, this paper proposes PU-MFA, a novel method to fuse multi-scale features and attention mechanisms. PU-MFA solves point cloud up-sampling through an attention mechanism that uses an adaptive feature for each layer. The contributions of this research are as follows:

- This paper proposes a point cloud up-sampling method of U-Net structure using Multi-scale Features (MFs) adaptively to Global Features (GFs).
- Global Context Refining Attention (GCRA), a structure for effectively combining MFs and attention mechanisms, is proposed. To the best of the authors' knowledge, this is the first MultiHead Cross-Attention (MCA) mechanism proposed in point cloud up-sampling.
- This study demonstrates the effect of MFs by visualizing the attention map of GCRA in ablation studies.

This method was compared with various state-of-the-art methods using the Chamfer Distance (CD), Hausdorff Distance (HD), and Point-to-Surface (P2F) evaluation metrics for the PU-GAN [10] and the KITTI [14] dataset. As a result, the effectiveness of this method was confirmed by showing better performace at generating dense point set.

2. Related Work

2.1. Optimization-Based Point Cloud Up-Sampling

Various optimization-based studies have been performed to generate a dense set of points from a sparse set. Alexa et al. solved up-sampling by inserting new points into the Voronoi diagram of the local tangential space computed based on the moving-least-squares error [8]. Lipman et al. explained the up-sampling using the Locally Optimal Projection (LOP) operator [9]. In this study, the points were re-sampled by using L_1 norm. Huang et al. up-sampled a noisy and non-uniform set of points using an improved LOP that is a weighted LOP [15]. Later, Huang et al. proposed an advanced method called Edge-Aware Re-sampling of a set of points (EAR). The EAR first re-samples the edges and then uses edge-aware up-sampling to resolve the up-sampling [16].

2.2. Learning-Based Point Cloud Up-Sampling

With the successful performance of learning-based image super-resolution, many studies have proposed a learning-based point cloud up-sampling method.

As with image analysis, many studies have used MFs in point cloud up-sampling. PU-Net [11], the first attempt at deep learning for point cloud up-sampling, showed good performance at generating high-quality point set by extracting MFs through hierarchical feature learning and interpolation based on the framework of PointNet++ [17]. 3PU [12] performed well using MFs via an Intra-Level Dense connection and Inter-Level Skip connection. PU-GCN [7] uses MFs extracted by Inception DenseGCN. In this study, Inception DenseGCN could effectively extract MFs with an InceptionNet-inspired structure [18].

Because of the advantages of learning the long-range dependency of the self-attention mechanism, it is used in various point cloud up-sampling studies. In PU-GAN [10], generators are trained using discriminators that apply a self-attention mechanism. Pugeo-Net [19] showed good performance at generating high-quality dense point set by using it in Feature Recalibration. PU-EVA [6] showed successfully generating up-sampled point set using an EVA Expansion Unit with the mechanism. Dis-PU [13] performed well using the Local Refinement Unit with self-attention applied to the generated point set. PU-Transformer [5], which applied the transformer structure for the first time in point cloud up-sampling, uses Shifted Channel MultiHead Self-Attention to show the state-of-the-art performance of generating high-quality point set.

3. Problem Description

Given an unordered sparse point set $S = \{s_i\}_{i=1}^{N}$ of N samples, we aim to generate low-noise, uniform, and dense point set $Q = \{q_i\}_{i=1}^{rN}$ using $D = \{d_i\}_{i=1}^{rN}$ as Ground Truth (GT), where N is the input patch size and r is the up-sampling ratio. Figure 1 shows the problem description of this study. Also, Table 1 summarizes the definitions of symbols to be used.

Table 1. Description of symbols.

Symbol	Description
S	Sparse point set
s_i	Element of S
S'^{Δ}	Offset of S
$s_i'^{\Delta}$	Element of S'^{Δ}
D	Ground truth point set
d_i	Element of D
Q'	Coarse point set
q_i'	Element of Q'
Q'^{Δ}	Offset of Q'
$q_i'^{\Delta}$	Element of Q'^{Δ}
Q	Dense point set
q_i	Element if Q
N	Input patch size
r	Up-sampling ratio
H	Depth of layer
F_h	Set of point wise feature extracted from h^{th} Point Transformer
f_i^h	Point-wise feature extracted from h^{th} Point Transformer
C	Channel
K, K'	Expansion rate
$patch_i$	Patch created through KNN based on s_i
$patch\_size$	Neighbor size of KNN

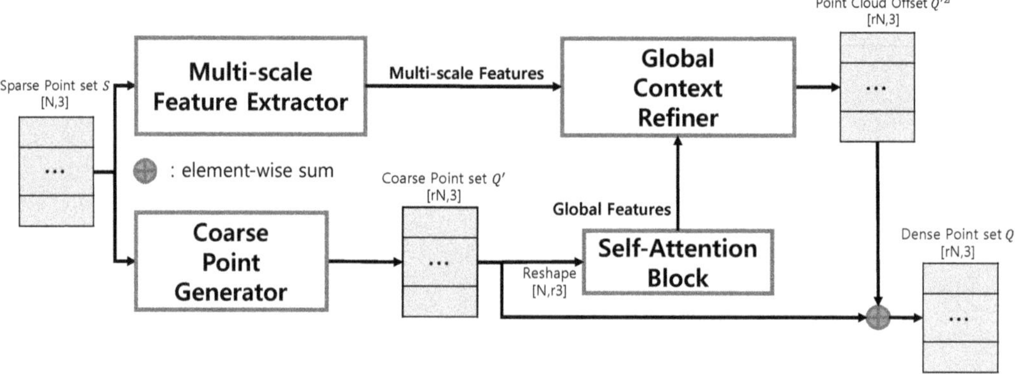

Figure 1. Illustration of an overview of the proposed method.

4. Method

This method consists of a Multi-scale Feature Extractor (MFE), Global Context Refiner (GCR), Coarse Point Generator (CPG), and Self-Attention Block (SAB). As shown in Figure 2, MFE extracts MFs, an adaptive feature for use in GCR. GCR uses MFs to refine GFs adaptively and finally produce Q'^{Δ}, where Q'^{Δ} is defined as $Q'^{\Delta} = \{q_i'^{\Delta}\}_{i=1}^{rN}$. CPG generates Q' from S and SAB extracts GFs from Q', where Q' is defined as $Q' = \{q'_i\}_{i=1}^{rN}$. Based on the definitions of Q' and Q'^{Δ}, Q is formulated as Equation (1), where $\oplus$ is an element-wise sum.

$$Q = Q' \oplus Q'^{\Delta} (\oplus : element\text{-}wise\ sum) \tag{1}$$

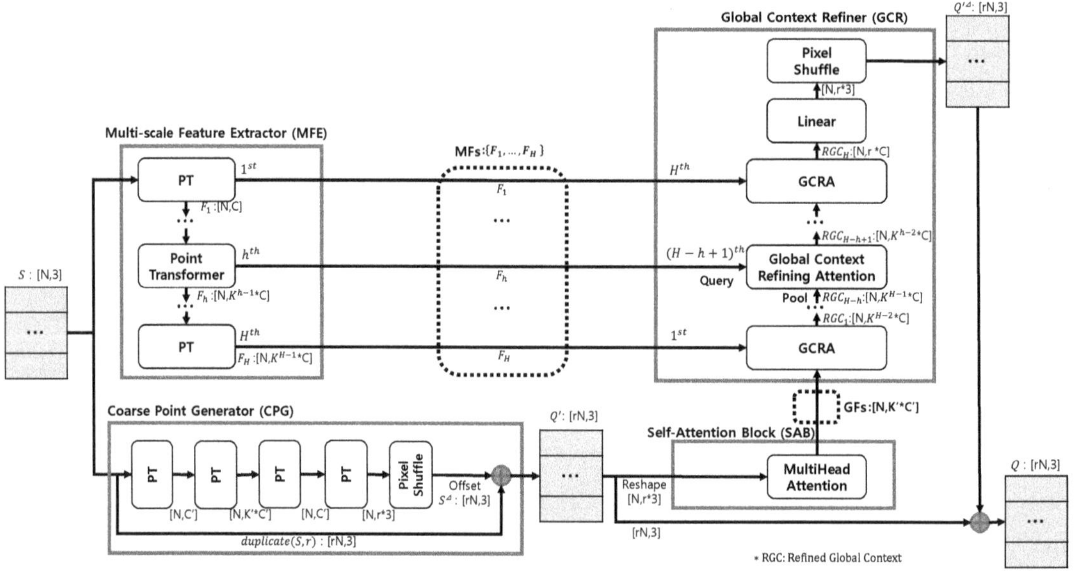

Figure 2. Illustration of the proposed framework. Here, 3 is the coordinate dimension, and H is the depth of the layer. In Multi-scale Feature Extractor (MFE) and Global Context Refiner (GCR), C is the channel, and K is the expansion ratio. In Coarse Point Generator (CPG) and Self-Attention Block (SAB), C' is the channel and K' is the expansion ratio.

4.1. Multi-Scale Feature Extractor

Because the GFs extracted from Q' via SAB is a feature extracted from a set of points in which geometric information about the original input S is diluted, MFE using Point Transformer (PT) [20], an advanced point cloud analysis technique, extracts MFs from S. As shown in Figure 2, the MFE consists of H PT, and the set of point-wise features extracted from the hth PT is $F_h \in \mathbb{R}^{N \times K^{h-1}C}$. MFs are the set of F_h extracted from all layers of the MFE. The extracted MFs and F_h are formulated as in Equation (2), where f_i^h is a point-wise feature extracted from the hth PT.

$$F_h = \left\{ f_i^h \right\}_{i=1}^{N}, MFs = \{F_h\}_{h=1}^{H} \tag{2}$$

Point Transformer

PT consists of two elements. The first is the K-Nearest Neighbor (KNN), and the second is the Vector Self-Attention (VSA) mechanism. At the hth PT, the point-wise feature f_i^{h-1} of point $s_i \in S$ is updated to f_i^h through VSA, which uses s_i and $patch_i$ as the inputs. The $patch_i$ is generated through KNN using s_i as the input. This operation works on all points in S, updating the point-wise feature of all points [20]. This is formulated in Equation (3), where $patch\_size$ is the size of KNN's neighbor size.

$$patch_i = KNN(s_i, patch\_size)$$
$$f_i^h = \sum_{p_k \in patch_i} VectorSelfAttention(s_i, p_k) \tag{3}$$
$$(s_i \in S, i \in \{1, 2, ..., N\})$$

Inspired by this operation, this considered $patch_i$ is equivalent to the CNN's kernel. In CNN, even if the kernel of the CNN is fixed, a deeper layer, means a wider receptive field. Therefore, even if the patch size of the KNN in PT is fixed, the deeper the layer, the more s_i can interact with a wider range of points. Figure 3 is an example with a KNN patch size of four. In Figure 3a, when the hth PT updates f_i^{h-1} to f_i^h, VSA is performed on $patch_i$, which is composed of s_i and incidental points, to update f_i^{h-1}. In Figure 3b, the hth PT updates each feature by performing a VSA for each patch in all incidental points. In Figure 3c, the $(h+1)^{th}$ PT updates f_i^h to f_i^{h+1} by performing VSA using $patch_i$ similar to the hth PT. However, the $(h+1)^{th}$ PT updates f_i^h to f_i^{h+1} using a wider receptive field than the receptive field of the hth PT because the features of the incidental points of the $(h+1)^{th}$ PT are updated by the hth PT. This operation allows the MFE to extract the MFs effectively.

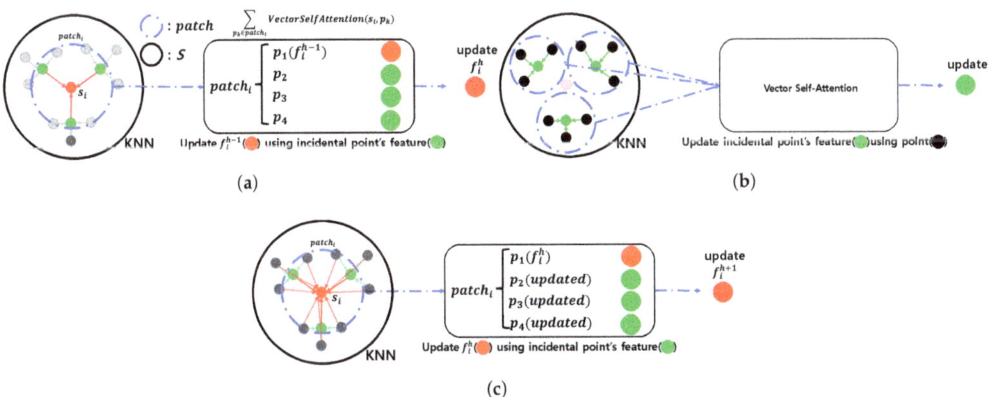

Figure 3. Illustration of KNN and VSA in PT. (**a**) h^{th} Point Transformer layer with s_i. (**b**) h^{th} Point Transformer layer with incidental points. (**c**) $(h+1)^{th}$ Point Transformer layer with s_i.

4.2. Global Context Refiner

Because GCR and MFE are U-Net [21] structures, they are composed of H GCRA. GCRA effectively refines GFs by querying MFs, which is adaptive geometric information applied to each layer. As shown in Figure 2, the $(H - h + 1)^{th}$ GCRA generates $RGC_{H-h+1} \in \mathbb{R}^{N \times K^{h-2}C}$ by using $F_h \in \mathbb{R}^{N \times K^{h-1}C}$ as a query and RGC_{H-h} as a pool. However, $\mathbb{R}^{N \times r3}$ was used instead to prevent RGC_H from becoming $\mathbb{R}^{N \times \frac{C}{K}}$. After refining the GFs, the linear layer was used to perform the transformation. PixelShuffle was then used to generate the $Q'^{\Delta} \in \mathbb{R}^{rN \times 3}$.

Global Context Refining Attention

Inspired by Skip-Attention [22], which acts as a communicator between the encoder and decoder, GCRA uses MFs and GFs to apply the MCA mechanism. In various studies, self-attention mechanisms are used to extract the features of point sets or to generate up-sampling point sets [5,10]. However, the self-attention mechanism is limited because it uses only limited information due to the structure in which key, query, and value are generated from the same input. With these limitations in mind, GCRA in the H hierarchy uses GFs$\in \mathbb{R}^{N \times K'C'}$ as the pool (key, values) and MFs as the queries, progressively refining the GFs through MCA. GCRA consists of MCA [23], Batch Normalization (BN) [24], and Feed Forward. As shown in Figure 4, the output shape of applying MCA using the query and pool is $\mathbb{R}^{N \times F_p}$. The pool was then refined by adding the pool and the MCA output. The BN was used for stable training after addition. Feed Forward transforms the output of the BN and produces a Refined Global Context (RGC)$\in \mathbb{R}^{N \times F_o}$.

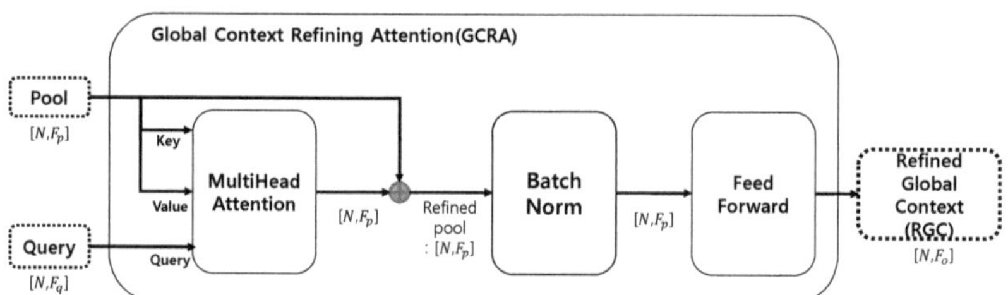

Figure 4. Illustration of Global Context Refining Attention (GCRA). F_p is the pool input channel, F_q is the query input channel, and F_o is the output channel.

4.3. Coarse Point Generator

CPG generates Q'. In CPG, PT [20] and PixelShuffle [5,25] generate S'^{Δ} from S, where, S'^{Δ} is defined as $S'^{\Delta} = \{s_i'^{\Delta}\}_{i=1}^{rN}$. The structure of CPG consists of four layers, such as the structure of the 3PU's Feature Extraction Unit [12]. As shown in Figure 2, to make the final output into 3D coordinates, first, PT was first used to expand the features, and then gradually reduce them. Subsequently, PixelShuffle generates 3D coordinates using those features. Q' is generated through the element-wise sum of the generated S'^{Δ} and $duplicate(S, r) \in \mathbb{R}^{rN \times 3}$. This process is formulated as Equation (4).

$$duplicate(S, r) = \left\{ \overbrace{s_i, ..., s_i}^{r \; times} \right\}_{i=1}^{N}$$

$$Q' = duplicate(S, r) \oplus S'^{\Delta}$$

(4)

4.4. Self-Attention Block

Inspired by self-attention, which learns long-range dependency [23], we use Multi-Head Self-Attention (MSA) was used to extract the GFs from Q'. As shown in Figure 2, the shape of Q' was changed from $\mathbb{R}^{rN \times 3}$ to $\mathbb{R}^{N \times 3r}$, and the coordinates of Q' were used as features of the original point set S. The GFs$\in \mathbb{R}^{N \times K'C'}$ was then extracted using the changed shape Q' as the input to the MSA.

5. Experimental Settings

5.1. Datasets

All methods were trained using the most popular PU-GAN [10] dataset in these experiments and evaluated using the PU-GAN dataset and the KITTI [14] dataset. The PU-GAN dataset was a synthetic point cloud dataset produced from 147 3D meshes, and the KITTI dataset was a real-scanned point cloud dataset collected using real LiDAR.

The training phase used 120 3D meshes from the PU-GAN dataset. All patches were generated via the Poisson disk sampling after converting the original mesh to a point cloud, just like the patch-based up-sampling approach. The sampling resulted in 24,000 input-output pairs.

In the evaluation phase, 27 3D meshes from the PU-GAN dataset were converted into point clouds to test the synthetic point up-sampling, and the real-scanned point up-sampling test was performed using the KITTI dataset. The generated patches should cover all point sets when evaluating the synthetic point cloud and real-scanned point cloud up-sampling. After merging each up-sampled patch, the up-sampled point set was reconstructed by farthest point sampling. More details can be found at study in PU-GAN [10]. This dataset was downloaded and used from https://github.com/liruihui/PU-GAN (accessed date: 12 July 2022).

5.2. Loss Function

In most point cloud reconstruction methods, CD is used as the loss function [22,26,27]. However, it was confirmed empirically that the Using Density-Aware Chamfer Distance as loss function showed good performance at point cloud reconstruction, considering the uniformity of the points set on the CD [28]. Therefore, the total loss was formulated as Equation (5), where α is linearly interpolated from 0.1 to 1 during training and $\|\cdot\|_2$ is L_2 norm.

$$Loss(Q', Q, D) = L_{CD}(Q', D) + \alpha \times L_{DCD}(Q, D)$$

$$L_{CD}(Q', D) = \frac{1}{|Q'|} \sum_{x \in Q'} \min_{y \in D} \|x - y\|_2 + \frac{1}{|D|} \sum_{y \in D} \min_{x \in Q'} \|y - x\|_2 \tag{5}$$

$$L_{DCD}(Q, D) = \frac{1}{|Q|} \sum_{x \in Q} \min_{y \in D} (1 - e^{-\|x-y\|_2}) + \frac{1}{|D|} \sum_{y \in D} \min_{x \in Q} (1 - e^{-\|y-x\|_2})$$

5.3. Metric

This study evaluated the method using CD, HD, and P2F metrics, as in prior studies [5,6,13]. CD is a metric that measures the similarity between a set of GT points and a set of predicted points for each point, and HD is an evaluation metric that measures the outliers in a set of predicted points based on a set of GT points. P2F is an index that measures the similarity between the original mesh and the predicted point set and measures the quality of the predicted point set. The parameter complexity was also measured by measuring the number of parameters. For all metrics, a lower the number, meant better performance.

5.4. Comparison Methods

The proposed method was compared with three state-of-the-art methods: Dis-pu [13], PU-EVA [6], and PU-Transformer [5] to validate the method. For an exact comparison, all methods were implemented using pytorch [29] version 1.7.0 on Ubuntu 20.04 and trained on the same Intel i9-10980XE CPU and NVIDIA TITAN RTX environment.

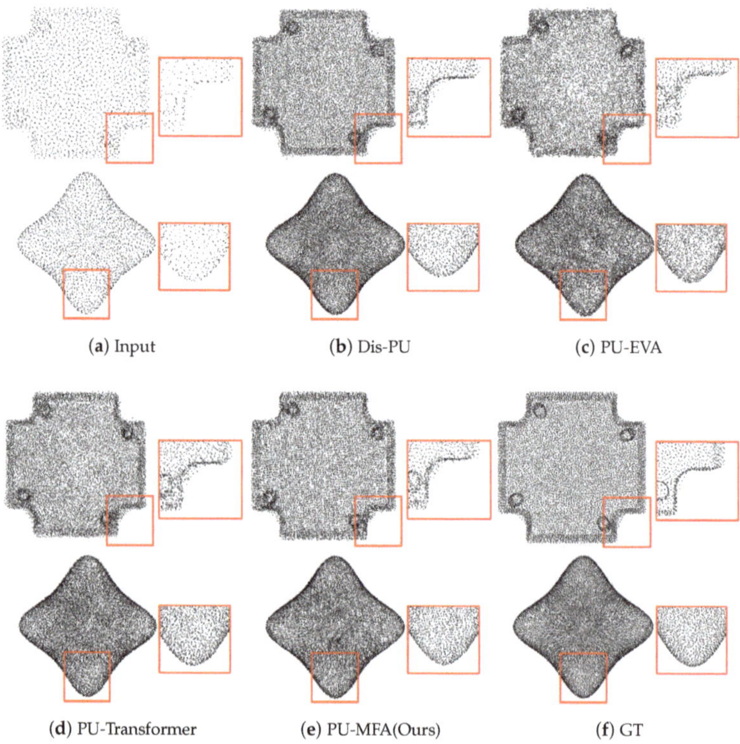

(a) Input   **(b)** Dis-PU   **(c)** PU-EVA

(d) PU-Transformer   **(e)** PU-MFA(Ours)   **(f)** GT

Figure 6. Visualization result of $\times 4$ up-sampling on PU-GAN dataset (non-tubular objects).

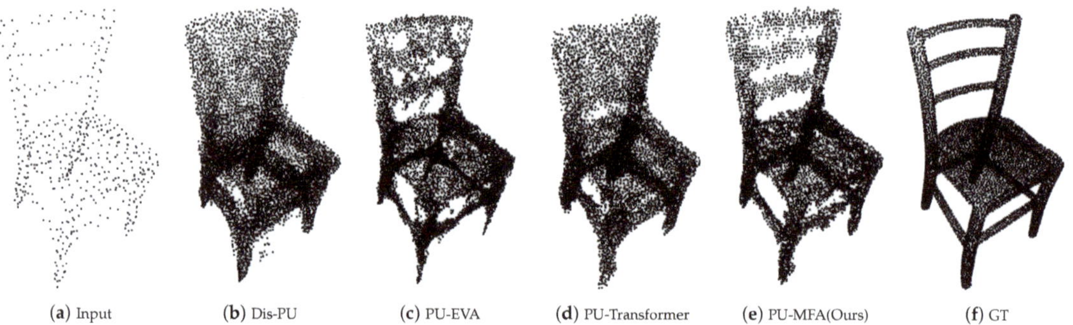

(a) Input   **(b)** Dis-PU   **(c)** PU-EVA   **(d)** PU-Transformer   **(e)** PU-MFA(Ours)   **(f)** GT

Figure 7. Visualization result of $\times 16$ up-sampling of PU-GAN dataset.

6.2. Results on Real-Scanned Datasets

Dis-PU, PU-EVA, PU-Transformer, and the present method were evaluated using the KITTI dataset for $\times 4$ up-sampling. Figure 8 shows $\times 4$ up-sampling. In Figure 8b–d, the boundary between the window and the door of the vehicle was unclear. However, Figure 8e generated by the present method, showed that the boundary was clearer.

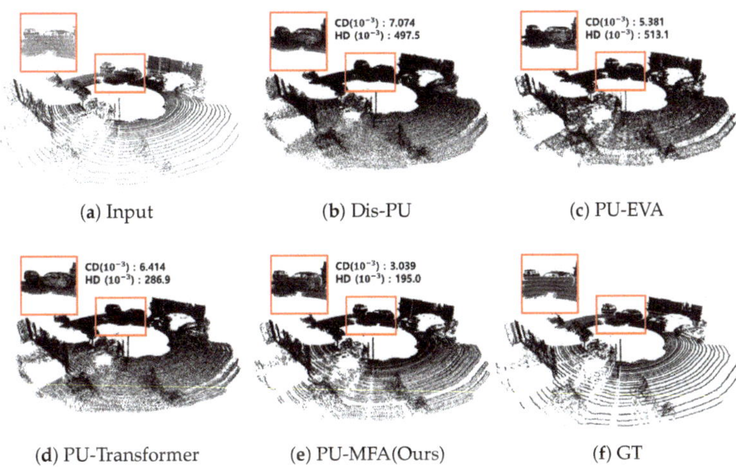

(**a**) Input   (**b**) Dis-PU   (**c**) PU-EVA

(**d**) PU-Transformer   (**e**) PU-MFA(Ours)   (**f**) GT

Figure 8. Visualization result of ×4 up-sampling of the KITTI dataset.

6.3. Ablation Study

This method, was evaluated by performing various ablation studies using the PU-GAN dataset.

6.3.1. Effect of Components

To demonstrate the effectiveness of the contribution, four cases were divided into ablation studies. The cases were as follows: Case 1 was a structure using GCR, CPG, and SAB, with the MultiHead Attention (MHA) of GCR and SAB consisting of self-attention with one head. Case 2 was a structure changed from Case 1 to eight heads. Case 3 was a structure using GCRA composed of MCA by adding MFE to Case 2, where the query of all GCRA becomes F_4, the final output of MFE. Case 4 was PU-MFA. As shown in Table 4, all contributions affected the method performance of generating point set.

Table 4. Ablation study results to analyze the effect of the present contribution.

Case	Contribution			Metric		
	MHA	**MFE**	**MFs**	**CD (10^{-3})**	**HD (10^{-3})**	**P2F (10^{-3})**
1				0.3349	4.461	4.926
2	√			0.2473	1.101	2.829
3	√	√		0.2500	2.735	2.737
4	√	√	√	0.2362	1.094	2.545

6.3.2. Multi-Scale Features Attention Analysis

By visualizing the attention maps of all GCRAs, it was confirmed that the GCRAs of GCR with $H = 4$ refined the GFs by adaptively using the MFs extracted from receptive fields of various sizes. Figure 9 shows the results visualized by choosing three attention heads in the GCRA and selecting 30 points, which had the highest attention score in S, from each head. The attention map was visualized using Case 3 in Table 4 without MFs in Figure 9b to compare that MFs operated adaptively. As shown in Figure 9a, in the low-layer GCRA, an attention map was formed for a wide range of points in a point set, and in high-layer GCRA, an attention map was formed for a relatively narrow range of points. On the other hand, in Figure 9b, a wide range of attention maps was formed regardless of the high and low levels of the hierarchy. This phenomenon confirmed that PU-MFA uses the adaptive point feature for each layer of the GCRA.

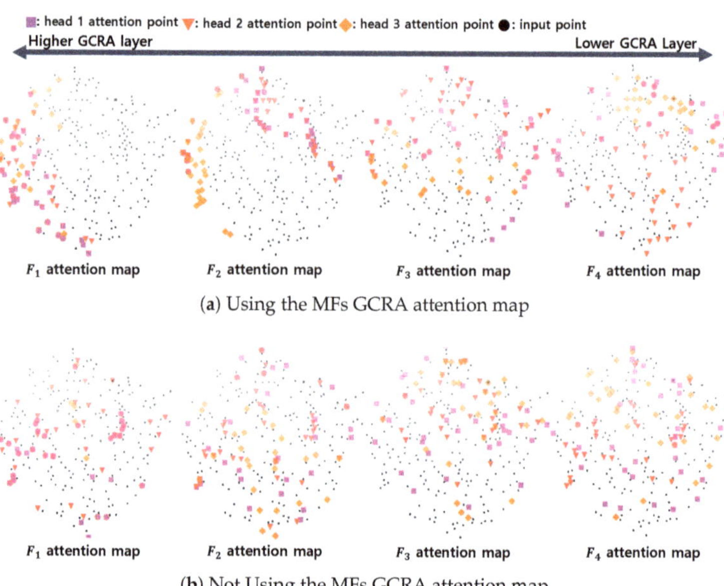

(**a**) Using the MFs GCRA attention map

(**b**) Not Using the MFs GCRA attention map

Figure 9. Visualization of attention map generated using MFs as a query in GCR with $H = 4$.

Method	Noise Level		
	0.001	0.01	0.02
Dis-PU			
PU-EVA			
PU-Transformer			
PU-MFA(Ours)			

Figure 10. Visualization result of the effect of noise.

6.3.3. Effect of Noise

Table 5 lists the $\times 4$ up-sampling results of Dis-PU [13], PU-EVA [6], PU-Transformer [5], and the present method using the PU-GAN dataset with various noises added. The noise effect evaluated the result obtained by adding different levels of Gaussian noise $\mathcal{N}(0,\text{noise level})$ to a set of input points. As shown in Table 5, the proposed method showed the most robustness to various noise levels. As shown in Figure 10, it can be seen that the boundary between the fingers blurred in the dense set of points generated by the state-of-the-art methods as the noise level was increased. On the other hand, the proposed method showed that the boundary between the fingers was maintained in the dense set of points generated by the present method.

Table 5. Quantitative evaluation results of the noise effects using the PU-GAN dataset.

Method	Various Noise Levels Test at $\times 4$ Up-Sampling (CD with 10^{-3})					
	0	0.001	0.005	0.01	0.015	0.02
Dis-PU	0.2703	0.2751	0.2975	0.3257	0.3466	0.3706
PU-EVA	0.2969	0.2991	0.3084	0.3167	0.3203	0.3268
PU-Transformer	0.2671	0.2717	0.2905	0.3134	0.3331	0.3585
PU-MFA (Ours)	0.2326	0.2376	0.2547	0.2764	0.2989	0.3195

7. Conclusions

In this paper, we proposed PU-MFA, a point cloud up-sampling method of U-Net structure that combines multi-scale features and attention mechanism. One of the most significant differences from the prior point cloud up-sampling methods was that PU-MFA used multi-scale features adaptively and effectively through fusion with the cross-attention mechanism. Also, the PU-MFA is the first method to apply the cross-attention mechanism to point cloud up-sampling to the best of the authors' knowledge. Various experiments were performed on PU-MFA and other state-of-the-art methods using the PU-GAN and the KITTI dataset. As a result, PU-MFA showed better performance of generating high-quality dense point set than other state-of-the-art methods in various experiments. In addition, ablation study showed that multi-scale features are very useful in PU-MFA for generating high-quality point sets by choosing receptive field size adaptively for each layer.

Despite the successful performance at generating high-quality dense point set of PU-MFA, PU-MFA cannot cope with an arbitrary up-sampling ratio. Because PU-MFA is a patch-based up-sampling of $\times 4$, up-sampling is only possible for 4 to the M power. A method that can respond to an arbitrary up-sampling ratio is planned in the future to overcome this limitation.

Author Contributions: Conceptualization, S.L.; methodology, H.L.; software, H.L.; validation, S.L.; formal analysis, H.L.; investigation, H.L. and S.L.; resources, H.L. and S.L.; data curation, H.L.; writing—original draft preparation, H.L.; writing—review and editing, S.L.; visualization, H.L.; supervision, S.L.; project administration, S.L.; funding acquisition, S.L. All authors have read and agreed to the published version of the manuscript.

Funding: This research was supported by Korea Institute of Police Technology (KIPoT) grant funded by the Korea government (KNPA)(No.092021C26S03000), Basic Science Research Program through the National Research Foundation of Korea(NRF) funded by the Ministry of Education(No. 2022R1F1A1072626), and the BK21 Program(5199990814084) through the National Research Foundation of Korea (NRF) funded by the Ministry of Education.

Institutional Review Board Statement: Not applicable.

Informed Consent Statement: Not applicable.

Data Availability Statement: Publicly available datasets were analyzed in this study. This data can be found here: https://github.com/liruihui/PU-GAN (accessed date: 12 July 2022).

Conflicts of Interest: The authors declare no conflict of interest.

References

1. Koide, K.; Miura, J.; Menegatti, E. A portable three-dimensional LIDAR-based system for long-term and wide-area people behavior measurement. *Int. J. Adv. Robot. Syst.* **2019**, *16*. [CrossRef]
2. Lim, H.; Yeon, S.; Ryu, S.; Lee, Y.; Kim, Y.; Yun, J.; Jung, E.; Lee, D.; Myung, H. A Single Correspondence Is Enough: Robust Global Registration to Avoid Degeneracy in Urban Environments. *arXiv* **2022**, arXiv:2203.06612.
3. Niu, B.; Wen, W.; Ren, W.; Zhang, X.; Yang, L.; Wang, S.; Zhang, K.; Cao, X.; Shen, H. Single image super-resolution via a holistic attention network. In *European Conference on Computer Vision*; Springer: Berlin/Heidelberg, Germany, 2020; pp. 191–207.
4. Liang, J.; Cao, J.; Sun, G.; Zhang, K.; Van Gool, L.; Timofte, R. Swinir: Image restoration using swin transformer. In Proceedings of the IEEE/CVF International Conference on Computer Vision, Montreal, QC, Canada, 10–17 October 2021; pp. 1833–1844.
5. Qiu, S.; Anwar, S.; Barnes, N. PU-Transformer: Point Cloud Upsampling Transformer. *arXiv* **2021**, arXiv:2111.12242.

6. Luo, L.; Tang, L.; Zhou, W.; Wang, S.; Yang, Z.X. PU-EVA: An Edge-Vector Based Approximation Solution for Flexible-Scale Point Cloud Upsampling. In Proceedings of the IEEE/CVF International Conference on Computer Vision, Montreal, QC, Canada, 10–17 October 2021; pp. 16208–16217.

7. Qian, G.; Abualshour, A.; Li, G.; Thabet, A.; Ghanem, B. Pu-gcn: Point cloud upsampling using graph convolutional networks. In Proceedings of the IEEE/CVF Conference on Computer Vision and Pattern Recognition, Nashville, TN, USA, 20–25 June 2021; pp. 11683–11692.

8. Alexa, M.; Behr, J.; Cohen-Or, D.; Fleishman, S.; Levin, D.; Silva, C.T. Computing and rendering point set surfaces. *IEEE Trans. Vis. Comput. Graph.* **2003**, *9*, 3–15. [CrossRef]

9. Lipman, Y.; Cohen-Or, D.; Levin, D.; Tal-Ezer, H. Parameterization-free projection for geometry reconstruction. *ACM Trans. Graph. (TOG)* **2007**, *26*, 22-es. [CrossRef]

10. Li, R.; Li, X.; Fu, C.W.; Cohen-Or, D.; Heng, P.A. Pu-gan: A point cloud upsampling adversarial network. In Proceedings of the IEEE/CVF International Conference on Computer Vision, Seoul, Republic of Korea, 27–28 October 2019; pp. 7203–7212.

11. Yu, L.; Li, X.; Fu, C.W.; Cohen-Or, D.; Heng, P.A. Pu-net: Point cloud upsampling network. In Proceedings of the IEEE Conference on Computer Vision and Pattern Recognition, Salt Lake City, UT, USA, 18–22 June 2018; pp. 2790–2799.

12. Yifan, W.; Wu, S.; Huang, H.; Cohen-Or, D.; Sorkine-Hornung, O. Patch-based progressive 3d point set upsampling. In Proceedings of the IEEE/CVF Conference on Computer Vision and Pattern Recognition, Long Beach, CA, USA, 15–20 June 2019; pp. 5958–5967.

13. Li, R.; Li, X.; Heng, P.A.; Fu, C.W. Point cloud upsampling via disentangled refinement. In Proceedings of the IEEE/CVF Conference on Computer Vision and Pattern Recognition, Nashville, TN, USA, 20–25 June 2021; pp. 344–353.

14. Geiger, A.; Lenz, P.; Stiller, C.; Urtasun, R. Vision meets robotics: The kitti dataset. *Int. J. Robot. Res.* **2013**, *32*, 1231–1237. [CrossRef]

15. Huang, H.; Li, D.; Zhang, H.; Ascher, U.; Cohen-Or, D. Consolidation of unorganized point clouds for surface reconstruction. *ACM Trans. Graph. (TOG)* **2009**, *28*, 1–7. [CrossRef]

16. Huang, H.; Wu, S.; Gong, M.; Cohen-Or, D.; Ascher, U.; Zhang, H. Edge-aware point set resampling. *ACM Trans. Graph. (TOG)* **2013**, *32*, 1–12. [CrossRef]

17. Qi, C.R.; Yi, L.; Su, H.; Guibas, L.J. Pointnet++: Deep hierarchical feature learning on point sets in a metric space. *Adv. Neural Inf. Process. Syst.* **2017**, *30* ; pp. 5105–5114.

18. Szegedy, C.; Liu, W.; Jia, Y.; Sermanet, P.; Reed, S.; Anguelov, D.; Erhan, D.; Vanhoucke, V.; Rabinovich, A. Going deeper with convolutions. In Proceedings of the IEEE Conference on Computer Vision and Pattern Recognition, Boston, MA, USA, 7–12 June 2015; pp. 1–9.

19. Qian, Y.; Hou, J.; Kwong, S.; He, Y. PUGeo-Net: A geometry-centric network for 3D point cloud upsampling. In *European Conference on Computer Vision*; Springer: Berlin/Heidelberg, Germany, 2020; pp. 752–769.

20. Zhao, H.; Jiang, L.; Jia, J.; Torr, P.H.; Koltun, V. Point transformer. In Proceedings of the IEEE/CVF International Conference on Computer Vision, Montreal, QC, Canada, 10–17 October 2021; pp. 16259–16268.

21. Ronneberger, O.; Fischer, P.; Brox, T. U-net: Convolutional networks for biomedical image segmentation. In *International Conference on Medical Image Computing and Computer-Assisted Intervention*; Springer: Berlin/Heidelberg, Germany, 2015; pp. 234–241.

22. Wen, X.; Li, T.; Han, Z.; Liu, Y.S. Point cloud completion by skip-attention network with hierarchical folding. In Proceedings of the IEEE/CVF Conference on Computer Vision and Pattern Recognition, Seattle, WA, USA, 14–19 June 2020; pp. 1939–1948.

23. Vaswani, A.; Shazeer, N.; Parmar, N.; Uszkoreit, J.; Jones, L.; Gomez, A.N.; Kaiser, Ł.; Polosukhin, I. Attention is all you need. *Adv. Neural Inf. Process. Syst.* **2017**, *30* ; pp. 1936-1945.

24. Ioffe, S.; Szegedy, C. Batch normalization: Accelerating deep network training by reducing internal covariate shift. In *International Conference on Machine Learning*; PMLR: Lille, France, 2015; pp. 448–456.

25. Shi, W.; Caballero, J.; Huszár, F.; Totz, J.; Aitken, A.P.; Bishop, R.; Rueckert, D.; Wang, Z. Real-time single image and video super-resolution using an efficient sub-pixel convolutional neural network. In Proceedings of the IEEE Conference on Computer Vision and Pattern Recognition, Las Vegas, NV, USA, 27–30 June 2016; pp. 1874–1883.

26. Nguyen, A.D.; Choi, S.; Kim, W.; Lee, S. Graphx-convolution for point cloud deformation in 2d-to-3d conversion. In Proceedings of the IEEE/CVF International Conference on Computer Vision, Seoul, Republic of Korea, 27 October–2 November 2019; pp. 8628–8637.

27. Yu, X.; Rao, Y.; Wang, Z.; Liu, Z.; Lu, J.; Zhou, J. Pointr: Diverse point cloud completion with geometry-aware transformers. In Proceedings of the IEEE/CVF International Conference on Computer Vision, Montreal, QC, Canada, 10–17 October 2021; pp. 12498–12507.

28. Wu, T.; Pan, L.; Zhang, J.; Wang, T.; Liu, Z.; Lin, D. Density-aware chamfer distance as a comprehensive metric for point cloud completion. *arXiv* **2021**, arXiv:2111.12702.

29. Paszke, A.; Gross, S.; Massa, F.; Lerer, A.; Bradbury, J.; Chanan, G.; Killeen, T.; Lin, Z.; Gimelshein, N.; Antiga, L.; et al. PyTorch: An Imperative Style, High-Performance Deep Learning Library. In *Advances in Neural Information Processing Systems 32*; Wallach, H., Larochelle, H., Beygelzimer, A., d'Alché-Buc, F., Fox, E., Garnett, R., Eds.; Curran Associates, Inc.: Vancouver, BC, Canada, 2019; pp. 8024–8035.

30. Kingma, D.P.; Ba, J. Adam: A method for stochastic optimization. *arXiv* **2014**, arXiv:1412.6980.

Article

MASPC_Transform: A Plant Point Cloud Segmentation Network Based on Multi-Head Attention Separation and Position Code

Bin Li [1,2,*] and Chenhua Guo [1]

1 School of Computer Science, Northeast Electric Power University, Jilin 132012, China
2 Gongqing Institute of Science and Technology, No. 1 Gongqing Road, Gongqing 332020, China
* Correspondence: libinjlu5765114@163.com

Abstract: Plant point cloud segmentation is an important step in 3D plant phenotype research. Because the stems, leaves, flowers, and other organs of plants are often intertwined and small in size, this makes plant point cloud segmentation more challenging than other segmentation tasks. In this paper, we propose MASPC_Transform, a novel plant point cloud segmentation network base on multi-head attention separation and position code. The proposed MASPC_Transform establishes connections for similar point clouds scattered in different areas of the point cloud space through multiple attention heads. In order to avoid the aggregation of multiple attention heads, we propose a multi-head attention separation loss based on spatial similarity, so that the attention positions of different attention heads can be dispersed as much as possible. In order to reduce the impact of point cloud disorder and irregularity on feature extraction, we propose a new point cloud position coding method, and use the position coding network based on this method in the local and global feature extraction modules of MASPC_Transform. We evaluate our MASPC_Transform on the ROSE_X dataset. Compared with the state-of-the-art approaches, the proposed MASPC_Transform achieved better segmentation results.

Keywords: point cloud; plant phenotyping; point cloud segmentation; multi-head attention; attention separation; position code

Citation: Li, B.; Guo, C. MASPC_Transform: A Plant Point Cloud Segmentation Network Based on Multi-Head Attention Separation and Position Code. *Sensors* **2022**, *22*, 9225. https://doi.org/10.3390/s22239225

Academic Editors: Miaohui Wang, Guanghui Yue, Jian Xiong and Sukun Tian

Received: 28 October 2022
Accepted: 25 November 2022
Published: 27 November 2022

Publisher's Note: MDPI stays neutral with regard to jurisdictional claims in published maps and institutional affiliations.

1. Introduction

Plant phenotype is to study how to measure the shape characteristics of plants, such as plant height, leaf organ size, root distribution, fruit weight, etc. These traits are closely related to the yield, quality, and stress resistance of plants. The study of plant phenotype has important value for agricultural modernization breeding [1], water and fertilizer managements of crop [2], and pest control [3].

In the process of plant phenotypic feature extraction, accurate segmentation of plant data according to different organs (stems, leaves, flowers, etc.,) is the premise of high-precision plant phenotype [4]. Plant organ segmentation technology based on 2D images has been very mature [5–8]. In recent years, with the development of LiDAR technology, more and more 3D spatial information of plants has been collected [9]. The plant point cloud contains the 3D spatial position, RGB color, normal vector, and other information of the collected object. Compared with 2D images, plant point cloud retains more spatial details and is not easily affected by occlusion, and can extract the plant structure more accurately.

By summarizing the existing segmentation methods of plant point clouds, we find that the existing methods have poor segmentation effect at the junction of different plant organs. For example, in the segmentation result in the fifth row and the first column in Figure 7, some stems are erroneously recognized as leaves. And this phenomenon is more obvious when the stem contacts the leaves. In the segmentation result in row 6 and column 2 of Figure 7, the part of the small calyx in the point cloud is erroneously divided into leaves. The reasons for the above segmentation errors are as follows: (1) for the plant segmentation task, the points belonging to the same plant organ are far from each other and are interwoven with the point clouds of other organs. For example, in Figures 5 and 6 the stems of the plants are almost distributed in the whole point cloud space and interweaved with other organs. The segmentation network often extracts the point cloud features of the whole plant without distinction, and does not mine the relationship of point clouds belonging to the same organ in the point cloud space. (2) Plant point clouds have the characteristics of disorder and irregularity, which will affect feature extraction.

To further improve the segmentation accuracy of plant point clouds, we use the Point Transformer [10] as the backbone of the proposed MASPC_Transform. The Point Transform uses the multi-head attentional mechanism in the process of local and global feature extraction. The multi-head attention mechanism can form associations between points of the same organ. And the multi-head attention mechanism is composed of multiple parallel self-attention mechanisms, which makes the whole feature generate multiple sub feature spaces and can extract feature information from multiple dimensions. However, the features of multi-head attention extraction may tend to be similar [11], that is, multiple attention heads establish connections for similar semantic point clouds at different positions in the point cloud space, but these point clouds may be located in the same area (for example, these point clouds may all be located on the same leaf of a plant). Therefore, we propose a multi-head attention separation loss based on spatial similarity, so that the attention positions of different attention heads can be separated from each other as much as possible, so as to establish a connection for point clouds that are distant in the point cloud space but belong to the same organ. In order to suppress the influence of point cloud disorder and irregularity on feature extraction, we added position coding network in the local and global feature extraction modules of MASPC_Transform.

The main contributions of this paper are summarized as follows:

1. We propose a plant point cloud segmentation network named MASPC_Transform, and evaluate its segmentation performance on the ROSE_X dataset.
2. We propose a loss function of multi-head attention separation based on spatial similarity. This loss can make the attention positions of different attention heads as dispersed as possible, and establish a connection for the point clouds that are far away but belong to the same organ, thus providing more semantic information for accurate segmentation.
3. In order to reduce the impact of point cloud disorder and irregularity on feature extraction, we propose a position coding method that can reflect the relative position of points, and use the position coding network in the local and global feature extraction modules of MASPC_Transform.

The rest of this paper is organized as follows. Section 2 introduces the related work of plant point cloud segmentation. Section 3 describes the detailed structure of MASPC_Transform. In Section 4, we evaluated the segmentation performance of MASPC_Transform on the ROSE_X dataset and analyzed the experimental results. The last part is the conclusion of this paper.

2. Related Work

Traditional methods achieve the segmentation of plant point clouds through geometric features [12]. These methods use geometric information such as point cloud edge points, smoothness, plane fitting residual [13], curvature gradient [14] to classify and aggregate each point. On this basis, the method of clustering and model fitting [15] is further applied to complete the segmentation of point cloud data. Lee et al. [16] developed an adaptive clustering method, which can segment the point cloud data of pine forest to manage individual pine trees. This method is suitable for different sizes of canopy, but it needs a lot of data for pre training. Tao et al. [17] completed the segmentation task of single tree by setting a reasonable spacing threshold by using the characteristics of different trees and combining the "growth" algorithm. Xu et al. [18] applied the traditional Dijkstra shortest path algorithm to the spatial point cloud to complete the separation of tree branches and leaves. Matheus et al. [19] fused a variety of algorithms to realize the recognition of geometric characteristics in tree point cloud, and combined with the shortest path algorithm to complete the segmentation of point cloud structure, which greatly improved the robustness of the algorithm. Li et al. [20] designed a new algorithm to more accurately estimate the inclination and azimuth of the blades in the point cloud, and constructed a new projection coefficient model. In the follow-up study, Li et al. [21] developed a new path discrimination method by improving Laplace's shrinkage skeletonization algorithm to obtain the relevant parameters of the branch architecture. Traditional algorithms are easily affected by outliers and noise, which reduces the segmentation accuracy. The design of such algorithms often depends on the empirical design of geometric features, which are only effective for specific segmentation tasks.

Compared with traditional algorithms, deep learning methods are data-driven, do not need too many artificial design features, and have better performance. Currently, the deep learning methods that have been applied to point cloud segmentation include methods based on multi-view [22], voxel [23], and point cloud [24–26]. The method based on point cloud has the characteristics of directly processing point cloud and greatly retaining data information, so it has gradually become the mainstream research direction. Qi et al. [24] first proposed the network structure pointnet for directly processing point cloud data. This network proposed to use multilayer perceptron (MLP) with shared parameters to learn features and use symmetric functions to obtain global features. However, it has the problem that it cannot make full use of local information of points to extract fine-grained features. In order to solve this problem, an improved pointnet++ network [25] is proposed, which performs hierarchical and progressive learning on points from a large local area to obtain accurate geometric features near each point. In order to better extract the features of point clouds, Lee et al. [27] proposed an attention network, which can deal with disordered sets by adjusting the internal parameters of the network and can be used to extract the features of point clouds. Engel et al. [10] designed the Point Transformer network for point cloud segmentation, used multi-head attention in the network, and designed the SortNet structure to ensure the permutation invariance of extracted features.

Although great progress has been made in the research of deep learning segmentation algorithms for point cloud data, there are still few research on the segmentation of plant point cloud using deep learning methods. Wu et al. [28] adjusted the pointnet architecture to make the framework more suitable for processing the segmentation task of branches and leaves, and proposed a contribution score evaluation method. Jin et al. [29] made corn point cloud voxelized and applied convolutional neural network to voxelized data to complete a series of research work such as corn population segmentation and individual segmentation. Dutagaci et al. [30] provided valuable rosette data sets and provided benchmarks. Turgut et al. [31] verified the segmentation accuracy of various point based deep learning methods based on the work of Dutagaci [30], and studied the feasibility of three-dimensional synthesis model for training networks. Compared with other field point cloud segmentation tasks, plant point cloud segmentation is more challenging. This is because the stems, leaves, flowers, and other parts of plants are intertwined, resulting in the segmentation effect of existing segmentation methods is not ideal. The particularity of plant point cloud is that each part (organ) of the plant is very small and interwoven. This study proposes MASPC_Transform for segmentation of complex point clouds such as plant point clouds. In addition to the plant point cloud segmentation task, it is also applicable to the segmentation task of other point clouds with complex interwoven structures, such as forest point clouds [32].

3. Approach

3.1. Architecture of MASPC_Transform

The architecture of MASPC_Transform is shown in Figure 1. We use Point Transformer [10] as the network framework of MASPC_Transform. The difference between the proposed MASPC_Transform and the Point Transformer is that the proposed position coding network is used in the PC_MSG and PC_SortNet modules, and the proposed multi-head attention separation loss based on spatial similarity is added to the loss function of the entire network. MASPC_Transform includes feature extraction part and detection head. The feature extraction network has two branches: location feature generation and global feature generation. These two branches are responsible for extracting local and global features of plant point clouds. The global features (F_{Globel}) and local features ($F_{Location}$) are aggregated in the detection head and the segmentation results are obtained.

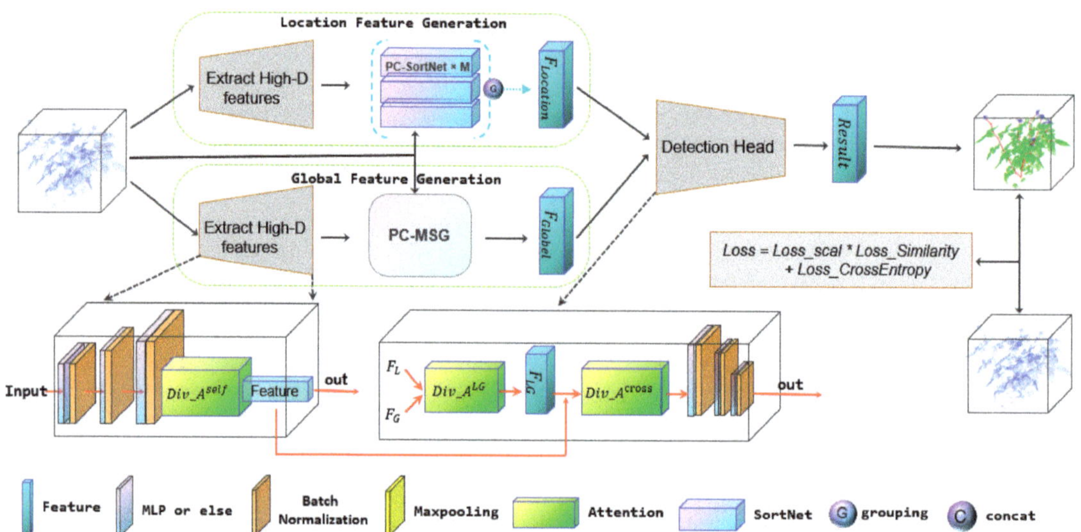

Figure 1. Architecture of MASPC_Transform.

First, the plant point cloud is input to the location feature generation and global feature generation branches for processing. Both branches first extract the feature of point cloud. In the location feature generation, the PC-SortNet module can evaluate the importance of the features of different areas of the point cloud, and select the important features as the local features of the plant point cloud. In the global feature generation branch, the multi-scale grouped (MSG) feature extraction network can obtain the point cloud features of three scales to adapt to different sizes of plant organs. The features of the three scales are fused as the global features of the whole plant point cloud. We use the position coding proposed in this paper in PC_MSG and PC_SortNet modules. Position coding is discussed in detail in Section 3.1. In detection head, the global feature F_{Globel} and the local feature $F_{Location}$ are associated and fused by the multi-head attention module. The multi-layer perceptron (MLP) in the detection head obtains the final segmentation result based on the fused features.

Multi-head attention [10] in MASPC_Transform is defined as follows:

$$Multihead(Q, K, V) = \left(F_{sa}^1 \bigoplus \cdots \bigoplus F_{sa}^i \right) W^O \tag{1}$$

$$A^{MH}(X, Y) = LayerNorm(S + \Phi(S)) \tag{2}$$

In Equation (1), Q, K, and V respectively represent the query matrix, key matrix, and value matrix of attention, and their matrix dimensions are d_k, d_k, and d_v. $F_{sa}^i = A\left(QW_i^Q, KW_i^K, VW_i^V \right)$ represents the features output by the ith attention head, $W_i^Q, W_i^K \in \mathbb{R}^{d_m \times d_k}$, $W_i^V \in \mathbb{R}^{d_m \times d_v}$, and $W^O \in \mathbb{R}^{hd_v \times d_m}$ are the learnable parameters. The symbol $\bigoplus$ indicates that the features outputted by different attention heads are concatenated together. In Equation (2), $LayerNorm$ is layer normalization [33]. S is defined as $S = LayerNorm(X + Multihead(X, Y, Y))$, Φ is a network module with multiple MLPs, which is responsible for further feature extraction of S. $A^{MH}(X, Y)$ is the prototype of all multi-head attention in the network.

$$A^{self}(P) = A^{MH}(P, P) \tag{3}$$

$$A^{LG}(P, Q) = A^{cross}(P, Q) = A^{MH}(P, Q) \tag{4}$$

In Equations (3) and (4), A^{self}, A^{LG}, and A^{cross} are derived from A^{MH}. A^{self} can perform the calculation of multi-head attention among all elements of P, while A^{LG} and A^{cross} can handle different sets P and Q, and perform the calculation of multi-head attention between the two sets.

We proposed a multi-head attention separation loss based on spatial similarity (loss in Figure 1). This loss acts on all the multi-head attention modules in MASPC_Transform. Therefore, we call the three attention modules that are affected by the proposed loss as Div_A^{self}, Div_A^{LG}, and Div_A^{cross}. These three multi-head attention modules are responsible for establishing connections for similar features at different positions in the point cloud space. We will discuss the loss function of multi-head attention separation based on spatial similarity in Section 3.3.

3.2. Position Code

Plant point cloud data are a collection of a series of points in space. Point sets have the characteristics of disordered and irregular distribution, so we propose a unique point cloud position coding method. The position code contains the relative position information of each point and its adjacent points, so as to avoid the interference of the disorder of the point cloud on the feature extraction. Position code function δ is defined as follows:

$$\delta = \theta\left(\bigcup_{i=1}^{n} (P_i, (P_i - P_{i1}), \dots\dots, (P_i - P_{ij}))\right) \bigoplus \theta(P_i, P_{i1}, \dots\dots, P_{ij}) \tag{5}$$

Suppose there are n points in the whole point cloud space. In Equation (5), P_i is a point in a subspace after the ball query, $P_i, P_{i1}, P_{i2}, P_{i3}, \dots, P_{ij} \in P$, P is the set of all points in the subspace. $P_i, (P_i - P_{i1}), \dots\dots, (P_i - P_{ij})$ are the relative position codes of point P_i, $\bigcup_{i=1}^{n}()$ represents the relative position code of all points in the space. Function θ is a multi-layer perceptron (MLP) used for feature extraction of position code. The symbol $\bigoplus$ indicates that the obtained two features are concatenated. Equation (5) indicates that the position coding δ of the point cloud space is composed of the relative position code (RPC) and the absolute position code (APC) of each point in the space. The absolute position code of a point is the coordinates of the point cloud. The relative position code of a point is the difference between the coordinates of the point and all points in its subspace. The relative position code keeps a certain invariance to the disorder of the point cloud, and it reflects the relationship between a point and its adjacent points, which can make the feature contain more local information. The position code network is shown in Figure 2.

3.3. MSG and SortNet Based on Position Code Network

In MASPC_Transform, we improved the MSG [10] in Point Transformer, and used the Position code-MSG(PC-MSG) module to extract global features. The structure of PC-MSG is shown in Figure 3. PC-MSG first takes the farthest point sampling (FPS), then the sampling point is taken as the center point, and three different radius are selected for ball query. According to the method in 3.1, the RPC of points is calculated in the subspace of each scale in PC-MSG. After that, the RPC features of each scale were extracted using MLP. In Figure 3, the orange rectangle represents the extracted RPC features of each scale, the blue rectangle represents the extracted APC features of each scale, and the high D features are the features extracted by the high-dimensional feature extraction network before the PC-MSG network.

Finally, the RPC features, APC features, and high D features are concatenated together. Because the network structures of different scales in the MSG are the same, the feature extraction process of the second scale of the network is omitted in Figure 3.

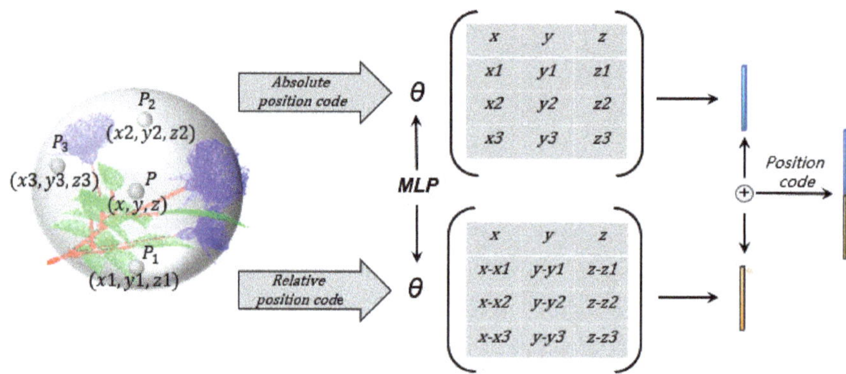

Figure 2. Position code network.

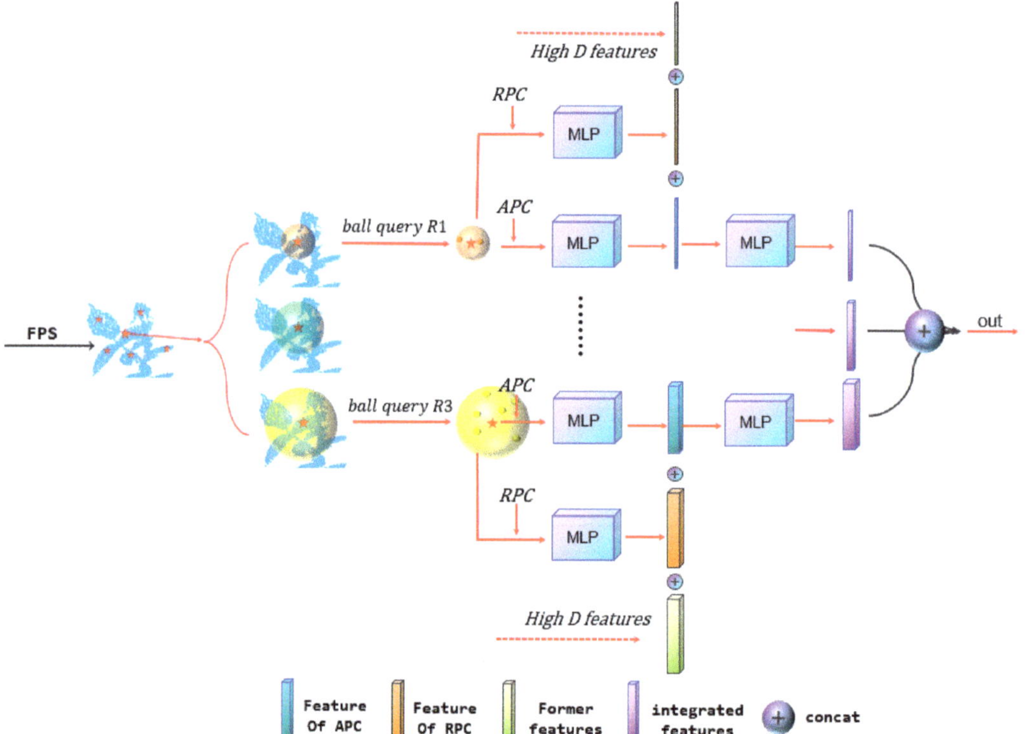

Figure 3. Position code in MSG.

We also improved SortNet in Point Transformer network [10], replacing SortNet with PC-SortNet with position code. As shown in Figure 4, in the PC-SortNet, the input features first pass through multiple MLPs, and its feature dimension is reduced to 1 dimension. This feature calculates a learnable importance score for each point in the point cloud space.

After that, k points with the highest score are selected through the Top-k module. We take k points as the center of the ball query and extract the features of the region within the ball. We use a method similar to skip connect to concatenate the features of different stages. As shown by the Red PC in Figure 4, we use the position code proposed in Section 3.1 when querying the ball and extracting features.

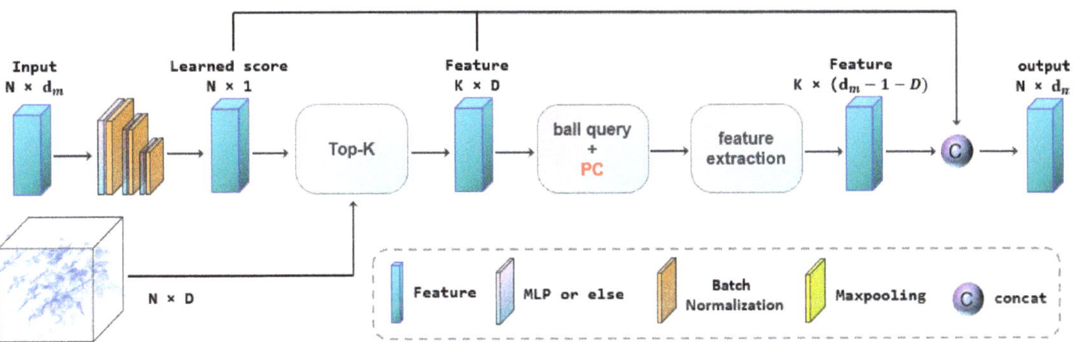

Figure 4. Position code in PC_SortNet.

3.4. Multi-Head Attention Separation Loss Based on Spatial Similarity

When multi-head attention is used for feature extraction, there is a possibility that the generated multiple attention spaces are similar [11], which will cause multiple attention spaces to overlap each other, resulting in repeated extraction in some areas and insufficient feature extraction in other areas. Therefore, we propose a multi-head attention separation loss based on spatial similarity, which makes each attention positions of the segmented network tend to be separated. Its definition is as follows:

$$Separation\_Loss = -\frac{1}{n^2} \sum_{F_{sa}^i, F_{sa}^j \in \mathcal{F}, i \neq j} \frac{\left| \sum F_{sa}^i F_{sa}^j \right|}{\|F_{sa}^i\|_2 \|F_{sa}^j\|_2} \tag{6}$$

In Equation (6), F_{sa}^i and F_{sa}^j are the different attention feature spaces of multi-head attention output, and $\mathcal{F}$ is the set of feature spaces output by the attention mechanism. The symbol $|\,.\,|$ represents the module of the matrix, $\|\,.\,\|_2$ denotes the 2-norm of the matrix. Equation (6) can calculate the average cosine distance of all output feature spaces. Cosine distance is an index to measure the difference of feature space in direction, so it can be used to evaluate the similarity of feature space. By dividing by n^2, we can make the calculated value tend to a reasonable range and avoid the difficulty of network training. Using a negative sign to indicate *Separation_Loss* penalizes network parameters that make F_{sa}^i and F_{sa}^j tend to be similar. We take the Separation Loss as a part of the loss function and train the network, so that the attention features tend to be diverse. The loss function of MASPC_Transform is as follows:

$$Loss\_CrossEntropy = -\sum_x (p(x)logq(x) + (1 - p(x))\log(1 - q(x))) \tag{7}$$

$$Loss = Loss_{CrossEntropy} + Loss_{scal} \times Separation\_Loss \tag{8}$$

In Equation (7), $p(x)$ is the real classification probability distribution of the input point cloud, and $q(x)$ is the prediction probability distribution actually given by the network. Equation (7) depicts the difference between the classification result and the real value. The smaller the value of *Loss_CrossEntropy*, the more realistic the prediction given by the network. As shown in Equation (8), we used the *Loss_CrossEntropy* and *Separation_Loss* as

MASPC_Transform's loss function. Where *Loss_scal* is the weight of *Separation_Loss* in the loss function. Using the new loss function to train MASPC_Transform can make multiple attention feature spaces specific.

4. Experiment

4.1. Data Set

We evaluated the performance of MASPC_Transform on the ROSE_X dataset [30]. The ROSE_X dataset contains a total of 11 rose point cloud data. The rose point cloud data contain three semantic tags, namely, flower, leaf, and stem. The petals, calyx, and bud of rose are all marked as "flower" label, and the stem and petiole are all marked as "stem" label. We use nine rose point clouds to train the network, and the other two rose point clouds to test the segmentation performance of the network after training. We denoted the two roses used for the test as test_R1 and test_R2. Because the volume of a single rose point cloud is large and the number of points is large, and the amount of data that can be processed at a single time is limited, it is necessary to divide the point cloud into smaller blocks. We adopt the same blocking method as in [30], that is, the size and number of points of each block are as consistent as possible, and the structure within the block is as complete as possible. With this method, we divided the nine rose point clouds used for training into 596 point clouds and the two point clouds used for testing into 143 point clouds.

4.2. Implementation Details

For the model training, the Adam optimizer is used to update and optimize the network parameters. The initial learning rate is set to 0.001 and the batch size is 16. The GPU model is NVIDIA GeForce RTX 2080Ti, operating system is Ubuntu 18.04 LTS, CUDA version is 11.0. The proposed model is implemented in PyTorch with Python version 3.6. When training MASPC_Transform network, the input point cloud only contains three-dimensional X-Y-Z coordinates, and the number of input points is 2048.

4.3. Evaluation Methodology

We use the Intersection over Union (IoU) and Mean Intersection over Union (MIoU) to evaluate the performance of all networks. Where IoU is equal to the ratio of intersection and union between the predicted point set and the real point set, and MIoU represents the average value of IOU of all categories. The higher the values of these two indicators, the better the segmentation effect of the point cloud. The mathematical definition is as follows:

$$IoU_c = \frac{TP_c}{TP_c + FP_c - FN_c} \tag{9}$$

$$MIoU = \frac{\sum_c IoU_c}{k} \tag{10}$$

where TP_c, FP_c, and FN_c are the number of positive samples of category C that have been correctly identified, the number of negative samples that have been misreported. and the number of positive samples that have been missed, $C \in \{Flower, stem, leaf\}$, k is the number of all categories.

4.4. Segmentation Results

In Table 1, we show the segmentation results of different segmentation networks on ROSE_X dataset, including PointNet [24], PointNet++ [25], DGCNN [34], PointCNN [35], ShellNet [36], RIConv [37], and the proposed MASPC_Transform.

Table 1. Comparison of network segmentation effect indicators (%).

Evaluation	Category	Pointnet	Pointnet++	DGCNN	PointCNN	ShellNet	RIConv	Point Transformer	Ours
IoU	Flower	15.83	74.12	8.34	53.56	49.36	54.12	80.93	**83.32**
	Leaf	82.56	**95.36**	84.17	91.76	89.69	88.96	91.76	94.36
	Stem	5.27	77.69	24.97	70.89	54.78	35.79	74.99	**78.96**
MIoU	MIou	34.55	82.39	39.16	72.14	64.61	60.79	82.56	**85.52**

In Table 1, we can see MASPC_Transform has the highest MIoU, and MASPC_Transform achieves the best segmentation results on both the flower and stem classes. As an improved version of PointNet, PointNet++ can flexibly extract local features by adjusting the neighborhood radius, and has the ability to extract the features of small organs of plants. So, it achieves the best segmentation results in leaf class. The IoU value of MASPC_Transform on the leaf class is slightly lower than that of PointNet++, but the MIoU value of MASPC_Transform is higher than that of PointNet++.

4.5. Visual Effects

Figures 5 and 6 respectively show the segmentation results of different segmentation networks on test_R1 and test_R2. Figures 5a and 6a are the ground truth of test_R1 and test_R2. In Figures 5a and 6a, we can see that the stems, leaves, and flowers of the two plants are interlaced and occluded each other, which creates great difficulties for the segmentation algorithm. In Figure 5d,f and Figure 6d,f, we can see that PointNet and DGCNN hardly segment different plant organs. It can be seen from the area within the dotted circle in Figures 5 and 6 that the segmentation ability of the comparison network (Point Transformer, PointNet++, DGCNN, PointCNN, ShellNet and RIConv) for details is inferior to that of MASPC_Transform. As shown in Figure 5c, Point Transformer mistakenly divides some petals into leaves. As shown in Figure 5e, PointNet++ mistakenly divided part of the calyx at the top into leaves and stems. As shown in Figure 5g, PointCNN mistakenly divided part of the calyx at the top into stems, and mistakenly divided the stems in the lowest red circle into leaves. As shown in Figure 5h, ShellNet mistakenly divided the calyx in the red circle into leaves. As shown in Figure 5i, RICov mistakenly divided some flowers in the top red circle into leaves. In Figure 6, there is also a case of false segmentation in the comparison network. The proposed MASPC_Transform has the best segmentation effect for the interlaced parts of different plant organs.

In order to show the segmentation effect of each method more clearly, we extracted some regions from the segmented plant point cloud and showed them more clearly in Figure 7. As can be seen from the first column in Figure 7, the objects to be segmented are leaves and stems. Among the segmentation results of all methods, the results corresponding to MASPC_Transform proposed by us are the most similar to ground truth. PointNet was failed to segment stems and leaves. DGCNN and PointCNN hardly segment the stem and leaf correctly. The stems segmented by PointNet++, ShellNet, and RIConv were shorter than those separated by MASPC_Transform, and they mistakenly divided the stems between two leaves into leaves. Point Transformer also mistakenly divides some stems into leaves at the intersection of leaves. In the segmentation results of the second and third columns of Figure 7, the MASPC_Transform also achieves the best segmentation effect.

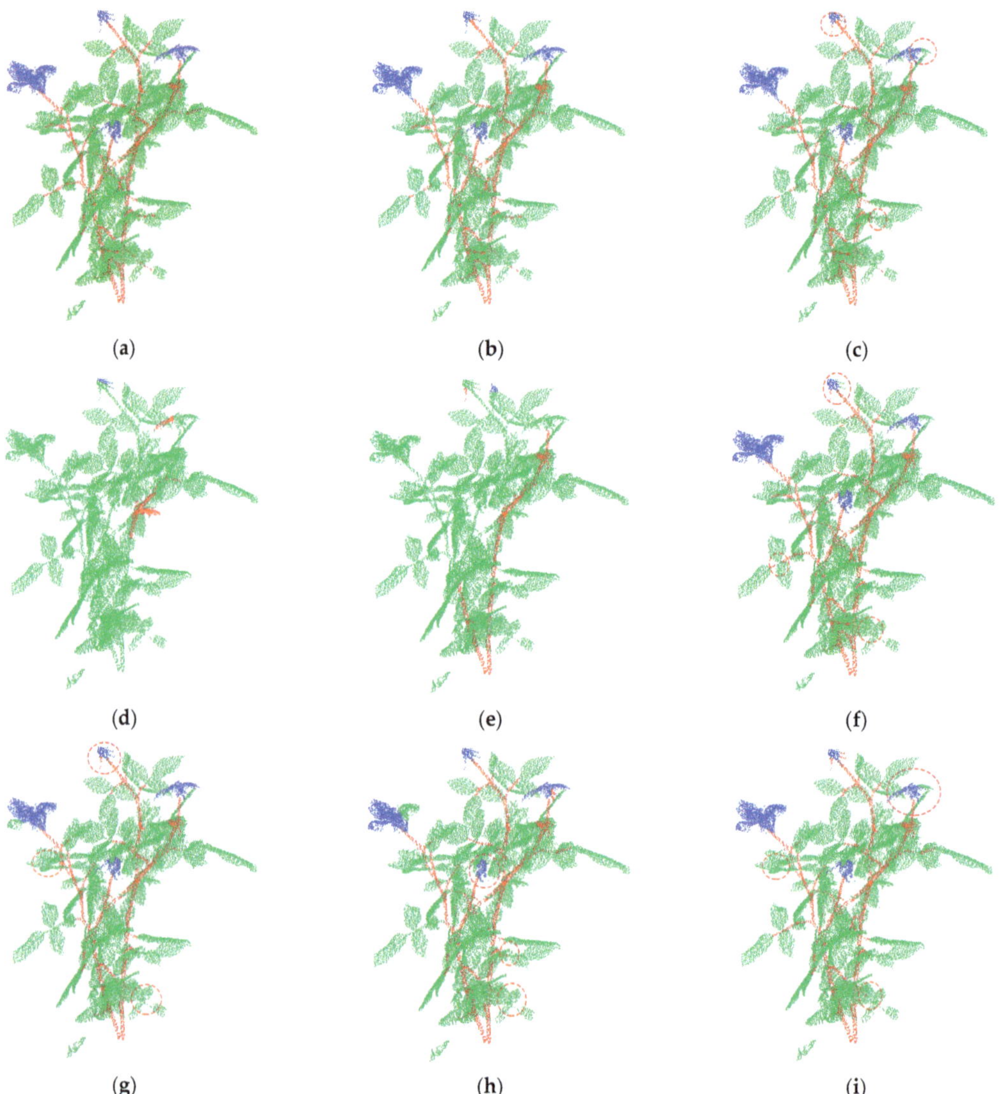

Figure 5. Segmentation result of each network on test_R1. (**a**) Ground Truth; (**b**) MASPC_Transform; (**c**) Point Transformer; (**d**) PointNet; (**e**) PointNet++; (**f**) DGCNN; (**g**) PointCNN; (**h**) ShellNet; (**i**) RICov.

It can be seen from the segmentation effect shown in Figures 5–7 that the MASPC_Transform has the best segmentation effect. This is because the multi-head attention and the multi-head attention separation loss based on spatial similarity in MASPC_Transform establish a connection for the same kind of point clouds (point clouds with similar semantics) scattered in different regions of the point cloud space. In areas where multiple categories are interlaced, this association can help MASPC_Transform achieve better segmentation effect in detail.

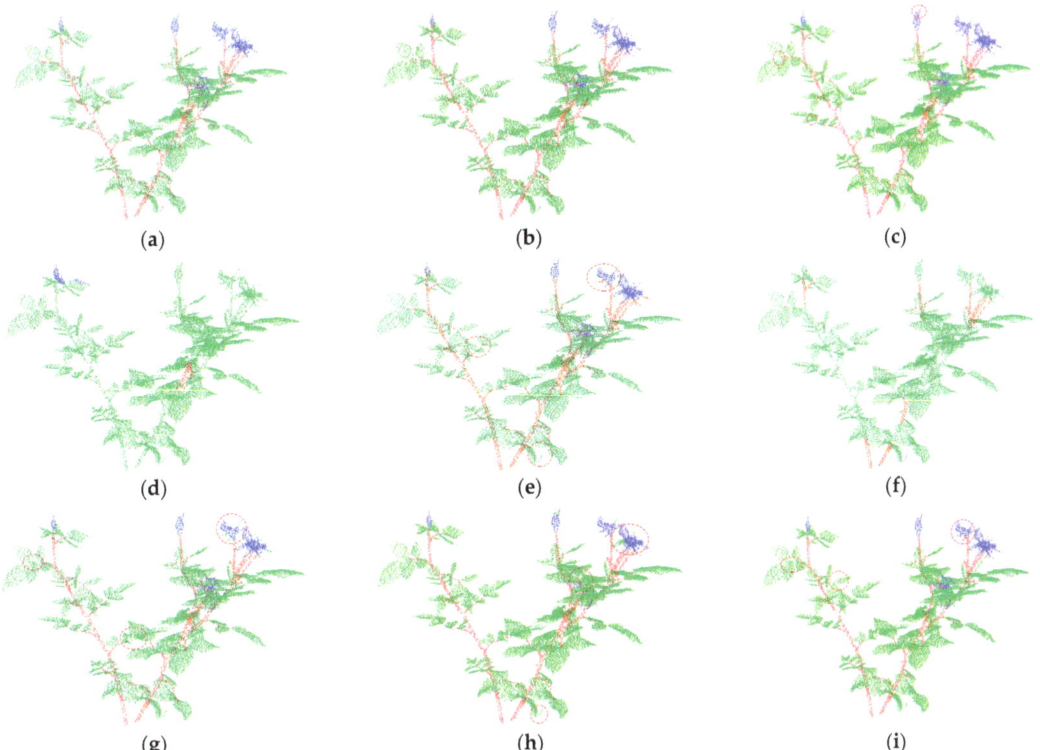

Figure 6. Segmentation result of each network on test_R2. (**a**) Ground Truth; (**b**) MASPC_Transform; (**c**) Point Transformer; (**d**) PointNet; (**e**) PointNet++; (**f**) DGCNN; (**g**) PointCNN; (**h**) ShellNet; (**i**) RICov.

4.6. Ablation Studies

Table 2 shows the results of our ablation studies on the ROSE_X dataset. In the ablation studies, we used the original Point Transformer [10] as the baseline. In Table 2, Without RPC represents a network that does not use RPC, but still uses our Equation (8) to train the network. Without Separation_Loss means that the proposed multi-head attention Separation_Loss is not used in the network, and only CrossEntropy is used to train the network. Note that RPC is used in the Without Separation_Loss network. The last column presents the experimental results of MASPC_Transform proposed by us. From the results shown in Table 2, we can see that the values of IoU and MIoU of MASPC_Transform are the highest. The IoU and MIoU of each category of MASPC_Transform without multi-head attention separation loss function and MASPC_Transform without relative position code are lower than those of MASPC_Transform, but better than the Point Transformer.

According to the experimental results in Section 4.4, the proposed MASPC_Transform outperforms the state-of-the-art approaches. The visualization results shown in Figures 6 and 7 confirm the experimental results in Section 4.4. The visualization results of these comparison approaches for rose point clouds with interlaced stems, leaves, and flowers is not as good as MASPC_Transform. This shows that our multi-head attention separation loss can distract the attention positions of different attention heads as much as possible, and establish connections for point clouds that are far away but belong to the same organ. However, these comparison approaches do not have this ability, so that these segmentation networks believe that two flowers (stems or leaves) far away belong to different categories.

The results of ablation studies verify the effectiveness of the multi-head attention separation loss (Separation_Loss) and position code (PC).

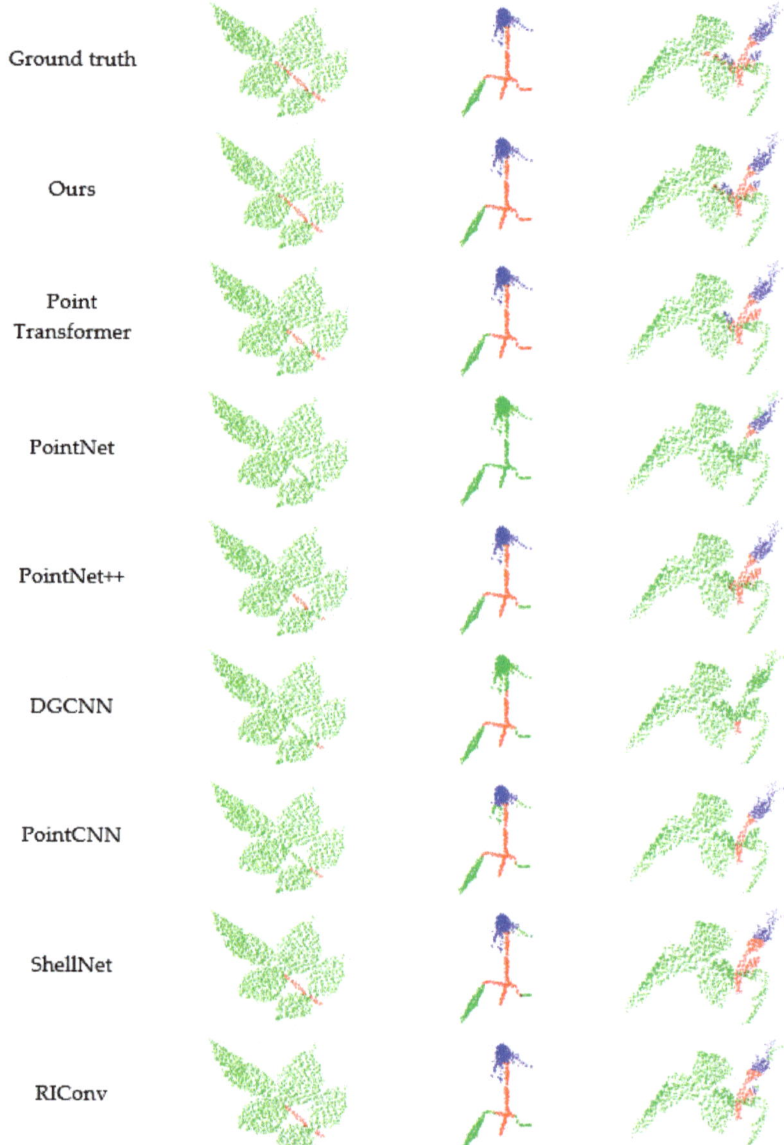

Figure 7. Segmentation rendering of different networks.

Table 2. Ablation study on ROSE_X dataset.

Evaluation	Category	Point Transformer	Without RPC	Without Separation_Loss	Ours
IoU	Flower	80.93	83.10	82.28	**83.32**
	Leaf	91.76	93.03	92.89	**94.36**
	Stem	74.99	77.64	76.71	**78.96**
MIoU	MIou	82.56	84.29	83.96	**85.52**

5. Conclusions

We propose a plant point cloud segmentation network named MASPC_Transform. In order to make the attention positions of different attention heads of MASPC_Transform as dispersed as possible, we propose a multi-head attention separation loss based on spatial similarity. In order to reduce the impact of point cloud disorder and irregularity on feature extraction, we use position coding in the local and global feature extraction modules of MARP_Transform. We evaluated the proposed MASPC_Transform on the ROSE_X dataset. The results of segmentation experiments show that MASPC_Transform network performs better than the state-of-the-art approaches. The results of ablation experiments demonstrate the effectiveness of the proposed position code and attention separation loss. Due to ROSE_X dataset is the only open source benchmark plant point cloud segmentation dataset, so the MASPC_Transform has only been tested on this dataset. If a new open source plant point cloud segmentation dataset appears, the MASPC_Transform should accept more tests.

Author Contributions: Conceptualization, methodology, resources, funding acquisition, B.L.; software, validation, visualization, C.G. All authors have read and agreed to the published version of the manuscript.

Funding: This work was supported by the Science and Technology Development Plan Project of Jilin Province under Grant 20200201165JC.

Institutional Review Board Statement: Not applicable.

Informed Consent Statement: This article is not about human research.

Data Availability Statement: Not applicable.

Conflicts of Interest: The authors declare no conflict of interest.

References

1. Crossa, J.; Fritsche-Neto, R.; Montesinos-Lopez, O.A.; Costa-Neto, G.; Dreisigacker, S.; Montesinos-Lopez, A.; Bentley, A.R. The modern plant breeding triangle: Optimizing the use of genomics, phenomics, and enviromics data. *Front. Plant Sci.* **2021**, *12*, 651480. [CrossRef] [PubMed]
2. Shi, Y.; Zhu, Y.; Wang, X.; Sun, X.; Ding, Y.; Cao, W.; Hu, Z. Progress and development on biological information of crop phenotype research applied to real-time variable-rate fertilization. *Plant Methods* **2020**, *16*, 1–15. [CrossRef] [PubMed]
3. Arbona, V.; Iglesias, D.J.; Talón, M.; Gómez-Cadenas, A. Plant phenotype demarcation using nontargeted LC-MS and GC-MS metabolite profiling. *J. Agric. Food Chem.* **2009**, *57*, 7338–7347. [CrossRef]
4. Sun, S.; Li, C.; Chee, P.W.; Paterson, A.H.; Jiang, Y.; Xu, R.; Robertson, J.S.; Adhikari, J.; Shehzad, T. Three-dimensional photogrammetric mapping of cotton bolls in situ based on point cloud segmentation and clustering. *ISPRS J. Photogramm. Remote Sens.* **2020**, *160*, 195–207. [CrossRef]
5. Sun, S.; Liang, N.; Zuo, Z.; Parsons, D.; Morel, J.; Shi, J.; Wang, Z.; Luo, L.; Zhao, L.; Fang, H.; et al. Estimation of botanical composition in mixed clover–grass fields using machine learning-based image analysis. *Front. Plant Sci.* **2021**, *12*, 622429. [CrossRef]
6. Aginako, N.; Lozano, J.; Quartulli, M.; Sierra, B.; Olaizola, I.G. Identification of plant species on large botanical image datasets. In Proceedings of the 1st International Workshop on Environnmental Multimedia Retrieval co-located with ACM International Conference on Multimedia Retrieval, EMR@ ICMR 2014, Glasgow, UK, 1 April 2014; pp. 38–44.
7. Grand-Brochier, M.; Vacavant, A.; Cerutti, G.; Kurtz, C.; Weber, J.; Tougne, L. Tree leaves extraction in natural images: Comparative study of preprocessing tools and segmentation methods. *IEEE Trans. Image Process.* **2015**, *24*, 1549–1560. [CrossRef] [PubMed]

8. Yogeswararao, G.; Malmathanraj, R.; Palanisamy, P. Fractional weighted nuclear norm based two dimensional linear discriminant features for cucumber leaf disease recognition. *Multimed. Tools Appl.* **2022**, *81*, 1–21. [CrossRef]

9. Li, Y.; Gao, J.; Wang, X.; Chen, Y.; He, Y. Depth camera based remote three-dimensional reconstruction using incremental point cloud compression. *Comput. Electr. Eng.* **2022**, *99*, 107767. [CrossRef]

10. Engel, N.; Belagiannis, V.; Dietmayer, K. Point transformer. *IEEE Access* **2021**, *9*, 134826–134840. [CrossRef]

11. Li, J.; Tu, Z.; Yang, B.; Lyu, M.R.; Zhang, T. Multi-head attention with disagreement regularization. *arXiv* **2018**, arXiv:1810.10183.

12. Perez-Perez, Y.; Golparvar-Fard, M.; El-Rayes, K. Segmentation of point clouds via joint semantic and geometric features for 3D modeling of the built environment. *Autom. Constr.* **2021**, *125*, 103584. [CrossRef]

13. Li, L.; Sung, M.; Dubrovina, A.; Yi, L.; Guibas, L.J. Supervised fitting of geometric primitives to 3d point clouds. In Proceedings of the IEEE/CVF Conference on Computer Vision and Pattern Recognition, Long Beach, CA, USA, 15–20 June 2019; pp. 2652–2660.

14. He, Y.; Kang, S.H.; Liu, H. Curvature regularized surface reconstruction from point clouds. *SIAM J. Imaging Sci.* **2020**, *13*, 1834–1859. [CrossRef]

15. Xia, S.; Chen, D.; Wang, R.; Li, J.; Zhang, X. Geometric primitives in LiDAR point clouds: A review. *IEEE J. Sel. Top. Appl. Earth Obs. Remote Sens.* **2020**, *13*, 685–707. [CrossRef]

16. Lee, H.; Slatton, K.C.; Roth, B.E.; Cropper, W.P., Jr. Adaptive clustering of airborne LiDAR data to segment individual tree crowns in managed pine forests. *Int. J. Remote Sens.* **2010**, *31*, 117–139. [CrossRef]

17. Tao, S.; Wu, F.; Guo, Q.; Wang, Y.; Li, W.; Xue, B.; Hu, X.; Li, P.; Tian, D.; Li, C.; et al. Segmenting tree crowns from terrestrial and mobile LiDAR data by exploring ecological theories. *ISPRS J. Photogramm. Remote Sens.* **2015**, *110*, 66–76. [CrossRef]

18. Xu, H.; Gossett, N.; Chen, B. Knowledge and heuristic-based modeling of laser-scanned trees. *ACM Trans. Graph.* **2007**, *26*, 19. [CrossRef]

19. Vicari, M.B.; Disney, M.; Wilkes, P.; Burt, A.; Calders, K.; Woodgate, W. Leaf and wood classification framework for terrestrial LiDAR point clouds. *Methods Ecol. Evol.* **2019**, *10*, 680–694. [CrossRef]

20. Li, Y.; Su, Y.; Hu, T.; Xu, G.; Guo, Q. Retrieving 2-D leaf angle distributions for deciduous trees from terrestrial laser scanner data. *IEEE Trans. Geosci. Remote Sens.* **2018**, *56*, 4945–4955. [CrossRef]

21. Li, Y.; Su, Y.; Zhao, X.; Yang, M.; Hu, T.; Zhang, J.; Liu, J.; Liu, M.; Guo, Q. Retrieval of tree branch architecture attributes from terrestrial laser scan data using a Laplacian algorithm. *Agric. For. Meteorol.* **2020**, *284*, 107874. [CrossRef]

22. Su, H.; Maji, S.; Kalogerakis, E.; Learned-Miller, E. Multi-view convolutional neural networks for 3d shape recognition. In Proceedings of the IEEE International Conference on Computer Vision, Santiago, Chile, 7–13 December 2015; pp. 945–953.

23. Maturana, D.; Scherer, S. Voxnet: A 3d convolutional neural network for real-time object recognition. In Proceedings of the 2015 IEEE/RSJ International Conference on Intelligent Robots and Systems (IROS), Hamburg, Germany, 28 September–3 October 2015; pp. 922–928.

24. Qi, C.R.; Su, H.; Mo, K.; Guibas, L.J. Pointnet: Deep learning on point sets for 3d classification and segmentation. In Proceedings of the IEEE Conference on Computer Vision and Pattern Recognition, Honolulu, Hawaii, 21–26 July 2017; pp. 652–660.

25. Qi, C.R.; Yi, L.; Su, H.; Guibas, L.J. Pointnet++: Deep hierarchical feature learning on point sets in a metric space. *Adv. Neural Inf. Process. Syst.* **2017**, *30*, 5099–5108.

26. Li, B.; Zhu, S.; Lu, Y. A single stage and single view 3D point cloud reconstruction network based on DetNet. *Sensors* **2022**, *22*, 8235. [CrossRef]

27. Kim, B.N.; Lee, J.S.; Shin, M.S.; Cho, S.C.; Lee, D.S. Regional cerebral perfusion abnormalities in attention deficit/hyperactivity disorder. *Eur. Arch. Psychiatry Clin. Neurosci.* **2002**, *252*, 219–225.

28. Wu, B.; Zheng, G.; Chen, Y. An improved convolution neural network-based model for classifying foliage and woody components from terrestrial laser scanning data. *Remote Sens.* **2020**, *12*, 1010. [CrossRef]

29. Jin, S.; Su, Y.; Gao, S.; Wu, F.; Hu, T.; Liu, J.; Li, W.; Wang, D.; Chen, S.; Jiang, Y.; et al. Deep learning: Individual maize segmentation from terrestrial lidar data using faster R-CNN and regional growth algorithms. *Front. Plant Sci.* **2018**, *9*, 866. [CrossRef]

30. Dutagaci, H.; Rasti, P.; Galopin, G.; Rousseau, D. ROSE-X: An annotated data set for evaluation of 3D plant organ segmentation methods. *Plant Methods* **2020**, *16*, 1–14. [CrossRef]

31. Turgut, K.; Dutagaci, H.; Galopin, G.; Rousseau, D. Segmentation of structural parts of rosebush plants with 3d point-based deep learning methods. *Plant Methods* **2022**, *18*, 1–23. [CrossRef]

32. Krisanski, S.; Taskhiri, M.S.; Aracil, S.G.; Herries, D.; Turner, P. Sensor agnostic semantic segmentation of structurally diverse and complex forest point clouds using deep learning. *Remote Sens.* **2021**, *12*, 1413. [CrossRef]

33. Ba, J.L.; Kiros, J.R.; Hinton, G.E. Layer normalization. *arXiv* **2016**, arXiv:1607.06450. Available online: https://doi.org/10.48550/arXiv.1607.06450 (accessed on 21 July 2022).

34. Zhang, K.; Hao, M.; Wang, J.; de Silva, C.W.; Fu, C. Linked dynamic graph cnn: Learning on point cloud via linking hierarchical features. *arXiv* **2019**, arXiv:1904.10014.

35. Li, Y.; Bu, R.; Sun, M.; Wu, W.; Di, X.; Chen, B. Pointcnn: Convolution on x-transformed points. *Adv. Neural Inf. Process. Syst.* **2018**, *31*, 820–830.

36. Zhang, Z.; Hua, B.S.; Yeung, S.K. Shellnet: Efficient point cloud convolutional neural networks using concentric shells statistics. In Proceedings of the IEEE/CVF International Conference on Computer Vision, Seoul, Republic of Korea, 27 October–2 November 2019; pp. 1607–1616.

37. Zhang, Z.; Hua, B.S.; Rosen, D.W.; Yeung, S.K. Rotation invariant convolutions for 3d point clouds deep learning. In Proceedings of the 2019 International Conference on 3d Vision (3DV), Quebec City, QU, Canada, 16 September 2019; pp. 204–213.

 *sensors*

Article

A Single Stage and Single View 3D Point Cloud Reconstruction Network Based on DetNet

Bin Li *, Shiao Zhu and Yi Lu

School of Computer Science, Northeast Electric Power University, Jilin 132011, China
* Correspondence: libinjlu5765114@163.com

Abstract: It is a challenging problem to infer objects with reasonable shapes and appearance from a single picture. Existing research often pays more attention to the structure of the point cloud generation network, while ignoring the feature extraction of 2D images and reducing the loss in the process of feature propagation in the network. In this paper, a single-stage and single-view 3D point cloud reconstruction network, 3D-SSRecNet, is proposed. The proposed 3D-SSRecNet is a simple single-stage network composed of a 2D image feature extraction network and a point cloud prediction network. The single-stage network structure can reduce the loss of the extracted 2D image features. The 2D image feature extraction network takes DetNet as the backbone. DetNet can extract more details from 2D images. In order to generate point clouds with better shape and appearance, in the point cloud prediction network, the exponential linear unit (ELU) is used as the activation function, and the joint function of chamfer distance (CD) and Earth mover's distance (EMD) is used as the loss function of 3DSSRecNet. In order to verify the effectiveness of 3D-SSRecNet, we conducted a series of experiments on ShapeNet and Pix3D datasets. The experimental results measured by CD and EMD have shown that 3D-SSRecNet outperforms the state-of-the-art reconstruction methods.

Keywords: 3D reconstruction; single view; single stage; point cloud

Citation: Li, B.; Zhu, S.; Lu, Y. A Single Stage and Single View 3D Point Cloud Reconstruction Network Based on DetNet. *Sensors* **2022**, *22*, 8235. https://doi.org/10.3390/s22218235

Academic Editors: Miaohui Wang, Guanghui Yue, Jian Xiong and Sukun Tian

Received: 22 September 2022
Accepted: 26 October 2022
Published: 27 October 2022

Publisher's Note: MDPI stays neutral with regard to jurisdictional claims in published maps and institutional affiliations.

1. Introduction

In many tasks, such as virtual reality [1], experimental assistance [2], and robot navigation [3], a detailed 3D model is required, however, the facilities required to sample 3D models from the real world are costly. Moreover, it is uneconomical to manually reconstruct 3D models from 2D maps on a large scale. Many researchers proposed methods to reconstruct 3D models from a single image [4–6].

There are mainly two structures to represent a 3D model, voxel and point cloud. The former is just like 2D pixels but fits 3D objects into grids, sometimes containing other information such as features. It is also a regular data structure so many successful 2D methods can be easily applied. Many approaches [7,8] have focused on the voxel grid as output. However, the computational cost increases cubically to perform better geometric information or apply convolution methods.

Point cloud represents geometric information by a set of data points, each point represented by (x, y, z). Fan et al. [4] firstly applied deep learning methods to generate point clouds and proposed the chamfer distance and Earth mover's distance, but there are many methods to improve its evaluation results. 3D-LMNet [5] used an autoencoder to design a two-stage point cloud construction network. Most of the existing works are similar to 3D-LMNet, which are multi-stage point cloud generation networks. These multi-stage networks inevitably suffered feature loss and were considered a waste of time. At the same time, the existing work often pays more attention to the structure of the point cloud generation network, while ignoring the feature extraction of 2D images.

The main tasks of point cloud reconstruction are: (1) To retain more details or small targets in the image when extracting 2D image features in order to obtain a better reconstruction effect. (2) To generate a point cloud through a simple network structure to reduce the loss of features in the process of transmission in different network stages. We propose an end-to-end point cloud reconstruction network called 3D-SSRecNet, which applies DetNet [9] as an image feature extractor and gains detailed features.

The key contributions of our work are as follows:

1. We propose a one-stage neural network for 3D reconstruction from a single image, namely, 3D-SSRecNet. 3D-SSRecNet takes an image as input and directly outputs the predicted point cloud without further processing.
2. 3D-SSRecNet includes feature extraction and 3D point cloud generation. The feature extraction network is better at extracting the detailed features of the 2D input. The point cloud generation network has a plain structure and uses a suitable activation function in its multi-layer perceptron, which reduces the loss of features during forwarding propagation to obtain an elaborate output.
3. Experiments on ShapeNet and pix3D dataset have shown that 3D-SSRecNet outperforms the state-of-art reconstruction methods for the task of single-view reconstruction. At the same time, we also proved the validity of the activation function of the point cloud generation network through experiments.

2. Related Work

The technology of reconstructing 3D models from 2D images has many practical applications. Therefore, 2D to 3D reconstruction technologies applicable to different application scenarios will be quite different. For example, in [2], a method is proposed to predict the liquid or solid in transparent vessels to XYZ maps. This method can be applied, for example, to the task of a robot arm taking containers and pouring liquid. Therefore, this study proposed a scale-invariant loss so that the predicted scale of XYZ map can conform to the original 2D input scale. Part-Wise AtlasNet proposed in [10] can output 3D reconstruction with a fine local structure. This is because each neural network of Part-Wise AtlasNet is only responsible for reconstructing a specific part of the 3D object. Part-Wise AtlasNet has achieved a very refined reconstruction effect, but the reconstruction process is very time consuming. Part-Wise AtlasNet is obviously more suitable for tasks such as high-precision reconstruction and the display of cultural relics. In order to obtain 3D data in real time and accurately, hardware assistance is required in addition to consideration of the camera parameters. For example, SLAM [11] constructs a 3D map by positioning the camera in real time. In virtual reality or game modeling tasks, a point cloud reconstruction network may be required to reconstruct point clouds with reasonable contour and shape without considering hardware parameters. The 3D-SSRecNet proposed in this paper is more suitable for virtual reality or game modeling scenarios. Here, we summarize the references similar to the application scenarios of 3D-SSRecNet. Due to the development of deep learning [12] and big data [13], great progress has been made in this field, and numerous valuable research has emerged. These studies can be roughly divided into 2D images to voxels and 2D images to point clouds.

From 2D images to voxels. Several approaches focus on generating voxelized output representations. V3DOR [14] applied autoencoders and variational autoencoders to generate a smoother and high-resolution 3D model. The encoder aims to learn latent representation from the image and the decoder tries to obtain corresponding 3D voxels. Xie et al. [15] applied ResNet [16] as a part of the autoencoder and proposed a multi-scale context-aware fusion module to gain better results with more views of the object. Han et al. [17] proposed a novel shape representation that enabled a tube-by-tube manner via discriminative neural networks. Based on the network, they proposed an RNN-based model to gain the 3D corresponding representation from the input image. TMVNet [18] applied the transformers to the encoder and proposed a 3D feature fusion layer to refine the predictions. Kniaz et al. [19] proposed an image-to-voxel translation model which applied a generative adversarial network. Sym3DNet [20] applied a symmetry fusion step and perceptual loss to apply symmetry prior. Yang et al. [21] designed a memory-based framework to obtain a heavily occluded 3D model to handle challenging situations. However, voxel reconstruction gains sparse space information and has a high costs. It is difficult to both predict higher resolution 3D models and process them efficiently.

From 2D images to point clouds. Fan et al. [1] designed a framework called PSGN, which firstly applied deep learning methods to the point sets generation problem. They proposed chamfer distance and Earth mover's distance to judge the distance between point sets. 3D-LMNet [2] trained a point cloud autoencoder, then try to map images to corresponding learned embedding and they proposed diversity loss for uncertain reconstruction. 3D-ARNet [22] combined an autoencoder and a point prediction network. After the image is input into the image encoder, a simple point cloud is obtained. Pumarola et al. [23] proposed a conditional flow-based generative model to generate a map from image to point cloud, which is different from other generative models such as VAEs or GAN. Hafiz et al. [24] proposed the SE-MD network. SE-MD uses an autoencoder network as the feature extraction network and multiple decoding networks as point cloud generation networks. The final result can be obtained by fusing all the outputs of all the point cloud generation networks. However, these multi-stage models may suffer more cost of computational resources and feature loss when the feature maps propagate across networks. 3D-ReConstnet [6] applied the residual network to extract the features from the input image and used MLPs to predict point sets and, meanwhile, learned Gaussian probability distribution to refine the self-occluded part of an object and then directly applied MLPs to predict the point cloud. Ping et al. [25] projected the predicted point cloud and tried to fit edge details with ground truth. In order to enhance the features of 2D images, 3D-FEGNet [26] adds an edge extraction module to the feature extraction network. After comparing the reconstruction results of the above networks, we found that the key to a better reconstruction network are: (1) the feature extraction part of the network reflects more detailed 2D image features; (2) the loss of the extracted two-dimensional image features in the network transmission is minimized. This paper designs a single-stage point cloud reconstruction network and uses DetNet, which can retain more detailed features, as the feature extraction network.

3. Approach

3.1. Architecture of 3D-SSRecNet

The architecture of 3D-SSRecNet is shown in Figure 1. 3D-SSRecNet has two main parts: a 2D image feature extraction network and a point cloud prediction network. These two parts constitute a simple single-stage point cloud reconstruction network. The single-stage network structure only transfers the features of 2D pictures within the network of 3D-SSRecNet. Compared with the two-stage reconstruction network that needs to transmit features across the network, this network structure reduces the loss of features.

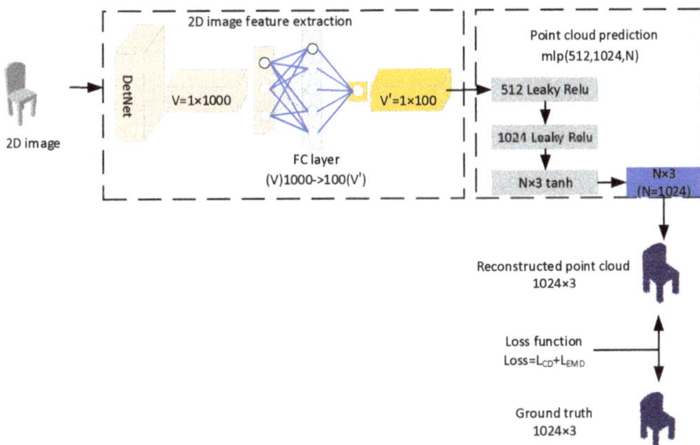

Figure 1. Architecture of 3D-SSRecNet.

Given a 2D image, firstly, we obtain a latent representation V by DetNet. Then, we map V to a low dimensionality feature V′ by a full connection (FC) layer. A multi-layer perceptron (MLP) is directly applied to predict a point set afterward. During training, chamfer distance and Earth mover's distance loss function are computed, and the update of trainable parameters is supervised.

3.2. 2D Image Feature Extraction

Many image feature networks applied downsampling, which brought a higher receptive field, but unavoidably caused the loss of image details. However, for reconstruction, image details are crucial for the recovery of geometric shape. This kind of network is more suitable for image classification tasks, but not for reconstruction tasks that require more detailed features.

DetNet [9] not only retains more details but also retains a large receptive field. Although DetNet was designed for object detection, its novel dilated bottleneck structure provides high-resolution feature maps and a large receptive field. We use DetNet as the backbone of image feature extraction. DetNet follows the same structure as ResNet-50 [16] until state 4, so DetNet also has the advantage of being easy to train and will not fall into gradient disappearance. Table 1 shows the parameters of the last two stages of DetNet, that is, the differences between DetNet and ResNet-50. After stage 4, DetNet keeps the size of the feature map at 16 × 16, which enables more details to be retained. The fifth and sixth stages of DetNet are composed of bottlenecks with dilated convolution, and some bottlenecks have 1 × 1 convolutions on their shortcut connections. Dilated convolution increases the receptive field. However, considering the amount of computation and memory, stage 5 and stage 6 set the same channel number of 256. At the end of the baseline, a fully connected layer is applied.

Table 1. The parameters of last two stages DetNet.

Stage	Feature Map Size	Parameters (Convolution Kernel Size, Number of Output Channels)
Stage 5	16 × 16	$\begin{bmatrix} 1 \times 1 kernel & 256 chanels \\ 3 \times 3 kernel, dilate\ 2 & 256 chanels \\ 1 \times 1 kernel & 256 chanels \end{bmatrix} + 1 \times 1 conv$ $\begin{bmatrix} 1 \times 1 kernel & 256 chanels \\ 3 \times 3 kernel, dilate\ 2 & 256 chanels \\ 1 \times 1 kernel & 256 chanels \end{bmatrix} \times 2$
Stage 6	16 × 16	$\begin{bmatrix} 1 \times 1 kernel & 256 chanels \\ 3 \times 3 kernel, dilate\ 2 & 256 chanels \\ 1 \times 1 kernel & 256 chanels \end{bmatrix} + 1 \times 1 conv$ $\begin{bmatrix} 1 \times 1 kernel & 256 chanels \\ 3 \times 3 kernel, dilate\ 2 & 256 chanels \\ 1 \times 1 kernel & 256 chanels \end{bmatrix} \times 2$
	1 × 1	16 × 16 average pool, 1000-d fully connected layer

As shown in Figure 1, after feature extraction of the input image, we obtain a 1000-dimensional latent feature V of the input image. After that, the full connection (FC) layer compresses the dimension of vector V from 1000 to 100 and obtains vector V′.

3.3. Point Could Prediction

We use three-layer MLP to predict the point sets directly. The dimensions of the outputs of the three MLP layers are 512, 1024, and N × 3, respectively. The output of the feature extraction network: vector V′ is fed into MLPs of the point cloud prediction network. On the first two layers, ELU [27] is introduced as an activation function which is defined as:

$$y = elu(x) = \begin{cases} x, x \geq 0 \\ \alpha(e^x - 1), x < 0 \end{cases} \tag{1}$$

where parameter α is set to 1. The curves of ELU activation function and its derivative are shown in Figure 2a,b, respectively. For common activation functions, such as ReLU, the value corresponding to the negative axis is 0. However, the normalized point cloud coordinate interval is $[-1, 1]$, which indicates that the point cloud coordinates will have negative values. As shown in Figure 2a, the value corresponding to the negative axis of the ELU activation function is non-zero. Therefore, using ELU as the activation function, the negative value information in the reconstructed network will not be lost in the forward propagation process. As shown in Figure 3b, the derivative of ELU is also non-zero on the negative axis. In the backpropagation process of the network, the negative gradient will not be lost, and it can help update the network weight.

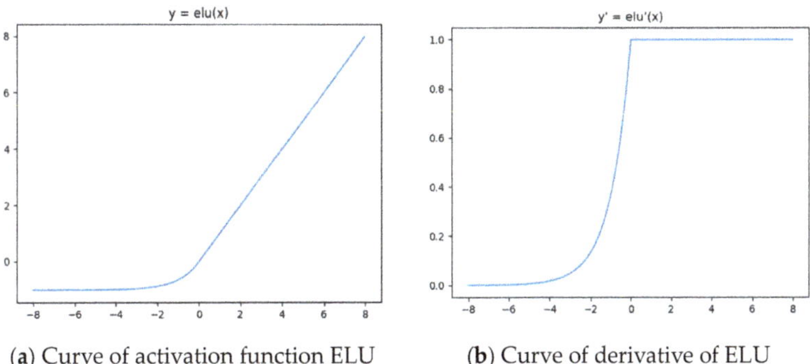

(**a**) Curve of activation function ELU   (**b**) Curve of derivative of ELU

Figure 2. ELU activation function and its derivative.

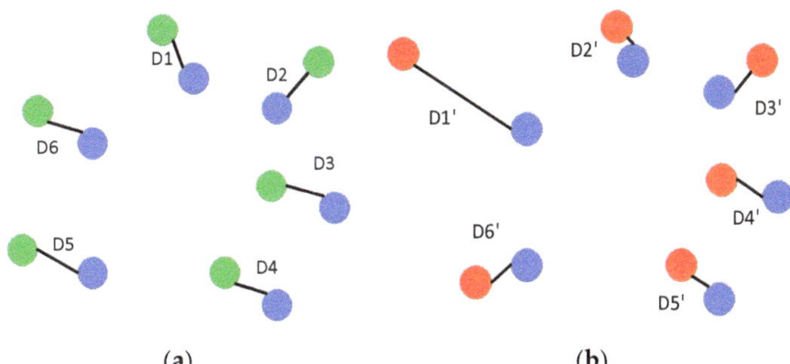

(**a**)   (**b**)

Figure 3. Different reconstructions with the same CD loss value.

In the experimental part, we prove that the reconstruction effect of using ELU as the activation function is better than that of using other activation functions. We directly output the predicted point sets after the last activation layer. It is actualized by the tanh function, whose outputs belong to $[-1, 1]$, which is the same as the required point set data.

3.4. Loss Function

To define the loss function of reconstruction of point clouds. We have to consider two important properties. (1) Point cloud is an unordered point set so we shall obtain the same data despite how we change the order of points. (2) The geometric feature of a real object shall not change significantly regardless of any rotation transformations. However, for instance, the point coordinates seem different as we take a rotation transformation. Fan et al. [4] proposed the use of the chamfer distance (CD) and Earth mover's distance (EMD) which satisfied the requirement. The chamfer distance is defined below.

$$L_{CD}(X_P^*, X_P) = \sum_{x \in X_P^*} \min_{y \in X_P} ||x - y||_2^2 + \sum_{y \in X_P^*} \min_{x \in X_P} ||x - y||_2^2 \tag{2}$$

It is a differentiable function with respect to point locations which are respectively calculated for each point, and it obtains the distance between it and its nearest point in the ground truth.

The Earth mover's distance is defined below.

$$L_{EMD}(X_P^*, X_P) = \min_{\varphi: X_P^* \to X_P} \sum_{x \in X_P^*} ||x - \varphi(x)||_2 \tag{3}$$

It is a measure between two distributions, so it can be considered as a "distance" between two point sets, and it constitutes the mapping between two point sets to guarantee the unique output.

Many research on point cloud reconstruction [5,17] point out the characteristics of CD and EMD: CD is related to the contour of the reconstructed point cloud. Lower CD values result in better point cloud contours. The 3D reconstruction network trained by CD can more easily capture the rough contour of objects in 2D images. However, the reconstructed network trained by CD can easily produce splash point clouds, but the visual effect is not good.

We use Figure 3 to explain the cause of point clouds. In Figure 3, the blue dot represents the ground truth, and the green dot and red dot represent two different reconstruction point clouds, respectively. D1~D6 respectively represent the distance between the six points obtained from the first reconstruction and their corresponding ground truth points. D1'~D6' respectively represent the distance between the six points obtained from the second reconstruction and their corresponding ground truth points. We can see that the reconstruction result represented by the green point in Figure 3a is better than the reconstruction result represented by the red point in Figure 3a because the distance D1' between one of the red points and its corresponding ground truth point is obviously greater than D2'~D6'. From Formula (2), if the sum of D1~D6 is equal to the sum of D1'~D6', the network trained with CD as the loss function cannot distinguish between the two reconstructions. Therefore, we can say that CD may confuse different reconstructions with similar chamfer distances. However, in the EMD loss function, φ represents the bijection relationship between the ground truth and the reconstructed point cloud, so EMD loss will not cause the above confusion.

EMD is related to the visual quality of the reconstructed point cloud. Lower EMD value always presumes higher visual quality. However, it is inclined to obtain a bad contour of the object. Synthesizing the pros and cons of CD and EMD, the loss function of our network is defined as:

$$Loss = L_{CD} + L_{EMD} \tag{4}$$

4. Experiment

We evaluated the proposed 3D-SSRecNet on ShapeNet [28] and Pix3D [29] datasets, respectively. ShapeNet is a big collection of textured CAD models which consists of 13 classes and 43,809 point cloud models for both training and testing. We used the 80–20% train/test split to perform our experiment. We performed the same experiment on the Pix3D database. The Pix3D database consists of three classes and 7595 point cloud models. This dataset is a CAD model of the real scene. Experiments on Pix3D can better evaluate the practicability of a point cloud reconstruction algorithm.

We used the gradient optimization algorithm Adam to optimize the proposed 3D-SSRecNet. In training, we set the learning rate to 0.0005 and the epoch to 50. The training environment is as follows: Ubuntu 18.04.6, CUDA 10.1, and the model of GPU is NVIDIA

Tesla T4 $\times$ 4. We used the CD and EMD values calculated on 1024 sampling points to evaluate the quality of the reconstructed point cloud.

4.1. Experiment on ShapeNet

To verify the advantage of ELU activation function, we fixed the structure of 3D-SSRecNet and replaced ELU with other activation functions such as Leaky ReLU, softsign, and softplus. Table 2 shows the reconstruction results using different activation functions. Among all the 13 classes, ELU performed better in 12 categories.

Table 2. Reconstruction results of different activation functions on ShapeNet dataset.

Class	Softsign	Softplus	Leaky ReLU	ELU	Softsign	Softplus	Leaky ReLU	ELU
	$CD \times 10^{-2}$				$EMD \times 10^{-2}$			
Airplane	2.60	2.38	2.37	2.38	3.05	2.72	2.68	2.72
Bench	3.69	3.58	3.55	3.51	3.34	3.27	3.18	3.15
Cabinet	5.19	5.09	4.9	4.77	4.28	4.21	4.03	3.91
Car	3.58	3.57	3.57	3.56	2.85	2.85	2.85	2.84
Chair	4.45	4.48	4.40	4.35	4.30	4.36	4.23	4.17
Lamp	5.04	5.21	4.97	4.99	6.36	6.75	6.23	6.19
Monitor	5.00	4.84	4.73	4.72	4.96	4.77	4.50	4.49
Rifle	2.52	2.58	2.49	2.45	3.69	3.98	3.48	3.48
Sofa	4.57	4.62	4.52	4.44	3.82	3.89	3.73	3.71
Speaker	6.26	6.26	5.99	5.94	5.58	5.65	5.27	5.23
Table	4.86	4.83	4.40	4.35	4.75	4.74	4.21	4.16
Telephone	3.75	3.71	3.57	3.52	3.41	3.52	3.06	3.05
Vessel	3.85	3.77	3.77	3.72	4.20	4.26	3.98	3.96

Figure 4 shows the point cloud reconstruction results obtained using different activation functions on the ShapeNet dataset. We can see that the network trained by ELU considers details better and preserves finer geometric information.

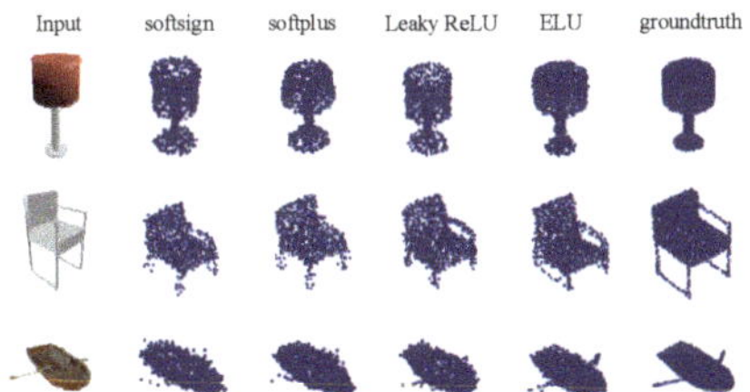

Figure 4. Visualization of 3D-SSRecNet's output with different activation functions on ShapeNet.

We compared our 3D-SSRecNet with PSGN [4], 3D-LMNet [5], SE-MD [24], 3D-VENet [25], 3D-ARNet [22], 3D-ReConstnet [6], and 3D-FEGNet [26]. All the experiments followed the same train/test split and used the same loss functions, CD and EMD. Tables 3 and 4 show the reconstruction results of different reconstruction methods on the ShapeNet dataset. The smaller the values of CD and EMD of a method, the better the reconstruction quality of the method. From the results shown in Table 3, the CD values of 3D-SSRecNet are slightly lower than that of 3D-Renstnet in the two categories of cabinet and monitor. In other categories, the reconstruction effect of 3D-SSRecNet is the best. In Table 3, PSGN, 3D-LMNet, SE-MD, and 3D-ARNet are two-stage networks, while other networks are single-stage networks. It can be seen from the CD values shown in Table 3 that the performance of single-stage networks in most categories is better than that of two-stage networks. This shows that the propagation of features between different stages of the network will cause feature loss. It can be seen from the EMD values shown in Table 4 that the reconstruction results of SSRecNet in most categories are better than those of other networks.

Table 3. Reconstruction results on ShapeNet evaluated by CD.

Class	PSGN	3D-LMNet	SE-MD	3D-VENet	3D-ARNet	3D-Reconstnet	3D-FEGNet	Ours
				$CD \times 10^{-2}$				
Airplane	3.74	3.34	3.11	3.09	2.98	2.42	2.36	2.38
Bench	4.63	4.55	4.34	4.26	4.44	3.57	3.60	3.51
Cabinet	6.98	6.09	5.89	5.49	6.01	4.66	4.84	4.77
Car	5.2	4.55	4.52	4.30	4.27	3.59	3.57	3.56
Chair	6.39	6.41	6.47	5.76	5.94	4.41	4.35	4.35
Lamp	6.33	7.10	7.08	6.07	6.47	5.03	5.13	4.99
Monitor	6.15	6.40	6.36	5.76	6.08	4.61	4.67	4.72
Rifle	2.91	2.75	2.81	2.67	2.65	2.51	2.45	2.45
Sofa	6.98	5.85	5.69	5.34	5.54	4.58	4.56	4.44
Speaker	8.75	8.10	7.92	7.28	7.65	5.94	6.00	5.94
Table	6.00	6.05	5.62	5.46	5.68	4.41	4.42	4.35
Telephone	4.56	4.63	4.51	4.20	4.10	3.59	3.50	3.52
Vessel	4.38	4.37	4.24	4.22	4.15	3.81	3.75	3.72
Mean	5.62	5.4	5.27	4.92	5.07	4.09	4.09	4.05

Table 4. Reconstruction results on ShapeNet evaluated by EMD.

Class	PSGN	3D-LMNet	SE-MD	3D-VENet	3D-ARNet	3D-Reconstnet	3D-FEGNet	Ours
				$EMD \times 10^{-2}$				
Airplane	6.38	4.77	4.78	3.56	3.12	2.80	2.67	2.72
Bench	5.88	4.99	4.61	4.09	3.93	3.22	3.75	3.15
Cabinet	6.04	6.35	6.37	4.69	4.81	3.84	4.75	3.91
Car	4.87	4.10	4.11	3.57	3.38	2.87	3.40	2.84
Chair	9.63	8.02	6.53	6.11	5.45	4.24	4.52	4.17
Lamp	16.17	15.80	12.11	9.97	7.60	6.40	6.11	6.19
Monitor	7.59	7.13	6.74	5.63	5.58	4.38	4.88	4.49
Rifle	8.48	6.08	5.89	4.06	3.39	3.63	2.91	3.48
Sofa	7.42	5.65	5.21	4.80	4.49	3.83	4.56	3.71
Speaker	8.70	9.15	7.86	6.78	6.59	5.26	6.24	5.23
Table	8.40	7.82	6.14	6.10	5.23	4.26	4.62	4.16
Telephone	5.07	5.43	5.11	3.61	3.25	3.06	3.39	3.05
Vessel	6.18	5.68	5.25	4.59	4.05	3.99	4.09	3.96
Mean	7.75	7.00	6.21	5.20	4.68	3.98	4.30	3.93

Figure 5 shows the reconstruction effect of 13 categories of 3D-SSRecNet on the ShapeNet dataset. The reconstructed point cloud resolution shown in Figure 5 is 2048.

4.2. Experiment on Pix3D

The Pix3D dataset consists of a large number of real indoor 2D images and their corresponding metadata (such as masks, ground truth CAD models, and attitudes). It can be seen from Figure 6 that the background of 2D images in Pix3D dataset is very complex, which poses a greater challenge to the feature extraction part of the reconstruction network. Therefore, the Pix3D dataset can be used to evaluate the generalization ability of the reconstructed network to real scenes.

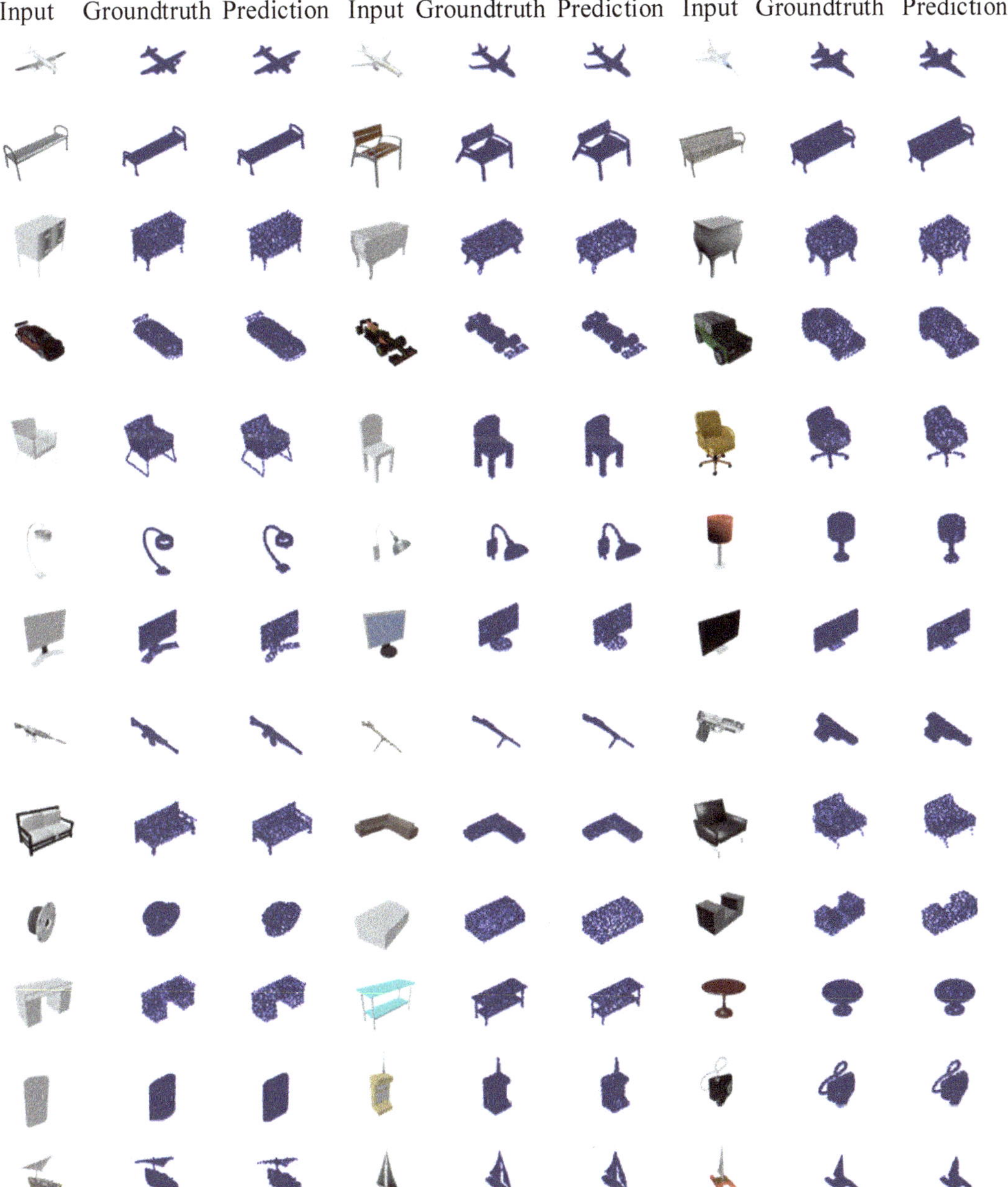

Figure 5. Visualization of our predictions on ShapeNet.

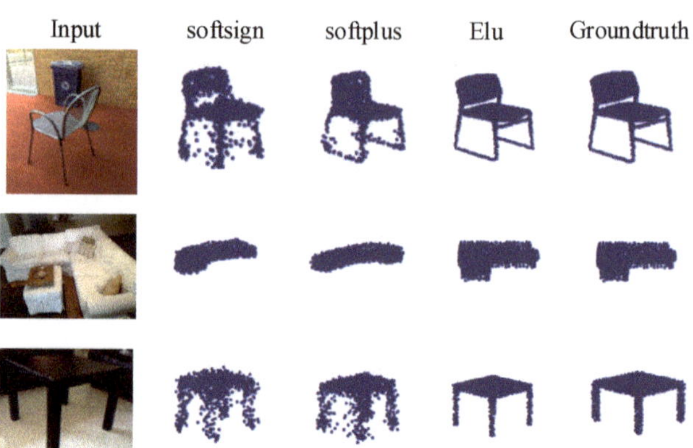

Figure 6. Visualization of 3D-SSRecNet's predictions with different activation functions on Pix3D.

Table 5 shows that ELU also provides better evaluation values on Pix3D. We exhibit the prediction outputs with different activation functions in Figure 6 to visualize how the activation function affects the prediction. Figure 6 shows that ELU also generates better qualitative results.

Table 5. Reconstruction results of different activation functions on Pix3D dataset.

Class	Softsign	Softplus	Leaky ReLU	ELU	Softsign	Softplus	Leaky ReLU	ELU
			$CD \times 10^{-2}$				$EMD \times 10^{-2}$	
Chair	5.70	5.57	5.58	5.52	6.32	6.13	6.08	6.05
Sofa	6.22	6.26	6.09	6.04	5.12	5.24	4.92	4.90
Table	7.73	7.32	7.03	6.88	8.66	8.20	7.60	7.59

We also compared our network with PSGN [4], 3D-LMNet [5], 3D-ARNet [22], 3D-Reconstnet [6], and 3D-FEGNet [26]. Tables 6 and 7 show the reconstruction results of different reconstruction methods on the Pix3D dataset. In all three categories, we can see that our 3D-SSRecNet has the lowest evaluation value (except the EMD value on chair category), which indicates that 3D-SSRecNet has a strong generalization ability for real scenes.

Table 6. Reconstruction results on Pix3D evaluated by CD.

Class	PSGN	3D-LMNet	3D-ARNet	3D-ReconstNet	3D-FEGNet	Ours
				$CD \times 10^{-2}$		
Chair	8.05	7.35	7.22	5.59	5.66	5.52
Sofa	8.45	8.18	8.13	6.14	6.23	6.04
Table	10.85	11.2	10.31	7.04	7.58	6.88
Mean	9.12	8.91	8.55	6.26	6.49	6.15

Table 7. Reconstruction results on Pix3D evaluated by EMD.

Class	PSGN	3D-LMNet	3D-ARNet	3D-ReconstNet	3D-FEGNet	Ours
			$EMD \times 10^{-2}$			
Chair	12.55	9.14	7.94	5.99	8.24	6.05
Sofa	9.16	7.22	6.69	5.02	6.77	4.90
Table	15.16	12.73	10.42	7.60	11.40	7.59
Mean	12.29	9.70	8.35	6.20	8.80	6.18

Figure 7 shows the reconstruction effect of 13 categories of 3D-SSRecNet on the Pix3D dataset. The reconstructed point cloud resolution shown in Figure 7 is 2048.

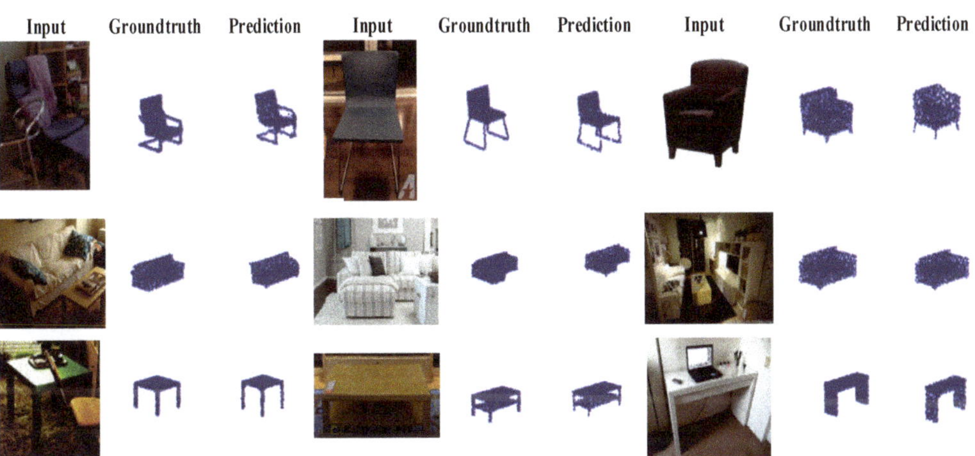

Figure 7. Visualization of our predictions on Pix3D.

5. Conclusions

In this paper, we proposed an efficient 3D point cloud reconstruction called 3D-SSRecNet. Given an image, it learns the latent representation and after dimensionality reduction we apply MLPs to predict the correspondent point cloud directly. We conducted several experiments on ShapeNet and Pix3D data sets. We proved that the reconstruction effect of using activation function ELU in the generation network is better than that of using other activation functions. That is, the CD and EMD values of the point cloud generated using the ELU are lower than those of the point cloud generated using other activation functions. Figures 4 and 6 also show that the shape and contour of the point cloud generated using ELU are better than those of the point cloud generated using other activation functions. By comparing CD and EMD values, it is proved that 3D-SSRecNet can outperform the state-of-the-art methods. The proposed 3D-SSRecNet is more suitable for tasks that need to obtain the contour and shape of a point cloud at a low cost. In future work, we will try to improve the local effect of the output point cloud with the Part-Wise AtlasNet on the premise of maintaining the existing low cost.

Author Contributions: Conceptualization, methodology, resources, funding acquisition, B.L.; software, validation, visualization, S.Z.; and writing—original draft preparation, writing—review and editing, Y.L. All authors have read and agreed to the published version of the manuscript.

Funding: This work was supported by the Science and Technology Development Plan Project of Jilin Province under grant 20200201165JC.

Institutional Review Board Statement: Not applicable.

Informed Consent Statement: Not applicable.

Data Availability Statement: Not applicable.

Conflicts of Interest: The authors declare no conflict of interest.

References

1. Garrido, D.; Rodrigues, R.; Augusto Sousa, A.; Jacob, J.; Castro Silva, D. Point Cloud Interaction and Manipulation in Virtual Reality. In Proceedings of the 2021 5th International Conference on Artificial Intelligence and Virtual Reality (AIVR), Kumamoto, Japan, 23–25 July 2021; pp. 15–20.
2. Eppel, S.; Xu, H.; Wang, Y.R.; Aspuru-Guzik, A. Predicting 3D shapes, masks, and properties of materials inside transparent containers, using the TransProteus CGI dataset. *Digit. Discov.* **2021**, *1*, 45–60. [CrossRef]
3. Xu, T.X.; Guo, Y.C.; Lai, Y.K.; Zhang, S.H. TransLoc3D: Point Cloud based Large-scale Place Recognition using Adaptive Receptive Fields. *arXiv* **2021**, arXiv:2105.11605v3.
4. Fan, H.; Su, H.; Guibas, L. A Point Set Generation Network for 3D Object Reconstruction from a Single Image. In Proceedings of the IEEE Conference on Computer Vision and Pattern Recognition 2017, Honolulu, HI, USA, 21–26 July 2017; pp. 2463–2471. [CrossRef]
5. Mandikal, P.; Navaneet, K.L.; Agarwal, M.; Babu, R.V. 3D-LMNet: Latent embedding matching for accurate and diverse 3D point cloud reconstruction from a single image. *arXiv* **2018**, arXiv:1807.07796.
6. Li, B.; Zhang, Y.; Zhao, B.; Shao, H. 3D-ReConstnet: A Single-View 3D-Object Point Cloud Reconstruction Network. *IEEE Access* **2020**, *1*, 99. [CrossRef]
7. Gwak, J.; Choy, C.B.; Chandraker, M.; Garg, A.; Savarese, S. Weakly supervised 3D reconstruction with adversarial con-straint. In Proceedings of the 2017 International Conference on 3D Vision (3DV), Qingdao, China, 10–12 October 2017; pp. 263–272.
8. Yang, B.; Wen, H.; Wang, S.; Clark, R.; Markham, A.; Trigoni, N. 3D Object Reconstruction from a Single Depth View with Adversarial Learning. In Proceedings of the IEEE International Conference on Computer Vision Workshops, Venice, Italy, 22–29 October 2017; pp. 679–688.
9. Li, Z.; Peng, C.; Yu, G.; Zhang, X.; Deng, Y.; Sun, J. DetNet: A Backbone network for Object Detection. *arXiv* **2018**, arXiv:1804.06215.
10. Yu, Q.; Yang, C.; Wei, H. Part-Wise AtlasNet for 3D point cloud reconstruction from a single image. *Knowl. Based Syst.* **2022**, *242*, 108395. [CrossRef]
11. Fuentes-Pacheco, J.; Ruiz-Ascencio, J.; Rendon-Mancha, J.M. Visual simultaneous localization and mapping: A survey. *Artif. Intell. Rev.* **2012**, *43*, 55–81. [CrossRef]
12. Cheikhrouhou, O.; Mahmud, R.; Zouari, R.; Ibrahim, M.; Zaguia, A.; Gia, T.N. One-Dimensional CNN Approach for ECG Arrhythmia Analysis in Fog-Cloud Environments. *IEEE Access* **2021**, *9*, 103513–103523. [CrossRef]
13. Kimothi, S.; Thapliyal, A.; Akram, S.V.; Singh, R.; Gehlot, A.; Mohamed, H.G.; Anand, D.; Ibrahim, M.; Noya, I.D. Big Data Analysis Framework for Water Quality Indicators with Assimilation of IoT and ML. *Electronics* **2022**, *11*, 1927. [CrossRef]
14. Tahir, R.; Sargano, A.B.; Habib, Z. Voxel-Based 3D Object Reconstruction from Single 2D Image Using Variational Autoencoders. *Mathematics* **2021**, *9*, 2288. [CrossRef]
15. Xie, H.; Yao, H.; Sun, X.; Zhou, S.; Zhang, S. Pix2Vox: Context-Aware 3D Reconstruction from Single and Multi-View Images. In Proceedings of the IEEE/CVF International Conference on Computer Vision, Seoul, Korea, 27 October–2 November 2019.
16. He, K.; Zhang, X.; Ren, S.; Sun, J. Deep residual learning for image recognition. In Proceedings of the 2016 IEEE Conference on Computer Vision and Pattern Recognition (CVPR), Las Vegas, NV, USA, 27–30 June 2016; pp. 770–778.
17. Han, Z.; Qiao, G.; Liu, Y.-S.; Zwicker, M. SeqXY2SeqZ: Structure Learning for 3D Shapes by Sequentially Predicting 1D Occupancy Segments from 2D Coordinates. In *European Conference on Computer Vision*; Springer: Cham, Switzerland, 2020; pp. 607–625.
18. Peng, K.; Islam, R.; Quarles, J.; Desai, K. TMVNet: Using Transformers for Multi-View Voxel-Based 3D Reconstruction. In Proceedings of the IEEE/CVF Conference on Computer Vision and Pattern Recognition, New Orleans, LA, USA, 19–20 June 2022; pp. 222–230.
19. Kniaz, V.V.; Knyaz, V.A.; Remondino, F.; Bordodymov, A.; Moshkantsev, P. Image-to-voxel model translation for 3d scene reconstruction and segmentation. In *European Conference on Computer Vision*; Springer: Cham, Switzerland, 2020; pp. 105–124.
20. Siddique, A.; Lee, S. Sym3DNet: Symmetric 3D Prior Network for Single-View 3D Reconstruction. *Sensors* **2022**, *22*, 518. [CrossRef] [PubMed]
21. Yang, S.; Xu, M.; Xie, H.; Perry, S.; Xia, J. Single-View 3D Object Reconstruction from Shape Priors in Memory. In Proceedings of the IEEE/CVF Conference on Computer Vision and Pattern Recognition, Nashville, TN, USA, 20–25 June 2021; pp. 3151–3160. [CrossRef]
22. Chen, H.; Zuo, Y. 3D-ARNet: An accurate 3D point cloud reconstruction network from a single-image. *Multimedia Tools Appl.* **2021**, *81*, 12127–12140. [CrossRef]
23. Pumarola, A.; Popov, S.; Moreno-Noguer, F.; Ferrari, V. C-flow: Conditional generative flow models for images and 3d point clouds. In Proceedings of the IEEE/CVF Conference on Computer Vision and Pattern Recognition, Seattle, WA, USA, 14–19 June 2020; pp. 7949–7958.
24. Hafiz, A.M.; Bhat, R.U.A.; Parah, S.A.; Hassaballah, M. SE-MD: A Single-encoder multiple-decoder deep network for point cloud generation from 2D images. *arXiv* **2021**, arXiv:2106.15325.
25. Ping, G.; Esfahani, M.A.; Chen, J.; Wang, H. Visual enhancement of single-view 3D point cloud reconstruction. *Comput. Graph.* **2022**, *102*, 112–119. [CrossRef]
26. Wang, E.; Sun, H.; Wang, B.; Cao, Z.; Liu, Z. 3D-FEGNet: A feature enhanced point cloud generation network from a single image. *IET Comput. Vis.* **2022**. [CrossRef]

27. Clevert, D.A.; Unterthiner, T.; Hochreiter, S. Fast and Accurate Deep Network Learning by Exponential Linear Units (ELUs). *arXiv* **2015**, arXiv:1511.07289.
28. Chang, A.X.; Funkhouser, T.; Guibas, L.; Hanrahan, P.; Huang, Q.; Li, Z.; Savarese, S.; Savva, M.; Song, S.; Su, H.; et al. ShapeNet: An information-rich 3D model repository. *arXiv* **2015**, arXiv:1512.03012.
29. Sun, X.; Wu, J.; Zhang, X.; Zhang, Z.; Zhang, C.; Xue, T.; Tenenbaum, J.B.; Freeman, W.T. Pix3D: Dataset and Methods for Single-Image 3D Shape Modeling. In Proceedings of the IEEE Conference on Computer Vision and Pattern Recognition 2018, Salt Lake City, UT, USA, 18–23 June 2018; pp. 2974–2983.

Review

A Survey on Deep-Learning-Based LiDAR 3D Object Detection for Autonomous Driving

Simegnew Yihunie Alaba and John E. Ball *

Department of Electrical and Computer Engineering, James Worth Bagley College of Engineering,
Mississippi State University, Starkville, MS 39762, USA
* Correspondence: jeball@ece.msstate.edu

Abstract: LiDAR is a commonly used sensor for autonomous driving to make accurate, robust, and fast decision-making when driving. The sensor is used in the perception system, especially object detection, to understand the driving environment. Although 2D object detection has succeeded during the deep-learning era, the lack of depth information limits understanding of the driving environment and object location. Three-dimensional sensors, such as LiDAR, give 3D information about the surrounding environment, which is essential for a 3D perception system. Despite the attention of the computer vision community to 3D object detection due to multiple applications in robotics and autonomous driving, there are challenges, such as scale change, sparsity, uneven distribution of LiDAR data, and occlusions. Different representations of LiDAR data and methods to minimize the effect of the sparsity of LiDAR data have been proposed. This survey presents the LiDAR-based 3D object detection and feature-extraction techniques for LiDAR data. The 3D coordinate systems differ in camera and LiDAR-based datasets and methods. Therefore, the commonly used 3D coordinate systems are summarized. Then, state-of-the-art LiDAR-based 3D object-detection methods are reviewed with a selected comparison among methods.

Keywords: autonomous vehicles; classification; deep learning; deep learning for point cloud processing; LiDAR; sparsity; 3D object detection

Citation: Alaba, S.Y.; Ball, J.E. A Survey on Deep-Learning-Based LiDAR 3D Object Detection for Autonomous Driving. *Sensors* **2022**, *22*, 9577. https://doi.org/10.3390/s22249577

Academic Editors: Miaohui Wang, Guanghui Yue, Jian Xiong and Sukun Tian

Received: 21 November 2022
Accepted: 5 December 2022
Published: 7 December 2022

Publisher's Note: MDPI stays neutral with regard to jurisdictional claims in published maps and institutional affiliations.

1. Introduction

Autonomous driving has succeeded since the 2007 DARPA urban challenge [1]. It has a high potential for decreasing traffic congestion, improving overall driving and road safety, fast and effective decision-making, and reducing carbon emission [2]. Autonomous vehicles need to perceive, predict, plan, decide, and execute decisions in an uncontrolled complex real world to achieve these goals, which is challenging. A small mistake in understanding the environment, plan, or decision causes fatal effects. A robust autonomous system is needed to avoid such mistakes. One branch of an autonomous system, the perception system, especially 3D object detection, helps to understand the driving environments, such as other vehicles and pedestrians. For the robust operation of the next branches in the autonomous system, the perception system should give precise information about the driving environment. It should also be robust enough to work in bad weather such as rain, snow, and fog and make fast and effective decisions to match high-speed driving [2,3].

The 3D sensors, such as Light Detection and Ranging (LiDAR) and Radio Detection and Ranging (radar), provide 3D information about the environment, including distance and speed estimation. Other sensors, such as depth sensors (RGB-D cameras), can also provide 3D information [4]. However, environmental variations, such as inclement weather, sensor limitations, and resolution differences, limit sensor performance. The LiDAR sensor is more robust to adverse weather than the camera but poorer in color and texture information. A solid-state LiDAR called flash LiDAR [5] uses optical flash operation to give texture information such as a standard camera. The major limitation of LiDAR sensors are the high price and struggle to detect close- or far-distance objects [5]. LiDAR data are sparse and unstructured, which makes processing LiDAR data challenging. On the other hand, a radar sensor is more robust in bad weather, better for long-range detection, and relatively cheaper. However, it has a lower resolution.

Due to the sparsity and unstructured nature, it is challenging to process LiDAR data. Different methods have been developed to process LiDAR point cloud data: projecting to 2D space, such as [6] MV3D [7] and AVOD [8], voxels, such as VoxelNet [9], F-PointNet [10], MVFP [11], and raw point clouds, such as PointNet [12], PointNet++ [13], and PointR-CNN [14]. The projection method converts the 3D LiDAR point cloud into a 2D plane representation, such as a range view or bird's-eye view (BEV). The projected data can be processed using the standard 2D detection models and reduces the computational burden due to 3D convolution; however, there is an information loss converting the LiDAR point cloud into 2D space representations. The voxel method discretizes the sparse point cloud data into a volumetric 3D grid of voxels and uses a series of voxels to represent the 3D point cloud. This representation avoids the unstructured nature of point clouds, but the 3D convolution is still the bottleneck and causes information loss from converting one representation to another [3]. On the other hand, the raw point cloud representation directly processes point cloud data to keep the information. However, the sparsity of the point cloud data and the high computational cost because of 3D convolution are challenges to this method (see Section 4 for details). Different compression techniques have been developed to remove spatial and temporal redundancies of point cloud data, to reduce the large volume, such as [15–17]. Although these compression techniques reduce redundancies, they can affect the performance of models (Details of compression techniques are out of the scope of this survey).

The major contributions of the paper can be summarized as follows:

1. An in-depth analysis of LiDAR-based 3D object detection, state-of-the-art (SOTA) methods, and a comparison of SOTA methods are presented.
2. The LiDAR processing and feature-extraction techniques are summarized.
3. The 3D coordinate systems commonly used in 3D detection are presented.
4. We categorize deep-learning-based LiDAR 3D detection methods based on LiDAR data processing techniques as projection, voxel, and raw point cloud.

The rest of the paper is organized as follows. Section 2 presents related work. Feature-extraction methods, stages of autonomous driving, and 3D coordinate systems representation are provided in Section 3. Section 4 summarizes the LiDAR 3D detection methods and compares the selected ones. The summary of the survey paper is presented in Section 5.

2. Related Work

Due to the rapid growth of deep learning (DL) in computer vision, object detection has become an extensively studied application. Three-dimensional detection draws much attention in robotics applications, such as autonomous driving, because the intention to know object driving environment and location has increased. Most review papers present general detection in 2D and 3D and for multiple sensors, mostly cameras and LiDARs. This work comprehensively reviews LiDAR 3D object-detection models for autonomous driving.

General object detection and semantic segmentation in 2D and 3D were presented in [2]. The available datasets and methods for autonomous driving were reviewed. Jiao et al. [18] presented DL-based methods, sensors, and datasets. However, this work focused mostly on 2D detection. Arnold et al. [4] reviewed 3D detection methods for autonomous driving. Rahman et al. [19] also reviewed 3D methods, datasets, and open challenges in 3D detections. Similarly, Li et al. [20] and Fernandes et al. [21] presented detection and segmentation in autonomous driving using LiDAR. DL-based 3D detection and segmentation methods in autonomous driving were also presented in [22]. Qian et al. [23] published a 3D detection method for autonomous driving. Alaba et al. [3] presented multisensor fusion-based 3D object-detection methods, 3D datasets, sensors, 3D detection challenges, and possible research directions for autonomous driving. Recently, Alaba and Ball [24] also reviewed an image 3D object detection for autonomous driving. The 3D object-detection evaluation techniques, 3D bounding box encoding techniques, and image-based (RGB and stereo) object detection works for autonomous driving were reviewed.

This survey presents point cloud encoding techniques, 3D coordinate systems in 3D object detection, and others not covered in the previous survey papers, including recently published works (For 3D bounding box encoding techniques, 3D detection evaluation methods, sensors in autonomous driving, radar-related 3D object-detection methods, 3D object-detection challenges, possible research directions, and datasets refer [3,24]). Most of the existing 3D object-detection methods reviewed general 2D and 3D detection methods and/or LiDAR and camera-based detections. This survey focuses on a detailed analysis of LiDAR-only methods. Few works reviewed LiDAR-only methods. However, we have included recently published works in addition to unique contributions, such as 3D coordinate systems in 3D object detection and stages of autonomous driving.

3. Background

This section presents feature-extraction techniques for point clouds, 3D coordinate systems for 3D object detection, and levels of autonomous driving.

3.1. Feature-Extraction Methods

Feature extraction plays a crucial role in detection and classification tasks. Before the DL era, image features were usually extracted using handcrafted feature extractors based on aspects such as texture, color, and shape. On the other hand, DL networks can extract features from images without prior engineered data. Optimal feature learning is important to achieve optimal performance. The Harris interest point detector [25], Shi-Tomasi corner detector [26], Scale Invariant Feature Transform (SIFT) [27], and Speed-Up Robust Features (SURF) [28] methods are commonly used handcrafted feature extractors. In different computer vision applications [29–31], convolutional neural networks can replace these traditional feature extractors because of their ability to extract complex features and learn features efficiently. The image feature-extraction networks use different backbone networks such as AlexNet [32], VGGNet [33], GoogleNet [34], ResNet [35] Inception-ResNet-V2 [36], MobileNet [37], and DarkNet-19 [38]. The point cloud representations differ from images because of the unstructured nature and sparsity of points, which makes their feature-extraction networks different from image feature extractors. We can categorize feature extractors of LiDAR point clouds as point-wise, segment-wise, object-wise, and CNN-based networks [21] (see the details in Section 4).

Point-wise feature extractors take the whole point cloud as an input and then process, analyze, and label each point individually, such as PointNet [12] and PointNet++ [13]. Segment-wise feature extractors segment point clouds into volumetric representations and stack the volumetric representations to form LiDAR point clouds such as voxels [9,39,40], pillars [41], and frustums [12]. The point-wise and segment-wise feature extractors extract features without prior knowledge of whether the point or voxel belongs to an object. The object-wise feature extractors project LiDAR point clouds into 2D representations and generate the object proposals using 2D networks. Then, they regress the region proposals to predict the 3D bounding boxes. MV3D [7] uses the object-wise feature extractor to generate object proposals from LiDAR point cloud and fuse multiview features. On the other hand, CNN-based feature extractors use CNN to learn features from LiDAR data, such as images. These CNN-based feature extractors can be 2D backbone [32–36], 3D backbone sparse convolutional networks [42,43], CNN networks based on voting scheme [44], and graph convolution network [45].

The 3D backbone feature extractors are directly applied to 3D space using 3D convolutions. However, these networks consume large amounts of memory and take a long time to process the points. The LiDAR point cloud data are sparse and unstructured. Applying a standard convolution to these data points is inefficient because most areas are empty, and time is wasted on areas that do not have relevant information. The sparse convolution solves this problem by reducing the number of points to be processed and exploiting the areas with point cloud data only, which facilitates feature extraction and saves memory. Graham [42,43] developed sparse convolution by processing only areas with relevant information to save computational power. Each hidden value with zero input is considered a ground state, but the ground state is nonzero because of the biased terms. Then, when the inputs are sparse, the hidden variables that differ from their ground state are considered for calculation. Sub-manifold sparse convolutional networks (SS-CNs) [46] developed as an efficient sparse convolution for sparse 3D operation. The work considers feature vectors active if the ground state is nonzero, which is similar to [42,43].

Wang et al. [44] proposed a voting-based sliding window feature-extraction technique. First, the 3D point clouds are discretized into a grid at a fixed resolution. Then, the feature vectors are extracted from occupied cells. This method applies filters only on the occupied grid rather than the entire grid. The 3D detection window of fixed size is placed on the feature grid and passed into the classifier, such as SVM. Finally, the classifier decides the presence of an object by returning the detection score. Then, Vote3deep [47] further improves voting-based feature extractions by proposing a feature-based feature extraction and L1 penalty on filter activation intermediate feature representations.

Similarly, the graph convolution network (GCN) operates similarly to CNN but encodes points as nodes and nodes connected through edges. Applying GCN operation to a set of neighborhood nodes extracts features. EdgeConv [45] applies dynamically updated graph convolutions on the edges of local geometric structure points of K-nearest neighboring pairs. Then, they apply a multilayer perceptron to extract edge features from each point. DeepMRGCN [48] proposed a memory-efficient GCN model for feature extraction of non-Euclidean data.

3.2. Coordinate Systems

Due to the variety of sensors used in 3D detection, there are different coordinate systems. Therefore, different 3D datasets follow different data formats. Although there are a variety of datasets and sensors, coordinate systems in 3D object detection can be categorized into three [49].

1. **Camera coordinate system:** In this coordinate system, the positive direction of the x-axis points to the right, the positive direction of the y-axis points to the ground, and the positive direction of the z-axis points to the front. Figure 1a shows a camera coordinate system.

2. **LiDAR coordinate system:** In the LiDAR coordinate system, the positive direction of the x-axis points to the front, the positive direction of the y-axis points to the left, and the negative direction of the z-axis points to the ground as shown in Figure 1b.

3. **Depth coordinate system:** In the depth coordinate system, the positive direction of the x-axis points to the right, the positive direction of the y-axis points to the front and the negative direction of the z-axis points to the ground. Three-dimensional object-detection networks, such as VoteNet [50], H3DNet [51], etc., used the depth coordinate system. The depth coordinate system is shown in Figure 1c.

The commonly used dataset, for example, KITTI [52] camera and LiDAR coordinate, are shown in Figure 1a,d, respectively. In multisensor fusion-based or point cloud-based methods, the coordinate transformation is necessary from the beginning for data preprocessing, such as data augmentation. We do not cover the detailed mathematical derivation of coordinate transformation, but we give the intuitions and basic concepts of coordinate transformations between LiDAR and the camera. For example, for a homogeneous 3D point, $p = (x, y, z, 1)^T$ in rectified (rotated) camera coordinates, the ith camera image corresponding point $y = (u, v, 1)^T$ can be expressed as:

$$y = p_{rect}^{(i)} p,$$
$$(1)$$

where $i \in 0, 1, 2, 3$ is the camera index (four cameras with centers aligned to the same x/y-plane in the KITTI dataset), and camera 0 is a reference camera.

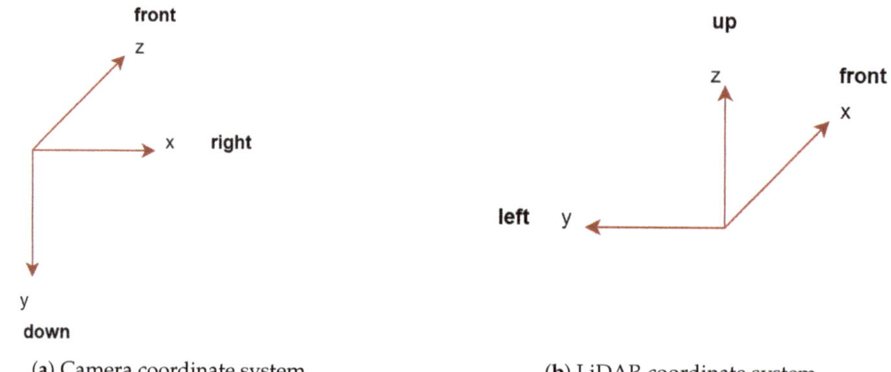

(**a**) Camera coordinate system

(**b**) LiDAR coordinate system

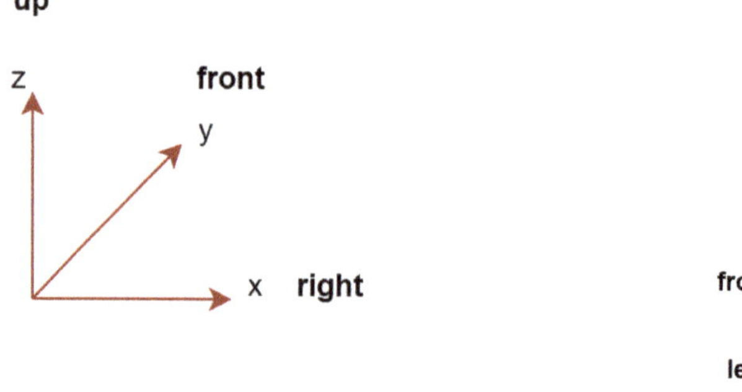

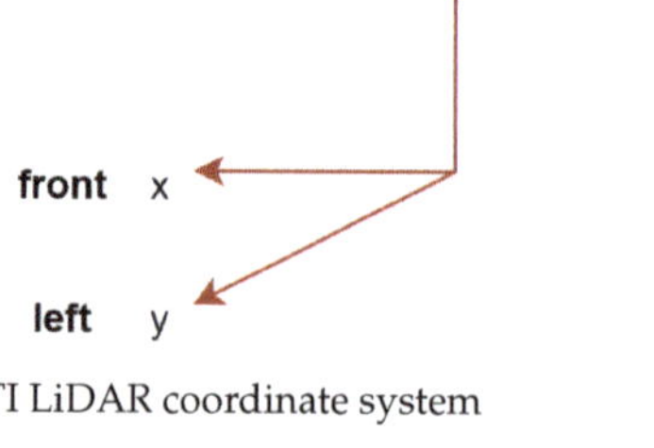

(**c**) Depth coordinate system

(**d**) KITTI LiDAR coordinate system

Figure 1. 3D object-detection coordinate systems. The KITTI camera coordinate system is similar to the camera coordinate system (**a**). However, its LiDAR coordinate system x-coordinate differs from the commonly used LiDAR x- coordinate system (compare (**b**,**d**)).

$p_{rect}^{(i)} \in R^{3\times4}$ is i^{th} the projection matrix after rectification, which can be given as:

$$p_{rect}^{(i)} = \begin{pmatrix} f_u^{(i)} & 0 & c_u^{(i)} & -f_u^{(i)} b_x^{(i)} \\ 0 & f_u^{(i)} & c_u^{(i)} & 0 \\ 0 & 0 & 1 & 0 \end{pmatrix}, \tag{2}$$

where $b_x^{(i)}$ is the baseline (in meters) with respect to a reference camera 0. To project the reference camera coordinates 3D points in p to the ith image plane in point y, the rectifying matrix of the reference camera R (0) should be considered. Thus:

$$y = p_{rect}^{(i)} R_{rect}^{(0)} p, \tag{3}$$

For 3D point, p in LiDAR coordinates the corresponding point y in the ith camera image can be given as:

$$y = p_{rect}^{(i)} R_{rect}^{(0)} T_{velo}^{cam} p, \tag{4}$$

$T_{velo}^{cam} \in R^{3\times4}$ is the rigid body transformation from Velodyne LiDAR to camera coordinates and $R_{rect}^{(0)} \in R^{3\times3}$ is the rectifying rotation matrix (Ref [52,53] for more details).

3.3. Stages of Autonomous Driving

Automated vehicles should have a safety-critical control function, such as steering, throttle, or braking, that can occur without direct driver intervention [54]. Vehicles that give safety warnings to drivers, such as forward crash warnings, are not considered automated unless they perform control functions even though the necessary data are received, processed, and the warning is given without driver intervention. When we think of autonomous driving, most people think of fully autonomous driving; however, autonomous driving has different stages. The vehicle automation levels range from level zero (no automation) to level five (fully autonomous). We present levels of autonomous driving for general understanding and to give insight into what should be needed for achieving fully autonomous driving. Vehicles should be equipped with multiple sensors to achieve robust driving, especially for levels four and five. However, this survey presents only LiDAR-related works (Refer [3] for different sensors used in autonomous driving). Therefore, the main goal of autonomous driving should be achieving level four and level five driving.

Level zero–No Automation: The driver is in complete control of driving and is responsible for monitoring the roadway. The vehicles with some driver support, such as steering, braking, etc., do not take action or do not have control authority over the vehicle. Therefore, it is considered no automation. Forward collision warning, lane departure warning, and blind spot monitoring driver support systems are considered level zero automation.

Level one–Function-specific Automation/Driver Assistance: In this level of automation, the driver has complete control and is solely responsible for the vehicle's safe operation. The system can perform a single automated system, such as steering or acceleration (as in adaptive cruise control), but not both. Drivers can take off their feet from the pedals so that it is sometimes called feet off level. Each system operates independently if a vehicle is equipped with multiple automated systems, such as lane keeping, steering, and acceleration. The automated system only supports the driver and does not replace the driver.

Level two–Partial Driving Automation: The driver is responsible for controlling the vehicle and monitoring the roadway. However, the system can perform two control functions: steering and acceleration. In this level of automation, the driver can be free from operating the steering wheel and pedal simultaneously. The driver engages whenever necessary to control both or one of the control functions. The driver can take off both feet and hands, called hand off level.

Level three–Conditional Automation: At this level of automation, vehicles can perform most of the tasks under specific traffic or environmental conditions (usually in urban areas with low-speed driving, such as 25 mph). The driver can override the system anytime. Therefore, the driver remains ready to take control when the system occasionally signals the driver to reengage in the driving task. The driver has sufficient transition time to control the vehicle. The driver can disengage from driving. Therefore, it is called eyes off level.

Level four–High Automation: At this level of automation, the vehicle can perform all the control functions under specific conditions. The driver does not engage in driving the vehicle but can override it. It is called mind off level. These technologies are not commercially available yet.

Level five–Full Automation: The vehicle performs all control operations and monitors roadway conditions without human intervention. The driver may provide navigation input, but all the controlling operations, including safe operation and roadway safety, rest on the vehicle. This level is also called mind off because vehicles can be without humans, or humans can do other activities. The diagrammatic representation of the levels of autonomous driving is shown in Figure 2.

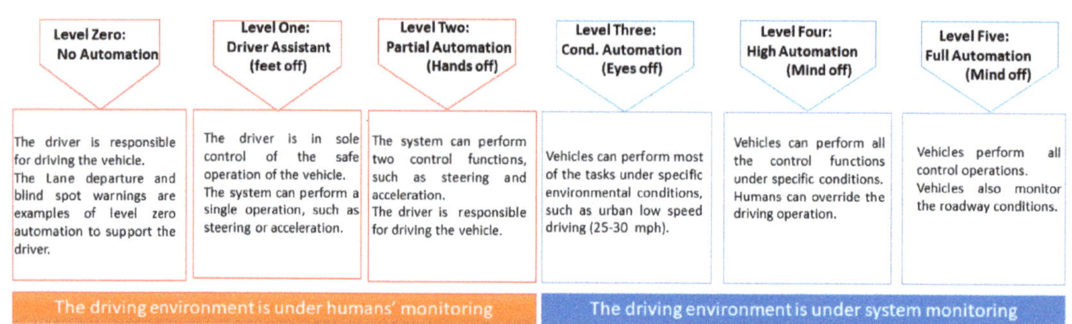

Figure 2. The levels of autonomous driving. In the first three levels (level zero to level two), the humans monitor the driving environment, whereas the system monitors the driving environment for the last three levels of driving.

The full automation level of autonomy replaces a human driver. The following four areas are vital to replacing human drivers in autonomous driving [55]:

1. **Vehicle Location and Environment:** For fully autonomous driving without human intervention, precise and accurate information about the driving environment must know the road signs, pedestrians, traffic, and others.
2. **Prediction and Decision Algorithms:** An efficient deep or machine learning algorithm is needed to detect, predict, and decide when interacting with other vehicles, pedestrians, and situations.
3. **High Accuracy and Real-time Maps:** Detailed, precise, and complete maps are needed to obtain information about the driving environment for path and trajectory planning.
4. **Vehicle Driver Interface:** Smooth and self-adaptive transition to/from the driver and an effective way to keep the driver alert and ready is needed, which increases customer satisfaction and confidence, especially at the beginning of the technology.

To replace human drivers, we need efficient sensors that can do a human eye's task, maps to do the human memory, ML algorithms to make the human brain decision, and vehicles to x communication such as human ears.

4. LiDAR 3D Object-Detection Methods

Although cameras are inexpensive, they lack the ability to create precise depth information for accurate 3D object detection. Additionally, cameras are vulnerable to adverse weather, such as snow, fog, and rain [3,24]. Point cloud-based methods provide solutions for such problems to improve performance significantly because sensors, such as LiDARs and radars, provide depth information, which is essential for accurate object size and precise location estimation. However, point cloud data are sparse, unordered, unstructured, and unevenly distributed. Therefore, they cannot fit directly into image-based deep-learning methods and need a different processing method for image-processing techniques. Different techniques have been proposed to reduce the sparsity of point cloud data, including dilated convolution [56] and flex convolution [57]. Yu et al. [56] put forth dilated convolution for dense prediction and enlarging receptive field. Although dilated convolution enlarges the receptive field and improves performance, it suffers from a gridding effect [58], which causes information discontinuity because of combination pixels from different neighbors [59]. Groh et al. [57] also developed flex-convolution to process irregular data, such as point clouds. We divide the point cloud representation methods into projection, voxel, and raw point cloud.

4.1. Projection Methods

The projection method transforms the 3D data points into a 2D space using plane (image) [60], spherical [61], cylindrical [62], or bird's-eye view (BEV) [7] projection techniques. The projected data can be processed using the standard 2D methods and regressed to obtain the 3D bounding boxes. It is hard to detect occluded objects in 2D plane representations. Additionally, the 2D plane representations cannot keep object length and width. Tian et al. [6] presented a range image-based 3D detection model. The network uses the projected range image to generate multilevel FPN features and feed the output to the detection head for classification and regression of 3D boxes. The model was trained and tested on the nuScenes [63] dataset. When a point cloud is projected to a range image, occlusion and scale change may reduce the detection performance, which is not the case when projected to BEV representation. Therefore, solving the occlusion and scale change problems would improve the model as with other representations, such as BEV.

The BEV representation, mostly encoded by height, intensity (reflectance value), and density, solves these problems because, in BEV (top view) representation, objects occupy separate spaces on the map. Therefore, it avoids occlusion and scale problems [7]. The BEV lies on the ground plane so that the variance in the vertical direction is small, which helps to obtain accurate 3D bounding boxes compared to other projection methods. Because of these reasons, BEV is mostly used as a LiDAR 2D representation for 3D detection. However, the BEV representation causes height compression at each position, which may cause semantic ambiguity.

Yu et al. [64] transformed point cloud data into BEV elevation images by encoding each pixel using maximal, median, and minimal height values as shown in Figure 3.

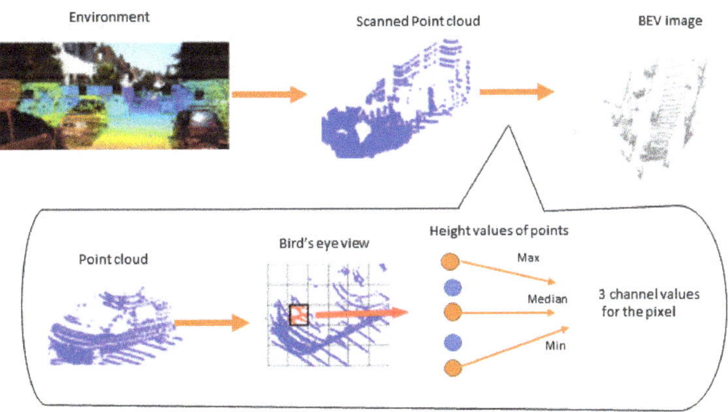

Figure 3. BEV image construction [64].

The BEV pixel of the image represents the ground coordinates so that it simultaneously detects and localizes vehicles in the ground coordinates. The classification of vehicle and localization is shown in Figure 4. Wirges et al. [65] encoded BEV using intensity, height, detections, observation, and decay rate. Similarly, BirdNet [66] and BirdNet+ [67] encoded BEV using height, intensity, and density information. The proposed density normalization map normalizes the number of points collected by different LiDAR sensors. BirdNet+ adds an improvised regression method to avoid the post-processing task in BirdNet to obtain the 3D boxes. Barrera et al. [68] modified BirdNet+ [67] and proposed a two-stage model using a Faster R-CNN [69] model. The LIDAR point cloud data are projected into BEV and normalized using the proposed density encoding method. The ResNet-50 [35] with FPN [70] is used as a backbone network and RPN as candidate proposals generation network. An ad-hoc set of nine anchors are employed to avoid the size constraints of objects in BEV. The category classification, 2D BEV rotated box regression, and 3D box regression is performed by the detection head's next stage. Even though the model shows a good performance on the KITTI [52] and nuScences [63] datasets, it is far from the real-time implementation due to slow speed detection. V and Pankaj proposed YOLO3D [71] model, an extension of YOLOv4 [72] 2D model. The point cloud data are projected onto the BEV domain before feeding to YOLOv4. Then, Euler-Region-Proposal Network is used to predict the 3D bounding box information.

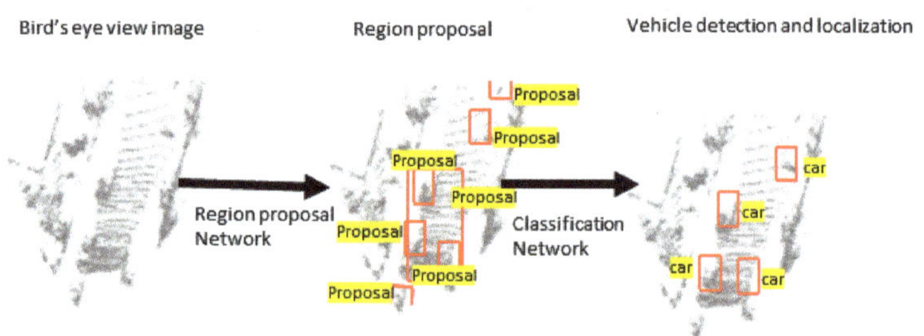

Figure 4. Two-stage vehicle detector [64].

Chen et al. [73] put forth a real-time multiclass 3D scene understanding two-stage model called MVLidarNet using multiple views. In the first stage, the point cloud data are projected into a perspective view to extract semantic information. In the second stage, the processed point cloud data are projected into BEV representation, which is important for classifying and detecting objects. An FPN-like encoder-decoder architecture is used in both stages. To detect individual object instances, DBSCAN [74] clustering algorithm is used. The model trained on the SemanticKITTI [75] dataset for semantic segmentation and KITTI [52] dataset for detection. The model detects objects and simultaneously determines the derivable space. Lu et al. [76] proposed the Range-Aware Attention Network (RAANet), which extracts more powerful BEV features and improves a 3D detection. Even though far-away objects have sparser LiDAR points, they do not appear smaller in the BEV representation. This weakens BEV feature extraction using shared-weight convolutional neural networks. The authors proposed RAANET to solve this challenge and extract more powerful features to enhance the overall 3D detection. They proposed the RAAConv layer instead of the Conv2D layer to extract more representative BEV features. They also developed an auxiliary loss of density estimation to enhance the detection of occlusion-related features. Even though the authors claimed the model is lightweight, the lite version runs a maximum of 22 Hz, which is far from the real-time implementation. We expect more lightweight models that can run at a higher frequency (Hz) for real-time implementation.

Du et al. [77] developed a detection network to provide an accurate 3D detection result. A two-stage CNN is presented for the final 3D box regression and classification based on the inputs fit into the 3D bounding box. PIXOR [78] is a proposal-free single-stage detector that balances high accuracy and real-time efficiency using a BEV representation of the point cloud. The network outputs a pixel-wise prediction of region proposals. Yang et al. put forth PIXOR++ [79], a single-stage 3D object-detection model. The model extracts geometric and semantic features from HD maps. A map prediction module is also proposed to estimate the map from raw LiDAR data. The model was trained and tested on the KITTI [52] dataset. Wenjie et al. [80] also proposed a deep NN model that jointly learns 3D detection, tracking, and motion forecasting by exploiting the BEV representation. The model is robust to occlusion and sparse data. Similarly, Complex-YOLO [81] extended the YOLOv2 [38] 2D image-based detection model for 3D object detection using a complex angle regression strategy for multiclass 3D box estimation. The introduced specific Euler-Region-Proposal Network (E-RPN) estimates the orientation of objects accurately by adding an imaginary and a real fraction for each bounding box. The model runs over 50 frames per second (fps), which is convenient for real-time operation. YOLO3D [82] also extended the loss function of YOLOv2 [38] using the yaw angle, 3D object bounding box center coordinates, and height of the box as a direct regression method.

4.2. Volumetric (Voxel) Methods

The volumetric method discretizes the unstructured and sparse 3D point clouds into a volumetric 3D grid of voxels (volume elements) and uses a series of voxels to represent a 3D point cloud. Voxels are similar to image pixels, but in 3D representation that explicitly provides depth information. Because of the 3D point clouds' sparsity nature, most volumes are empty. Consequently, processing the empty cells reduces performance. Another limitation of the voxel representation is that it uses 3D convolution for CNN models, increasing the computational cost and involving a trade-off between resolution and memory. In this section, we reviewed voxel- and pillar-based methods. Wang et al. [44] proposed a voting scheme-based sliding window approach, vote3D, for 3D detection to minimize the effect of the sparsity of point clouds.

Similarly, Sedaghat et al. [83] introduced a category-level classification network that estimates both the orientation and the class label. The orientation estimation during training improves the classification results during test time. Vote3Deep [47] presented the first sparse convolutional layers based on a feature-centric voting scheme [44] to leverage the sparsity point cloud. The computational cost is proportional to the number of occupied cells rather than the total number of cells in a 3D grid.

Likewise, Li put forth a 3DFCN [84] model for 3D object detection. They extended the 2DFCN [85] into a 3D operation on voxelized 3D data. The point clouds are discretized on a square grid with a 4D array representation using dimensions of length, width, height, and channels. Zhou and Tuzel created a VoxelNet [9] network for 3D object detection. The introduced voxel feature-encoding (VFE) layer learns a unified representation for each group of points within voxels instead of using handcrafted features as shown in Figure 5. Then, the region proposal network (RPN) takes the output of the volumetric representation and VFE layer output convolved with 3D convolutions to generate detections. Converting point clouds into a dense tensor structure help to implement stacked VFE operations in parallel across points efficiently. Processing only the nonempty voxels rather than the whole voxels in a grid improves the performance.

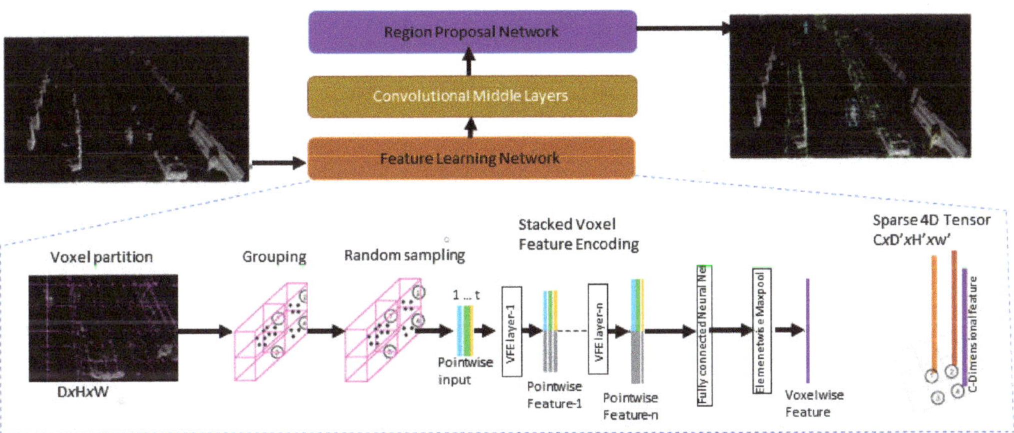

Figure 5. VoxelNet Architecture [9]. The raw point cloud is partitioned into voxels and transformed into vector representation by the feature learning network. The convolutional middle layers process the 4D tensor vector before the region proposal network generates 3D detection.

Furthermore, SECOND [39], a spatially sparse convolution network to extract information from LiDAR data, is introduced. The sparse convolution network uses a GPU-based rule-generation algorithm to increase the speed of operation. The work outperforms the previous results, such as VoxelNet. HVNet [86] is a hybrid single-stage LiDAR-based 3D detection network. The network fuses a multiscale VFE at a point-wise level using FPN [70]. Then, the result is projected into multiple pseudo-image feature maps using attentive VFE (AVFE) to solve the performance issues because of the size of the voxels. A small voxel improves performance, but the detection speed is slow. Large voxel sizes cannot capture the features of small objects. It uses a multiscale fusion network to solve these problems. The performance result shows a state-of-the-art mAP performance on the KITTI dataset.

Du et al. put forth Associate-3Ddet [87] model to learn the association between perceptual features extracted from real scenes using a perceptual voxel feature extractor (PFE) and conceptual features generated from augmented scenes using a conceptual feature generator (CFG) based on domain adaptation. First, CFG is separately trained on the conceptual scene and integrated with complete object models. The perceptual-to-conceptual module (P2C) uses an incompletion-aware re-weighting map to build associations between perceptual and conceptual features. Once the perceptual and conceptual domains are well aligned after training, the network can adaptively generate the conceptual features without CFG. The models show promising results on the benchmarks of the KITTI [52] dataset. Liu et al. proposed TANet [88], a robust network comprised of Stacked Triple Attention (STA) and Coarse-to-Fine Regression (CFA) modules. The STA module obtains multilevel feature attention by performing a stack operation on channel-wise, point-wise, and voxel-wise attention. The CFA module helps to achieve more accurate detection boxes without more outrageous computational costs. Then, the pyramid sampling aggregation (PSA) module takes the output of CFA and provides cross-layer feature maps. The cross-layer feature maps capture multilevel information, which has larger receptive fields with richer semantic information from the high-level features and larger resolution from low-level features. This helps to obtain more robust representative features for objects and enhances the detection capability.

Deng et al. put forth Voxel R-CNN [89], a voxel-based 3D detection model comprising a 3D backbone network, a 2D BEV RPN, and a detection head. The point cloud data are voxelized and fed into the 3D backbone network for feature extraction. Then, the features are converted into BEV representation before applying the 2D backbone and RPN for region proposal generation. ROI features are extracted from the 3D features by applying voxel RoI pooling. Finally, the detection head uses RoI features for box refinement. The experiments are conducted on the KITTI [52] and Waymo Open [90] datasets. Li et al. proposed SIENet [91], a two-stage spatial information enhancement network for 3D object detection. The authors designed a hybrid-paradigm region proposal network (HP-RPN), which includes the SPConv branch, the auxiliary branch, and the key-point branch, to learn discriminative features and generate accurate proposals for the spatial information enhancement (SIE) module. The SPConv branch learns more abstract voxel features from the voxelized point cloud. Concurrently, the corresponding voxel features are dynamically encoded by the key-point branch. The auxiliary branch is used to learn object structures. The spatial shapes of the foreground points in the candidate boxes are predicted using the proposed SIE module. The SIE module learns structure information to enhance the features for box refinement. The model was trained and tested on the KITTI [52] dataset.

Liu et al. [92] put forth a single-stage sparse multiscale voxel feature aggregation network (SMS-Net). The model comprises sparse multiscale-fusion (SMSF) and shallow-to-deep regression (SDR) modules. The SMSF module fuses point-wise and multiscale features at the 3D sparse feature-map level to achieve more fine-grained shape information. The SDR improves the localization and 3D box estimation accuracy through multiple aggregations at the feature-map level with less computational overhead. The model shows the comparable result on the KITTI [52] dataset. Sun et al. [93] proposed a semantic-aware 3D object-detection model. The proposed voxel-wise class-aware segmentation module learns the fine-grained semantic features. Additionally, the semantic-aware refinement module generates coarse proposals. The model shows a competitive result on the KITTI [52] dataset, especially for small objects.

Liu et al. proposed MA-MFFC [94], attention and multiscale feature fusion network with ConvNeXt module for 3D detection network. The multi-attention module comprises point-channel and voxel attention to enhance key point information in voxels and obtain more robust and discriminative voxel features in the 3D backbone. The convolutional layer is replaced with a ConvNeXt module to extract richer features more accurately. The experimental result on the KITTI [52] dataset shows an improvement over Voxel R-CNN [89] baseline network. Wang et al. put forth a self-attention graph convolutional network (SAT-GCN) [95] for 3D object detection. The proposed model consists of three modules. The first vertex feature-extraction (VFE) module with GCN decodes point cloud data and extracts local relationships between features. The second module, self-attention with dimension reduction, uses self-attention to further enhance the neighboring relationship between features. The final module, far-distance feature suppression, generates global features by suppressing far-away features. The experimental result on the KITTI [52] and nuScenes [63] datasets show the effectiveness of the proposed model. Fan et al. [96] proposed voxel-based fully sparse 3D object detector. The sparse instance recognition module generates instance features by grouping points before the prediction is employed. Grouping points into instances reduces the missing center feature, such as center-based detection and neighbor queries. The model shows competitive performance on the Waymo [90] dataset.

Li et al. [91] presented a two-stage model with a spatial information enhancement network. The first stage of the hybrid-paradigm RPN consists of the SPConv branch, the auxiliary branch, and the key-point branch to extract features and generate proposals. The second stage predicts the 3D bounding boxes via the spatial information enhancement module. The number of LiDAR points of an object decreases with distance. The spatial information enhancement module helps predict the spatial shapes of objects. It extracts the structure information and the representative features for further box refinement, which increases the detection accuracy of far-away objects. Hu et al. [97] proposed a two-stage point density-aware voxel (PDV) network to account the point density variations. The PDV with voxel point centroids localizes voxel features from the 3D sparse convolution backbone. The density-aware ROI-grid pooling module using kernel density estimation (KDE) and self-attention with point density positional encoding aggregates the localized features. The model was trained and tested on the Waymo [90] and KITTI [52] datasets.

Another work, Frustum PointNets (F-PointNet) [10], is a cascaded fusion network for 3D detection and semantic segmentation. The 2D region proposals are generated from RGB images using CNN and extruded to 3D region proposals called Frustum point clouds. Then, the Frustum PointNet performs 3D object instance segmentation and amodal 3D bounding box regression. Although the network shows incredible performance, there are failures because of inaccurate pose and size estimation in the sparse point cloud. The network also has a limitation when there are multiple instances from the same category and dark lighting or strong occlusion. Cao et al. [11] proposed multiview Frustum PointNet (MVFP), which is an extension of the F-PointNet [10] to reduce the rate of miss detection. F-PointNet [10] misses detection when the RGB feature detector does not capture all the information. MVFP added an auxiliary BEV detection to handle the miss detection. The authors use an IOU [98] to match the F-PointNet [10] 3D and BEV 2D bounding box. If there is a matching for the BEV box in the F-PointNet box, the object is detected successfully by F-PointNet [10]. Therefore, the F-PointNet result is the output. On the other hand, if there is no match for the BEV box, the object is possibly miss-detected. Thus, the 2D box in BEV will be fed back to the raw point cloud. The experimental result on the KITTI [52] dataset shows the model outperforms the F-PointNet [10].

Wang and Jia put forth Frustum-ConvNet [99], an amodal 3D model by sliding frustums to aggregate local point-wise features. The authors proposed a method to obtain a sequence of frustums by sliding a pair of planes along the Frustum axis. The pair of planes are perpendicular to the Frustum axis as well as the optical axis of the camera is perpendicular to the 2D region proposal. For each 2D region proposal, a sequence of frustums is generated, and all points inside the Frustum are grouped. Point-wise features inside each Frustum are aggregated as a Frustum-level feature vector. Then, Frustum-level feature extraction is undertaken using PointNet [12]. The Frustum-level feature vectors array as a 2D feature map and feed into a fully convolutional network before feeding to the detection head, which estimates oriented 3D bounding boxes. Finally, the proposed refinement network applies to ensure the predicted box precisely bounds object instances. The network was trained on the KITTI [52] and indoor SUN-RGBD [100] datasets.

Some works discretize point clouds into vertical columns called pillars instead of voxels. Unlike voxels, pillars are suitable for 2D convolutions. Pointpillars [41] is a single-stage 3D object-detection method by learning the features on pillars (vertical columns) to predict 3D oriented boxes for objects. It is a point cloud encoding technique by learning the features rather than using fixed encoders. It can run at 105 Hz and outperform previous detectors, such as SECOND [39], in BEV and 3D KITTI detection benchmarks. McCrae and Zakhor [101] modified PointPillars [41] as a recurrent network using fewer LiDAR frames per forward pass. The ConvLSTM layer is inserted between PointPillars [41] backbone and detection head to propagate information through time. The network takes fewer LiDAR frames than PointPillars [41], which reduces complexity by processing fewer data. The model outperforms the PointPillars [41] network, which uses 10 LiDAR frames per forwarding pass, with only three LiDAR frames per forwarding pass. However, there was a performance decline in the vehicle class. Detecting small objects is one of the main challenges in autonomous driving, and decreasing the LiDAR frames will further hurt performance.

Wang et al. [102] proposed an anchor-free pillar-based model for autonomous driving. As in other methods, the network predicts bounding box parameters per pillar. The authors also include an interpolation method in pillar-to-point projection to improve the final prediction. The model is evaluated on the Waymo [90] dataset. Fan et al. [103] proposed a single-stride sparse transformer network to avoid information loss because of multiple strides using PointPillars as a base network. The model was trained and tested on the Waymo [90] dataset. Tong et al. put forth ASCNet [104], a two-stage network. Pillar-wise spatial-context feature-encoding and length-adaptive RNN-based modules are proposed to learn features from point clouds and to solve the inhomogeneity in point clouds, such as a varying number of points in the pillars, the diverse size of Regions of Interest (RoI), respectively. The model shows competitive performance on the KITTI [52] dataset.

Zhang et al. [105] put forth a semisupervised pillar-based single-stage model by adopting a teacher-student framework. The authors used the same loss function as SECOND [39] and PointPillars [41]. The exponential moving average (EMA) teacher and asymmetric data augmentation method are used to improve the efficiency of a teacher-student network. Even though the performance of this model is not good enough for real-time application, it is a suitable model to start with by developing a semisupervised model to reduce the time and cost of dataset labeling and annotation. Caine et al. [106] put forth a pseudo-labeling domain adaptation method using PointPillars [41] as a student network. The teacher network pseudo-labels all unseen source data, and the student network trains with a union of labeled and pseudo-labeled data. Finally, the student and teacher networks were evaluated on the Waymo Open Dataset [90] and Kirkland validation split [107].

Bai et al. proposed PillarGrid [108], a cooperative perception model that fuses information from multiple 3D on-board and roadside 3D LiDARs. The PillarGrid model consists of four main components: (1) cooperative preprocessing for point cloud data transformation, (2) pillar-wise voxelization and feature extraction, (3) grid-wise deep fusion to fuse deep features, and (4) CNN-based augmented 3D object detection. The model was trained and tested on a dataset collected using CARLA [109]. Lin et al. [110] put forth a pointpillar-based 3D object-detection module to improve the performance for snow weather conditions. The proposed double-attention module is used to reweight the input features of pillars feature extraction. The feature refinement extraction module captures context information to reduce the noise of local features. Finally, the proposed maximum mean discrepancy module is employed to obtain the domain feature representation distribution. The module is trained and tested on the Canadian Adverse Driving Condition [111] and KITTI [52] datasets. Recently, Alaba and Ball proposed WCNN3D [59], a wavelet-based 3D object-detection network. The model comprises discrete wavelet transform (DWT) and inverse wavelet transform (IWT) with skip connection between the contrasting layers, expanding layers, and the previous layers in the model. The model is designed without the pooling operation to reduce the information loss during downsampling. The DWT is used as a downsampling operator, whereas IWT is an upsampling operator. The wavelet's lossless property helped recover the lost details during the downsampling operation. The experimental result on the KITTI [52] dataset shows the model outperforms pillar-based models, such as PointPillars [41], and PV-RCNN [112], and is more suitable for the detection of small objects such as pedestrians and cyclists.

4.3. Raw Point Cloud Methods

The LiDAR data projection and volumetric methods cause spatial information loss during conversion to another domain, so processing point clouds directly are important to keep this spatial information. However, the raw point cloud methods have high sparsity and computational costs due to 3D convolutions. PointNet [12] is a unified architecture for 3D object classification, part segmentation, and semantic segmentation that directly uses raw point cloud data, as shown in Figure 6. The classification network transforms n inputs into aggregate point features using feature transformation. It outputs k class classification scores. Then, the segmentation network concatenates the local and global features and generates per-point scores.

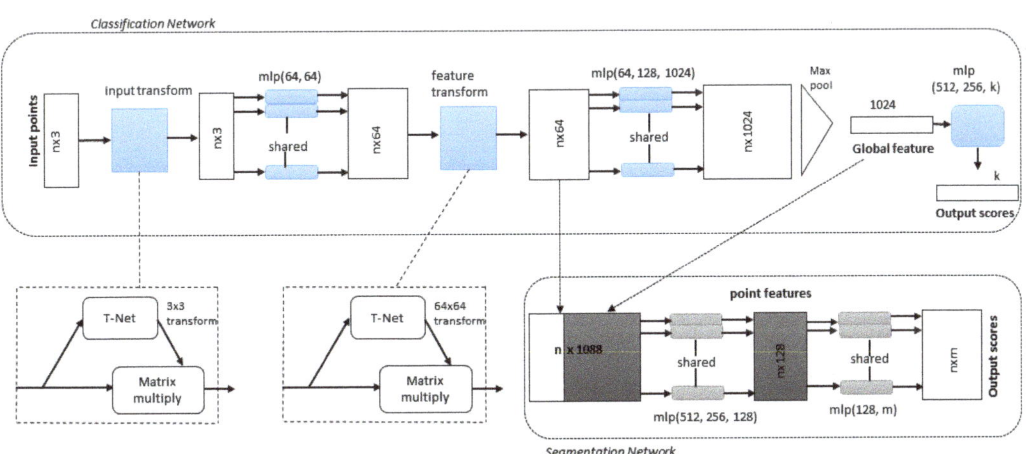

Figure 6. PointNet Architecture [12]. The classification network transforms n inputs into aggregate point features using feature transformation. It outputs k class classification scores. Then, the segmentation network concatenates the local and global features and generates per-point scores. The multilayer perceptrons (mlp) with layer size in brackets are also given.

The network applies feature transformation to each input and aggregates point features. The process generates classification scores for each class. Then, global features, local features, and outputs-per-point scores are aggregated for segmentation. Although the PointNet model is promising, it does not capture local structures, which is crucial for fine-grain patterns and better generalizability for unseen cases.

Qi et al. [13] developed PointNet++ to solve the PointNet architecture limitations by processing a set of points hierarchically sampled in a metric space. Distance metrics are used to partition the set of points into overlapping local regions by leveraging neighborhoods at multiple scales and centroid locations using the farthest point sampling (FPS) algorithm and then learning the local features using PointNet. Finally, multiscale grouping forms multiscale features by concatenating features at different scales. Multiresolution grouping adaptively aggregates information based on the distributional properties of points. It minimizes computational costs more than multiscale grouping.

Likewise, Shi et al. presented PointRCNN [14], a two-stage model using raw point cloud data. The first stage generates 3D proposals by segmenting the point clouds into foreground points and background bottom-up. The second stage refines the 3D bounding box proposals in the canonical coordinates to achieve better detection results. In the second stage, the authors adopted a pooling operation to pool learned point representations from proposal generation and transformed them into canonical coordinates. The canonical coordinates are combined with the pooled point features and the segmentation mask from the first stage to learn relative coordinate refinement and use all information from the first stage segmentation and proposal sub-network. They also proposed bin-based loss for efficient and effective 3D bounding box regression. The model was trained and tested on the KITTI [52] dataset.

Shi et al. [113] extended the pointRCNN [14] model by proposing a part-aware and aggregation neural network (Part-A2) as shown in Figure 7. The part-aware network learns to estimate the intra-object part locations of foreground points and generate 3D proposals simultaneously by extracting discriminative features from the point cloud. Intra-object part locations are the relative locations of the 3D foreground points regarding their corresponding ground-truth boxes.

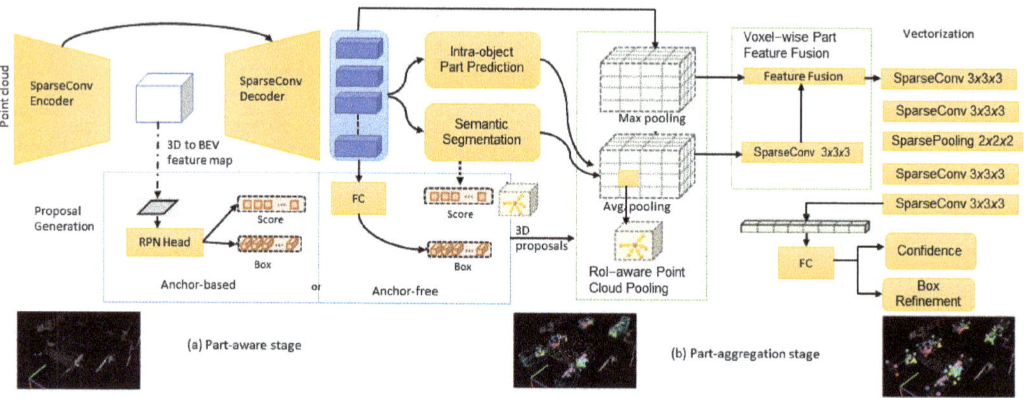

Figure 7. The part-aware and aggregation proposed 3D object-detection network architecture [113]. The model consists of two parts: (**a**) The intra-object part locations are predicted by the part-aware network. Then, 3D proposals are generated before feeding to the encoder-decoder network. (**b**) ROI-aware pooling is undertaken in the part-aggregation stage.

The discriminative point-wise features for foreground point segmentation and intra-object part location estimation are extracted using an encoder-decoder network with sparse convolution and deconvolution [46,114]. The model acquires the potential to infer the shape and pose of objects by learning to estimate the foreground segmentation mask and the intra-object part location of each point. The authors used anchor-free and anchor-based methods for 3D proposal generation. The anchor-free method is more memory efficient, whereas the anchor-based method gives better recall with a more GPU memory cost. Once the intra-object part locations and 3D proposals are generated, box scoring and proposal refinement is completed by aggregating the part information and learning point-wise features of all the points within the same proposal. The canonical transformation reduces effects due to the rotation and location variations of 3D proposals. Then, the RoI-aware point cloud feature pooling module removes the ambiguity of the previous point cloud pooling operation. The box proposal scoring and refinement information is fused for fine detection results. The model outperforms PointRCNN on the KITTI [52] dataset.

Yang et al. [115] put forth a sparse-to-dense (STD) two-stage 3D object-detection framework. In the first stage, the bottom-up proposal generation network uses a raw point cloud as input and seeds each point with a new spherical anchor to generate proposals. Then a PointNet++ [13] backbone extracts semantic context features for each point and generates objectness scores to filter anchors. The authors proposed a PointsPool layer to generate features for each proposal and transform sparse, unstructured, and unordered point-wise proposals into more compact features. In the second stage, a prediction is made. To reduce inappropriate removal during post-processing, they introduced augmenting a 3D IOU branch for predicting 3D IOU between predictions and ground-truth bounding boxes. The result on the KITTI dataset [52] outperforms models, such as PointPillars [41] and PointRCNN [14].

Yu et al. [116] proposed an equivariant network with a rotation equivariance suspension design to achieve object-level equivariance for 3D detection. This method helps the bounding box independent of object pose and scene motion. The proposed method tested on different models, such as VoteNet [117] on the ScanNetV2 [118] dataset and transfer-based network [119] on the SUN RGB-D [100] dataset for indoor scenes and PointRCNN [14] on the KITTI [52] dataset for outdoor scenes. Furthermore, 3DSSD [120] is a lightweight and efficient 3D single-stage framework using point clouds. The feature propagation upsampling layers and refinement module, mostly common for point cloud-based models, were removed to reduce the computational cost. A fusion set of abstraction downsampling layers were proposed to keep important information for regression and classification tasks. Finally, a box prediction network and anchor-free regression head with a 3D center label were introduced to enhance the final performance. He et al. [121] developed a structure-aware single-stage 3D detection (SASSD) network. A detachable auxiliary network with point-level supervision was designed for better localization performance through learning the structure information and an efficient feature-map part-sensitive wrapping operation, PSWarp, to correct the misalignment between the predicted bounding boxes and corresponding confidence maps. This design solves the spatial loss due to downsampling in a fully convolutional operation. Gustafsson et al. [122] proposed conditional energy-based models by extending SASSD [121] model. The authors designed a differentiable pooling operator to regress 3D bounding boxes accurately. They integrated the differentiable pooling operation into the SASSD model, and the experimental result on the KITTI dataset [52] shows the model outperforms SASSD [121] model.

Similarly, Zheng et al. [123] proposed a confident IOU-aware single-stage object detector (CIA-SSD) network to solve the localization accuracy and classification confidence misalignment. The authors proposed a spatial-semantic feature aggregation module to adaptively fuse high-level abstract semantic and low-level spatial features. This fusion helps to predict bounding boxes and classification confidence accurately. An IOU-aware confidence rectification module further rectifies the predicted confidence for more consistent confidence with localization accuracy. The IOU-aware confidence rectification module solves the complexity of SASSD [121] because of the interpolation operation. A distance-variant IOU-weighted NMS module uses rectified confidence to obtain smoother regressions and avoid redundant predictions. The network shows a comparable performance for 3D car detection of the KITTI [52] dataset. Shi and Rajkumar proposed Point-GNN [124], a graph neural network-based model. A graph was constructed using vertices and connecting neighboring points within a fixed radius. Voxel downsampling was used to reduce the point cloud density, but the representation is still a graph. After the graph is constructed, a graph neural network is developed to refine the vertex features by aggregating features along the edges. An autoregistration mechanism was proposed to align neighbor coordinates and reduce translation variance. A box merging and scoring operation were also proposed to combine detection from multiple vertices with confidence scores. The model outperforms others, such as PointRCNN [14]and STD [115].

Zhou et al. [125] put forth a two-stage joint 3D semantic segmentation and detection model for autonomous driving. The model consists of two parts: spatial embedding (SE) learning-based object proposal and the refinement of local bounding boxes (BBoxes). Point-wise features (local features and global context information) are extracted using PointNet++ [13] as a backbone network, along with sampling and grouping. A SE method was proposed to assemble all foreground points into the corresponding object centers. Based on the SE results, the object proposals, instance segmentation, and BBox can be generated using a simple clustering strategy (K-means) [126]. Non-maximal suppression (NMS) is not employed in this model because only one proposal is generated for each cluster. Finally, the proposed instance-aware ROI pooling outputs refined 3D BBoxes and instance masks. The model was trained on the KIITI [52] dataset.

Mao et al. [127] put forth a two-stage model, pyramid R-CNN, which is compatible with voxels, points, and other representations of the LiDAR region of interest (ROI). A pyramid ROI head was proposed, which comprises an ROI-grid pyramid, ROI-grid attention, and Density-Aware Radius Prediction (DARP), to learn the features adaptively from the sparse points of interest. An ROI-grid pyramid collects points of interest for each ROI in the pyramid, which helps to mitigate the sparsity problem. The ROI-grid attention component incorporates conventional attention-based and graph-based points into a unified form to encode richer information from sparse points. Finally, the DARP module is dynamically adjusting the focusing range of ROIs to adapt to different point density levels. This model is robust to the sparse data and imbalanced classes. Yang et al. presented ST3D [128], a self-training 3D domain adaptive network. A 3D object augmentation technique, random object scaling (ROS), was developed to overcome the bias in object size in the labeled source domain. A quality-aware triplet memory bank (QTMB) was also proposed for pseudo-label generation and assessing pseudo-boxes' quality. Finally, a curriculum data augmentation (CDA) strategy was developed using pseudo-labels by escalating the intensity of augmentation and simulating hard examples during training to overcome overfitting. This network trained with four datasets, namely KITTI [52], Waymo [90], nuScenes [63], and Lyft [129].

Hegde and Patel [130] proposed a source-free unsupervised domain adaptive model, which uses class prototypes to mitigate the effect of pseudo-label noise. The model performance dropped when tested with a dataset different from what was trained. This model minimizes the effect of domain change. During self-training, a transformer module was used to identify incorrect and overconfident annotation outliers and compute an attentive class prototype. Zheng et al. presented SE-SSD [131], a self-assembling single-stage 3D object detector. The CIA-SSD [123] model structure was used by removing the confidence function and DI-NMS. The SE-SSD framework consists of teacher SSD and student SSD. The teacher SSD receives the point of cloud input and produces a bounding box with confidence predictions. Then, the teacher supervises the student using the soft targets, which are the predictions after global transformations, with consistency loss to align the student predictions with soft targets. The teacher also supervises the student with hard targets using orientation-aware distance-IOU loss, focusing more on the alignment of box centers and the distance between the 3D centers of the predicted and ground-truth bounding boxes. The student parameters were updated using consistency loss and orientation-aware distance-IOU loss. In contrast, the teacher parameters were updated based on the student parameters with the standard exponential moving average (EMA) method. The model shows performance improvement on the KITTI [52] dataset over other methods, such as CIA-SSD [123].

Wang et al. [132] proposed a semisupervised network via temporal graph neural network. The teacher network takes a single point cloud frame input to generate candidate detections, which are input to the graph neural network to generate further refined detection scores. The generated pseudo-labels are then combined with the labeled point clouds to train the student's model. The teacher network is updated in each step using the exponential moving average method. This semisupervised method is important to leverage abundant unlabeled data. The model was trained and tested on the datasets of the nuScenes [63] and H3D [133] datasets. Zhang et al. proposed PointDistiller [134] knowledge distillation method for point cloud data. The model includes the local distillation to extract and distills the local geometric structure of point clouds to the student network using dynamic graph convolution with a reweighted learning strategy to handle the sparsity and noise in point clouds. Transferring the teacher knowledge to the students on the point cloud data trained and tested both on voxel and raw point cloud detectors, such as PointPillars [41], SECOND [39], and PointRCNN [14].

Wang et al. put forth POAT-Net [135], a parallel offset-attention assisted transformer model. The parallel offset-attention method helps to capture the distinguishing local features at different scales. The normalized multiresolution grouping (NMRG) helps the system to adapt the non-uniform density distribution of the 3D object point cloud from the input embedding. NMRG fuses features from the downsampling pyramid and upsampling pyramid of different scales and normalizes them. By leveraging the encoder and decoder structure of the transformer and incorporating T-net, POAT-Net is insensitive to the permutations of the point cloud and tolerates any initial rigid translation or rotation of the raw point cloud. The normalized NMRG and parallel offset-attention help POAT-Net improve the occluded object-detection rate. Ren et al. proposed a dynamic graph transformer 3D object-detection network (DGT-Det3D) [136]. The model consists of a dynamic graph transformer (DGT) and proposal-aware fusion (PAF) modules. The DGT module is a backbone network to encode long-range features and extracts spatial information. The proposed PAF module combines and enhances the spatial and point-wise semantic features for performance improvement. The model was trained and tested on the KITTI [52] and Waymo [90] datasets.

Theodose et al. [137] proposed a DL-based LiDAR resolution-agnostic model to minimize the effect of point cloud variation / distribution for 3D models. Two methods were proposed to improve the performance of a model for unknown data. The first one is increasing the data variability. The data variability increased by randomly discarding layers from the training. Each target was represented as a Gaussian function, and the corresponding loss function focused on the obstacle occupancy rather than regressing parameters as the other solution. The model has exclusively trained on the subset of KITTI [52] dataset and then evaluated on the nuScenes [63] and Pandaset [138] datasets. The model was trained only for car detection. Nagesh et al. [139] proposed an auxiliary network [121] to improve the localization accuracy of class imbalance on the nuScenes [63] dataset using the structure information of the 3D point cloud. The class-balanced grouping and sampling [140] was proposed to solve the classification loss problem due to class imbalance. The auxiliary network was combined with class-balanced grouping and sampling to improve the localization loss. The auxiliary network can be detached during the test time to reduce the computational burden. Wang et al. put forth a single-stage network, LSNet [141], which comprises a learned sampling (LS) module to sample important points. The LS sampling outperforms other techniques, such as farthest point sampling on the KITTI [52] dataset.

Hahner et al. [142] put forth a snowfall LiDAR simulation for 3D object detection. The snow particles were sampled for each LiDAR line in 2D space and used the induced geometry to modify each LiDAR beam measurement. Partially synthetic snow LiDAR data are generated and used these data to train several 3D networks, such as PV-RCNN [112], PointRCNN [14], SECOND [39], and PointPillars [41]. This technique is vital to simulate models when collecting data in such weather conditions is challenging.

Some methods use more than one LiDAR data representation to improve performance. Chen et al. developed Fast Point R-CNN [143], a two-stage hybrid 3D detection network using both voxel representations and raw point cloud data. In the first stage, VoxelRPN uses the voxel representations to make a few initial predictions. Each predicted bounding box in the first stage is projected to BEV. The second stage, RefinerNet, further improves the output of the first stage detection by directly processing the raw point cloud. The high-dimensional coordinate feature is fused with the convolutional feature to preserve accurate localization and context information. Shi et al. also proposed PV-RCNN [112] that uses both voxel representation and raw point clouds. The voxel-based part of the network encodes a multiscale representation of the features using sparse convolution. In contrast, the PointNet-based part of the network set abstraction to learn more discriminative features with small key-points. Each proposal introduces a multiscale ROI feature abstraction layer for points to preserve rich context information for accurate box refinement and confidence prediction. However, 3D convolution is still a bottleneck because of the high computational cost.

Likewise, Xu et al. presented Spg [144], an unsupervised domain adaptation model via semantic point generation. The model comprises three modules: the Voxel feature-encoding (VFE) module, the information propagation module, and the point generation module. The point cloud data are voxelized, and a prediction for each voxel, including occupied and empty, is generated. The VFE module from VoxelNet [9] aggregates points inside each voxel, and then the voxel features are stacked into pillars and projected onto a BEV feature space similar to PointPillars [41] and PV-RCNN [112]. The pillar features feed into the information propagation module for 2D convolutions operation. Then, the point generation module maps the pillar features to the corresponding voxels. These features are fed to the two detectors: Pointpillars [41] and PV-RCNN [112]. The model is evaluated on the Waymo open [90] and KITTI [52] datasets.

Table 1 gives the selected LiDAR-based 3D object-detection methods for the KITTI [52] dataset to show the performance improvement over time. The BEV and 3D KITTI eval-uation benchmarks are commonly used to compare the performance of models in the KITTI dataset.

Table 1. BEV and 3D performance comparison (%) of selected LiDAR-based 3D object-detection methods on the KITTI [52] test benchmark. r40 shows the mAP is calculated for 40 recall points instead of 11. E stands for easy, M for moderate, and H for hard.

Methods	AP_{BEV} Car			Pedestrians			Cyclists			AP_{3D} Car			Pedestrians			Cyclists		
	E	M	H	E	M	H	E	M	H	E	M	H	E	M	H	E	M	H
BirdNet [66]	75.5	50.8	50.0	26.1	21.4	20.0	39.0	27.2	25.5	14.8	13.4	12.0	14.3	11.8	10.6	18.4	12.4	11.9
BirdNet+ [67]	84.8	63.3	61.2	45.5	38.3	35.4	72.5	52.2	46.6	70.1	51.9	50.0	38.0	31.5	29.5	67.4	47.7	42.9
VoxelNet [9]	89.4	79.3	77.4	46.1	40.7	38.1	66.7	57.7	50.6	77.5	65.1	57.7	39.5	33.7	31.5	61.2	48.4	44.4
SECOND [39]	88.1	79.4	78.0	55.1	46.3	44.8	73.7	56.0	48.8	83.1	73.7	66.2	51.1	42.6	37.3	70.5	53.9	46.9
Fast point R-CNN [143]	88.0	86.1	78.2	-	-	-	-	-	-	84.3	75.7	67.4	-	-	-	-	-	-
PointPillars [41]	88.4	86.1	79.8	58.7	50.2	47.2	79.1	62.3	56.0	79.1	75.0	68.3	52.1	43.5	41.5	75.8	59.1	53.0
3DSSD [120]	88.4	79.6	74.6	-	-	-	-	-	-	-	-	-	-	-	-	-	-	-
SASSD [121]	88.8	79.8	74.2	-	-	-	-	-	-	-	-	-	-	-	-	-	-	-
CIA-SSD [123]	89.6	80.3	72.9	-	-	-	-	-	-	-	-	-	-	-	-	-	-	-
PIXOR++ [79]	89.4	83.7	78.0	-	-	-	-	-	-	-	-	-	-	-	-	-	-	-
TANet [88,141]	91.6	86.5	81.2	-	-	-	-	-	-	84.4	75.9	68.8	-	-	-	-	-	-
LSNet [141]	92.1	85.9	80.8	-	-	-	-	-	-	86.1	73.6	68.6	-	-	-	-	-	-
Associate-3Ddet [87]	91.4	88.1	83.0	-	-	-	-	-	-	86.0	77.4	70.5	-	-	-	-	-	-
HVNet [86] (r40)	92.8	88.8	83.4	54.8	48.9	46.3	84.0	71.2	63.7	-	-	-	-	-	-	-	-	-
Part-A2 [113]	91.7	87.8	84.6	-	-	-	-	-	-	87.81	78.49	73.51	-	-	-	-	-	-
PV-RCNN [112] (r40)	95.0	90.7	86.1	59.9	50.6	46.7	82.5	68.9	62.1	90.3	81.4	76.8	52.2	43.3	40.3	78.6	63.7	57.7
WCNN3D [59]	90.1	88.0	86.5	68.4	63.2	59.4	82.78	64.3	60.3	87.8	77.6	75.4	62.0	57.7	52.1	82.7	61.0	57.7
Point-GNN [124]	93.1	89.2	83.9	-	-	-	-	-	-	88.3	79.5	72.3	-	-	-	-	-	-
SE-SSD [131] (r40)	95.7	91.8	86.7	-	-	-	-	-	-	91.5	82.5	77.2	-	-	-	-	-	-

We recently addressed the challenges in 3D object detection, especially challenges related to processing point cloud data, LiDAR point encoding techniques, robust models, representative datasets, best fusion techniques, and others. We also indicated the possible research directions to solve the challenges in 3D object detection (Read [3,24] for more details). Different methods use different LiDAR encoding techniques and datasets. The comparison of LiDAR 3D object-detection methods based on the LiDAR encoding tech-niques, datasets, and publication years are summarized in Table 2 to show the performance improvement over time.

Table 2. Comparison of LiDAR-based 3D object-detection methods based on LiDAR encoding techniques, the dataset used, and year of publication. We categorized pillar representation in the volumetric encoding. Some methods use multiple datasets, but we report only datasets related to autonomous driving.

Method	LiDAR Encoding Technique	Dataset Used	Year of Publication
Yu et al. [64]	projection	KITTI	2017
BirdNet [66]	projection	KITTI	2017
Wirges et al. [65]	projection	KITTI	2018
PIXOR [78]	projection	KITTI	2018
Complex-YOLO [81]	projection	KITTI	2018
Birdnet+ [67]	projection	KITTI	2020
MVLidarNet [73]	projection	KITTI	2020
BirdNet+ [68]	projection	KITTI & Nuscenes	2021
YOLO3D [71]	projection	KITTI	2021
RAANet [76]	projection	KITTI & nuScenes	2021
Tian et al. [6]	Projection	nuScenes	2022
3DFCN [84]	voxel	KITTI	2017
VoxelNet [9]	voxel	KITTI	2018
SECOND [39]	voxel	KITTI	2018
F-PointNet [10]	voxel	KITTI & SUN RGB-D	2018
MVFP [11]	voxel	KITTI	2019
Frustum-ConvNet [99]	voxel	KITTI & SUN RGB-D	2019
Pointpillars [41]	pillar	KITTI	2019
McCrae and Zakhor [101]	pillar	KITTI	2020
Wang et al. [102]	pillar	Waymo	2020
HVNet [86]	voxel	KITTI	2020
Associate-3Ddet [87]	voxel	KITTI	2020
TANet [88]	voxel	KITTI	2020
Voxel R-CNN [89]	voxel	KITTI & Waymo	2021
SIENet [91]	voxel	KITTI	2021
Zhang et al. [105]	pillar	KITTI	2021
SMS-Net [92]	voxel	KITTI	2022
Sun et al. [93]	voxel	KITTI	2022
MA-MFFC [94]	voxel	KITTI	2022
SAT-GCN [95]	voxel	KITTI & Nuscenes	2022
Fan et al. [96]	voxel	Waymo	2022
Li et al. [91]	voxel	KITTI	2022
PDV [97]	voxel	KITTI & Waymo	2022
Fan et al. [103]	pillar	Waymo	2022
ASCNet [104]	pillar	KITTI	2022
PillarGrid [108]	pillar	synthetic data	2022
Lin et al. [110]	pillar	KITTI & CADC	2022
WCNN3D [59]	pillar	KITTI	2022
PointNet [12]	raw point cloud	ScanNet	2017
PointNet++ [13]	raw point cloud	ScanNet	2017
PointRCNN [14]	raw point cloud	KITTI	2019
STD [115]	raw point cloud	KITTI	2019
Part-A2 [113]	raw point cloud	KITTI	2020
3DSSD [120]	raw point cloud	KITTI & nuScenes	2020
SASSD [121]	raw point cloud	KITTI	2020
CIA-SSD [123]	raw point cloud	KITTI	2020
Point-GNN [124]	raw point cloud	KITTI	2020
Zhou et al. [125]	raw point cloud	KITTI	2020
Auxiliary network [121]	raw point cloud	nuScenes	2020
LSNet [141]	raw point cloud	KITTI	2021
Pyramid R-CNN [127]	raw point cloud	KITTI & Waymo	2021
ST3D [128]	raw point cloud	KITTI, Waymo, nuScenes & Lyft	2021
SE-SSD [131]	raw point cloud	KITTI	2021
Wang et al. [132]	raw point cloud	nuScenes & H3D	2021
POAT-Net [135]	raw point cloud	KITTI	2021
Theodose et al. [137]	raw point cloud	KITTI, nuScenes, & Pandaset	2021
DGT-Det3D [136]	raw point cloud	KITTI & Waymo	2022
Yu et al. [116]	raw point cloud	KITTI, ScanNetv2 & SUN RGB-D	2022
Hahner et al. [142]	raw point cloud	STF	2022
PointDistiller [134]	raw point cloud	KITTI	2022
Fast Point R-CNN [143]	voxel & raw point cloud	KITTI	2019
Pv-Rcnn [112]	voxel &raw point cloud	KITTI &Waymo	2020
Spg [144]	voxel & raw point cloud	KITTI &Waymo	2021

5. Conclusions

This survey presented state-of-the-art LiDAR-based 3D object detection for autonomous driving. The LiDAR feature-extraction methods and LiDAR encoding techniques were also summarized. The 3D coordinate systems are different for different sensors and datasets. Therefore, the commonly used 3D coordinate systems were reviewed. The stages of autonomous driving were also summarized. We categorized the 3D LiDAR perception systems methods based on the encoding technique as projection, voxel, and raw point cloud, with the pros and cons of each method. Generally, 3D object-detection methods show significant performance improvement for autonomous driving. However, several open issues exist in improving model speed and accuracy for real-time processing and level four and five driving. Some works, such as [145], proposed a computationally efficient algorithm for the quality of training and resource allocation, such as bandwidth and power allocation in multi-modal systems. We expect more works that efficiently train the complete system in autonomous driving.

Author Contributions: Conceptualization and methodology, S.Y.A.; formal analysis, S.Y.A.; writing—original draft preparation, S.Y.A.; writing—review and editing, J.E.B.; supervision, J.E.B. All authors have read and agreed to the published version of the manuscript.

Funding: This research received no external funding.

Institutional Review Board Statement: Not applicable.

Informed Consent Statement: Not applicable.

Data Availability Statement: Not applicable.

Conflicts of Interest: The authors declare no conflict of interest.

Abbreviations

The following abbreviations are used in this manuscript:

BEV	Bird's-Eye View
CNN	Convolutional Neural Network
DL	Deep Learning
GCN	Graph Convolution Network
IOU	Intersection Over Union
LiDAR	Light Detection and Ranging
mAP	Mean Average Precision
NMS	Non-maximal Suppression
RPN	Region Proposal Network
SOTA	State-of-the-art
VFE	Voxel Feature Encoding
WCNN3D	Wavelet Convolutional Neural Network for 3D Detection
YOLO	You Only Look Once

References

1. Urmson, C.; Anhalt, J.; Bagnell, D.; Baker, C.; Bittner, R.; Clark, M.; Dolan, J.; Duggins, D.; Galatali, T.; Geyer, C.; others. Autonomous driving in urban environments: Boss and the urban challenge. *J. Field Robot.* **2008**, *25*, 425–466. [CrossRef]
2. Feng, D.; Haase-Schuetz, C.; Rosenbaum, L.; Hertlein, H.; Duffhauss, F.; Gläser, C.; Wiesbeck, W.; Dietmayer, K.C. Deep Multi-Modal Object Detection and Semantic Segmentation for Autonomous Driving: Datasets, Methods, and Challenges. *IEEE Trans. Intell. Transp. Syst.* **2021**, *22*, 1341–1360. [CrossRef]
3. Alaba, S.; Gurbuz, A.; Ball, J. A Comprehensive Survey of Deep Learning Multisensor Fusion-based 3D Object Detection for Autonomous Driving: Methods, Challenges, Open Issues, and Future Directions. *TechRxiv* **2022**. [CrossRef]
4. Arnold, E.; Al-Jarrah, O.Y.; Dianati, M.; Fallah, S.; Oxtoby, D.; Mouzakitis, A. A survey on 3d object detection methods for autonomous driving applications. *IEEE Trans. Intell. Transp. Syst.* **2019**, *20*, 3782–3795. [CrossRef]
5. Khader, M.; Cherian, S. An Introduction to Automotive LIDAR. *Texas Instruments.* 2020. Available online: https://www.ti.com/lit/wp/slyy150a/slyy150a.pdf (accessed on 15 November 2022).

6. Tian, Z.; Chu, X.; Wang, X.; Wei, X.; Shen, C. Fully Convolutional One-Stage 3D Object Detection on LiDAR Range Images. *arXiv* **2022**, arXiv:2205.13764.

7. Chen, X.; Ma, H.; Wan, J.; Li, B.; Xia, T. Multi-view 3d object detection network for autonomous driving. In Proceedings of the IEEE Conference on Computer Vision and Pattern Recognition, Honolulu, HI, USA, 21–26 July 2017; pp. 1907–1915.

8. Ku, J.; Mozifian, M.; Lee, J.; Harakeh, A.; Waslander, S.L. Joint 3d proposal generation and object detection from view aggregation. In Proceedings of the 2018 IEEE/RSJ International Conference on Intelligent Robots and Systems (IROS), Madrid, Spain, 1–5 October 2018; pp. 1–8.

9. Zhou, Y.; Tuzel, O. Voxelnet: End-to-end learning for point cloud based 3d object detection. In Proceedings of the IEEE Conference on Computer Vision and Pattern Recognition, Salt Lake City, UT, USA, 18–23 June 2018; pp. 4490–4499.

10. Qi, C.R.; Liu, W.; Wu, C.; Su, H.; Guibas, L.J. Frustum pointnets for 3d object detection from rgb-d data. In Proceedings of the IEEE Conference on Computer Vision and Pattern Recognition, Salt Lake City, UT, USA, 18–23 June 2018; pp. 918–927.

11. Cao, P.; Chen, H.; Zhang, Y.; Wang, G. Multi-view frustum pointnet for object detection in autonomous driving. In Proceedings of the 2019 IEEE International Conference on Image Processing (ICIP), Taipei, Taiwan, 22–25 September 2019; pp. 3896–3899.

12. Qi, C.R.; Su, H.; Mo, K.; Guibas, L.J. Pointnet: Deep learning on point sets for 3d classification and segmentation. In Proceedings of the IEEE Conference on Computer Vision and Pattern Recognition, Honolulu, HI, USA, 21–26 July 2017; pp. 652–660.

13. Qi, C.R.; Yi, L.; Su, H.; Guibas, L.J. Pointnet++: Deep hierarchical feature learning on point sets in a metric space. *arXiv* **2017**, arXiv:1706.02413.

14. Shi, S.; Wang, X.; Li, H. Pointrcnn: 3d object proposal generation and detection from point cloud. In Proceedings of the IEEE/CVF Conference on Computer Vision and Pattern Recognition, Long Beach, CA, USA, 15–20 June 2019; pp. 770–779.

15. Sun, X.; Wang, S.; Wang, M.; Cheng, S.S.; Liu, M. An advanced LiDAR point cloud sequence coding scheme for autonomous driving. In Proceedings of the 28th ACM International Conference on Multimedia Seattle, Washington, DC, USA, 12–16 Octber 2020; pp. 2793–2801.

16. Sun, X.; Wang, M.; Du, J.; Sun, Y.; Cheng, S.S.; Xie, W. A Task-Driven Scene-Aware LiDAR Point Cloud Coding Framework for Autonomous Vehicles. *IEEE Trans. Ind. Inform.* **2022**, 1–11. [CrossRef]

17. Wang, Q.; Jiang, L.; Sun, X.; Zhao, J.; Deng, Z.; Yang, S. An Efficient LiDAR Point Cloud Map Coding Scheme Based on Segmentation and Frame-Inserting Network. *Sensors* **2022**, *22*, 5108. [CrossRef]

18. Jiao, L.; Zhang, F.; Liu, F.; Yang, S.; Li, L.; Feng, Z.; Qu, R. A survey of deep learning-based object detection. *IEEE Access* **2019**, *7*, 128837–128868. [CrossRef]

19. Rahman, M.M.; Tan, Y.; Xue, J.; Lu, K. Recent advances in 3D object detection in the era of deep neural networks: A survey. *IEEE Trans. Image Process.* **2019**, *29*, 2947–2962. [CrossRef]

20. Li, Y.; Ma, L.; Zhong, Z.; Liu, F.; Chapman, M.A.; Cao, D.; Li, J. Deep Learning for LiDAR Point Clouds in Autonomous Driving: A Review. *IEEE Trans. Neural Netw. Learn. Syst.* **2021**, *32*, 3412–3432. [CrossRef] [PubMed]

21. Fernandes, D.; Silva, A.; Névoa, R.; Simões, C.; Gonzalez, D.; Guevara, M.; Novais, P.; Monteiro, J.; Melo-Pinto, P. Point-cloud based 3D object detection and classification methods for self-driving applications: A survey and taxonomy. *Inf. Fusion* **2021**, *68*, 161–191. [CrossRef]

22. Guo, Y.; Wang, H.; Hu, Q.; Liu, H.; Liu, L.; Bennamoun, M. Deep learning for 3d point clouds: A survey. *IEEE Trans. Pattern Anal. Mach. Intell.* **2020**, *43*, 4338–4364. [CrossRef] [PubMed]

23. Qian, R.; Lai, X.; Li, X. 3D Object Detection for Autonomous Driving: A Survey. *arXiv* **2021**, arXiv:2106.10823.

24. Alaba, S.; Ball, J. Deep Learning-based Image 3D Object Detection for Autonomous Driving: Review. *TechRxiv* **2022**. [CrossRef]

25. Harris, C.G.; Stephens, M. A combined corner and edge detector. In Proceedings of the 4th Alvey Vision Conference, Manchester, UK, 31 August–2 September 1988; pp. 10–5244.

26. Shi, J.; Tomasi. Good features to track. In Proceedings of the 1994 IEEE Conference on Computer Vision and Pattern Recognition, Seattle, WA, USA, 21–23 June 1994; pp. 593–600.

27. Lowe, D.G. Distinctive image features from scale-invariant keypoints. *Int. J. Comput. Vis.* **2004**, *60*, 91–110. [CrossRef]

28. Bay, H.; Ess, A.; Tuytelaars, T.; Van Gool, L. Speeded-up robust features (SURF). *Comput. Vis. Image Underst.* **2008**, *110*, 346–359. [CrossRef]

29. Alaba, S.Y.; Nabi, M.; Shah, C.; Prior, J.; Campbell, M.D.; Wallace, F.; Ball, J.E.; Moorhead, R. Class-Aware Fish Species Recognition Using Deep Learning for an Imbalanced Dataset. *Sensors* **2022**, *22*, 8268. [CrossRef]

30. Islam, F.; Nabi, M.; Ball, J.E. Off-Road Detection Analysis for Autonomous Ground Vehicles: A Review. *Sensors* **2022**, *22*, 8463. [CrossRef]

31. Nabi, M.; Senyurek, V.; Gurbuz, A.C.; Kurum, M. Deep Learning-Based Soil Moisture Retrieval in CONUS Using CYGNSS Delay–Doppler Maps. *IEEE J. Sel. Top. Appl. Earth Obs. Remote Sens.* **2022**, *15*, 6867–6881. [CrossRef]

32. Krizhevsky, A.; Sutskever, I.; Hinton, G.E. Imagenet classification with deep convolutional neural networks. *Adv. Neural Inf. Process. Syst.* **2012**, *25*, 1097–1105. [CrossRef]

33. Simonyan, K.; Zisserman, A. Very deep convolutional networks for large-scale image recognition. *arXiv* **2014**, arXiv:1409.1556.

34. Szegedy, C.; Liu, W.; Jia, Y.; Sermanet, P.; Reed, S.; Anguelov, D.; Erhan, D.; Vanhoucke, V.; Rabinovich, A. Going deeper with convolutions. In Proceedings of the IEEE Conference on Computer Vision and Pattern Recognition, Boston, MA, USA, 7–12 June 2015; pp. 1–9.

35. He, K.; Zhang, X.; Ren, S.; Sun, J. Deep residual learning for image recognition. In Proceedings of the IEEE Conference on Computer Vision and Pattern Recognition, Las Vegas, NV, USA, 27–30 June 2016; pp. 770–778.
36. Szegedy, C.; Ioffe, S.; Vanhoucke, V.; Alemi, A. Inception-v4, inception-resnet and the impact of residual connections on learning. In Proceedings of the AAAI Conference on Artificial Intelligence, San Francisco, CA, USA, 4–9 February 2017.
37. Howard, A.G.; Zhu, M.; Chen, B.; Kalenichenko, D.; Wang, W.; Weyand, T.; Andreetto, M.; Adam, H. Mobilenets: Efficient convolutional neural networks for mobile vision applications. *arXiv* **2017**, arXiv:1704.04861.
38. Redmon, J.; Farhadi, A. YOLO9000: Better, faster, stronger. In Proceedings of the IEEE Conference on Computer Vision and Pattern Recognition, Honolulu, HI, USA, 21–26 July 2017; pp. 7263–7271.
39. Yan, Y.; Mao, Y.; Li, B. Second: Sparsely embedded convolutional detection. *Sensors* **2018**, *18*, 3337. [CrossRef]
40. Kuang, H.; Wang, B.; An, J.; Zhang, M.; Zhang, Z. Voxel-FPN: Multi-scale voxel feature aggregation for 3D object detection from LIDAR point clouds. *Sensors* **2020**, *20*, 704. [CrossRef]
41. Lang, A.H.; Vora, S.; Caesar, H.; Zhou, L.; Yang, J.; Beijbom, O. Pointpillars: Fast encoders for object detection from point clouds. In Proceedings of the IEEE/CVF Conference on Computer Vision and Pattern Recognition, Long Beach, CA, USA, 15–20 June 2019; pp. 12697–12705.
42. Graham, B. Spatially-sparse convolutional neural networks. *arXiv* **2014**, arXiv:1409.6070.
43. Graham, B. Sparse 3D convolutional neural networks. *arXiv* **2015**, arXiv:1505.02890.
44. Wang, D.Z.; Posner, I. Voting for voting in online point cloud object detection. In Proceedings of the Robotics: Science and Systems, Rome, Italy, 13–17 July 2015; pp. 10–15607.
45. Wang, Y.; Sun, Y.; Liu, Z.; Sarma, S.E.; Bronstein, M.M.; Solomon, J.M. Dynamic graph cnn for learning on point clouds. *Acm Trans. Graph. (Tog)* **2019**, *38*, 1–12. [CrossRef]
46. Graham, B.; Engelcke, M.; Van Der Maaten, L. 3d semantic segmentation with submanifold sparse convolutional networks. In Proceedings of the IEEE Conference on Computer Vision and Pattern Recognition, Salt Lake City, UT, USA, 18–23 June 2018; pp. 9224–9232.
47. Engelcke, M.; Rao, D.; Wang, D.Z.; Tong, C.H.; Posner, I. Vote3deep: Fast object detection in 3d point clouds using efficient convolutional neural networks. In Proceedings of the 2017 IEEE International Conference on Robotics and Automation (ICRA), Marina Bay Sands Convention Centre, Singapore, 29 May–3 June 2017; pp. 1355–1361.
48. Li, G.; Müller, M.; Qian, G.; Perez, I.C.D.; Abualshour, A.; Thabet, A.K.; Ghanem, B. Deepgcns: Making gcns go as deep as cnns. *IEEE Trans. Pattern Anal. Mach. Intell.* **2021**. [CrossRef]
49. MMDetection3D Contributors. MMDetection3D: OpenMMLab Next-Generation Platform for General 3D Object Detection. 2020. Available online: https://github.com/openmmla/mmdetection3d (accessed on 10 November 2022).
50. Ding, Z.; Han, X.; Niethammer, M. Votenet: A deep learning label fusion method for multi-atlas segmentation. In Proceedings of the International Conference on Medical Image Computing and Computer-Assisted Intervention, Shenzhen, China, 13–17 October 2019; pp. 202–210.
51. Zhang, Z.; Sun, B.; Yang, H.; Huang, Q. H3dnet: 3d object detection using hybrid geometric primitives. In Proceedings of the European Conference on Computer Vision, Glasgow, UK, 23–28 August 2020; pp. 311–329.
52. Geiger, A.; Lenz, P.; Urtasun, R. Are we ready for Autonomous Driving? The KITTI Vision Benchmark Suite. In Proceedings of the Conference on Computer Vision and Pattern Recognition (CVPR), Providence, RI, USA, 16–21 June 2012.
53. Geiger, A.; Lenz, P.; Stiller, C.; Urtasun, R. Vision meets robotics: The kitti dataset. *Int. J. Robot. Res.* **2013**, *32*, 1231–1237. [CrossRef]
54. US National Highway Traffic Safety Administration. *The Evolution of Automated Safety Technologies*; Technical Report; US National Highway Traffic Safety Administration: Washington, DC, USA, 2022.
55. Berger, R. *Think Act: Autonomous Driving*; Technical Report; Roland Berger Strategy Consultants GMBH: München, Germany, 2014.
56. Yu, F.; Koltun, V. Multi-scale context aggregation by dilated convolutions. *arXiv* **2015**, arXiv:1511.07122.
57. Groh, F.; Wieschollek, P.; Lensch, H. Flex-convolution (million-scale point-cloud learning beyond grid-worlds). *arXiv* **2018**, arXiv:1803.07289.
58. Wang, P.; Chen, P.; Yuan, Y.; Liu, D.; Huang, Z.; Hou, X.; Cottrell, G. Understanding convolution for semantic segmentation. In Proceedings of the 2018 IEEE Winter Conference on Applications of Computer Vision (WACV), Lake Tahoe, NV, USA, 12–15 March 2018; pp. 1451–1460.
59. Alaba, S.Y.; Ball, J.E. WCNN3D: Wavelet Convolutional Neural Network-Based 3D Object Detection for Autonomous Driving. *Sensors* **2022**, *22*, 7010. [CrossRef] [PubMed]
60. Su, H.; Maji, S.; Kalogerakis, E.; Learned-Miller, E. Multi-view convolutional neural networks for 3d shape recognition. In Proceedings of the IEEE International Conference on Computer Vision, Santiago, Chile, 7–13 December 2015; pp. 945–953.
61. Milioto, A.; Vizzo, I.; Behley, J.; Stachniss, C. Rangenet++: Fast and accurate lidar semantic segmentation. In Proceedings of the 2019 IEEE/RSJ International Conference on Intelligent Robots and Systems (IROS), Venetian Macao, Macau, 3–8 November 2019; pp. 4213–4220.
62. Li, B.; Zhang, T.; Xia, T. Vehicle detection from 3d lidar using fully convolutional network. *arXiv* **2016**, arXiv:1608.07916.
63. Caesar, H.; Bankiti, V.; Lang, A.H.; Vora, S.; Liong, V.E.; Xu, Q.; Krishnan, A.; Pan, Y.; Baldan, G.; Beijbom, O. nuscenes: A multimodal dataset for autonomous driving. In Proceedings of the IEEE/CVF Conference on Computer Vision and Pattern Recognition, Seattle, WA, USA, 13–19 June 2020; pp. 11621–11631.

64. Yu, S.L.; Westfechtel, T.; Hamada, R.; Ohno, K.; Tadokoro, S. Vehicle detection and localization on bird's eye view elevation images using convolutional neural network. In Proceedings of the 2017 IEEE International Symposium on Safety, Security and Rescue Robotics (SSRR), Shanghai, China, 11–13 October 2017; pp. 102–109.

65. Wirges, S.; Fischer, T.; Stiller, C.; Frias, J.B. Object detection and classification in occupancy grid maps using deep convolutional networks. In Proceedings of the 2018 21st International Conference on Intelligent Transportation Systems (ITSC), Maui, HI, USA, 4–7 November 2018; pp. 3530–3535.

66. Beltrán, J.; Guindel, C.; Moreno, F.M.; Cruzado, D.; Garcia, F.; De La Escalera, A. Birdnet: A 3d object detection framework from lidar information. In Proceedings of the 2018 21st International Conference on Intelligent Transportation Systems (ITSC), Maui, HI, USA, 4–7 November 2018; pp. 3517–3523.

67. Barrera, A.; Guindel, C.; Beltrán, J.; García, F. Birdnet+: End-to-end 3d object detection in lidar bird's eye view. In Proceedings of the 2020 IEEE 23rd International Conference on Intelligent Transportation Systems (ITSC), Rhodes, Greece, 20–23 September 2020; pp. 1–6.

68. Barrera, A.; Beltrán, J.; Guindel, C.; Iglesias, J.A.; García, F. BirdNet+: Two-Stage 3D Object Detection in LiDAR through a Sparsity-Invariant Bird's Eye View. *IEEE Access* **2021**, *9*, 160299–160316. [CrossRef]

69. Ren, S.; He, K.; Girshick, R.; Sun, J. Faster R-CNN: Towards Real-Time Object Detection with Region Proposal Networks. In Proceedings of the Neural Information Processing Systems (NIPS), Montreal, QC, Canada, 11–12 December 2015.

70. Lin, T.Y.; Dollár, P.; Girshick, R.; He, K.; Hariharan, B.; Belongie, S. Feature pyramid networks for object detection. In Proceedings of the IEEE Conference on Computer Vision and Pattern Recognition, Honolulu, HI, USA, 21–26 July 2017; pp. 2117–2125.

71. Priya, M.; Pankaj, D.S. 3DYOLO: Real-time 3D Object Detection in 3D Point Clouds for Autonomous Driving. In Proceedings of the 2021 IEEE International India Geoscience and Remote Sensing Symposium (InGARSS), Virtual Conference, 6–10 December 2021; pp. 41–44.

72. Bochkovskiy, A.; Wang, C.Y.; Liao, H.Y.M. Yolov4: Optimal speed and accuracy of object detection. *arXiv* **2020**, arXiv:2004.10934.

73. Chen, K.; Oldja, R.; Smolyanskiy, N.; Birchfield, S.; Popov, A.; Wehr, D.; Eden, I.; Pehserl, J. Mvlidarnet: Real-time multi-class scene understanding for autonomous driving using multiple views. In Proceedings of the 2020 IEEE/RSJ International Conference on Intelligent Robots and Systems (IROS), Las Vegas, NV, USA, 25–29 October 2020; pp. 2288–2294.

74. Ester, M.; Kriegel, H.P.; Sander, J.; Xu, X.; et al. A density-based algorithm for discovering clusters in large spatial databases with noise. In Proceedings of the Kdd, Portland, OR, USA, 2–4 August 1996; pp. 226–231.

75. Behley, J.; Garbade, M.; Milioto, A.; Quenzel, J.; Behnke, S.; Stachniss, C.; Gall, J. Semantickitti: A dataset for semantic scene understanding of lidar sequences. In Proceedings of the IEEE/CVF International Conference on Computer Vision, Seoul, Republic of Korea, 27 October–2 November 2019; pp. 9297–9307.

76. Lu, Y.; Hao, X.; Sun, S.; Chai, W.; Tong, M.; Velipasalar, S. RAANet: Range-Aware Attention Network for LiDAR-based 3D Object Detection with Auxiliary Density Level Estimation. *arXiv* **2021**, arXiv:2111.09515.

77. Du, X.; Ang, M.H.; Karaman, S.; Rus, D. A general pipeline for 3d detection of vehicles. In Proceedings of the 2018 IEEE International Conference on Robotics and Automation (ICRA), Brisbane, Australia, 21–25 May 2018; pp. 3194–3200.

78. Yang, B.; Luo, W.; Urtasun, R. Pixor: Real-time 3d object detection from point clouds. In Proceedings of the IEEE conference on Computer Vision and Pattern Recognition, Salt Lake City, UT, USA, 18–22 June 2018; pp. 7652–7660.

79. Yang, B.; Liang, M.; Urtasun, R. Hdnet: Exploiting hd maps for 3d object detection. In Proceedings of the Conference on Robot Learning, Zürich, Switzerland, 29–31 October 2018; pp. 146–155.

80. Luo, W.; Yang, B.; Urtasun, R. Fast and furious: Real time end-to-end 3d detection, tracking and motion forecasting with a single convolutional net. In Proceedings of the IEEE Conference on Computer Vision and Pattern Recognition, alt Lake City, UT, USA, 18–22 June 2018; pp. 3569–3577.

81. Simony, M.; Milzy, S.; Amendey, K.; Gross, H.M. Complex-yolo: An euler-region-proposal for real-time 3d object detection on point clouds. In Proceedings of the European Conference on Computer Vision (ECCV) Workshops, Munich, Germany, 8–14 September 2018.

82. Ali, W.; Abdelkarim, S.; Zidan, M.; Zahran, M.; El Sallab, A. Yolo3d: End-to-end real-time 3d oriented object bounding box detection from lidar point cloud. In Proceedings of the European Conference on Computer Vision (ECCV) Workshops, Munich, Germany, 8–14 September 2018.

83. Sedaghat, N.; Zolfaghari, M.; Amiri, E.; Brox, T. Orientation-boosted voxel nets for 3d object recognition. *arXiv* **2016**, arXiv:1604.03351.

84. Li, B. 3d fully convolutional network for vehicle detection in point cloud. In Proceedings of the 2017 IEEE/RSJ International Conference on Intelligent Robots and Systems (IROS), Vancouver, BC, Canada, September 24–28 2017; pp. 1513–1518.

85. Long, J.; Shelhamer, E.; Darrell, T. Fully convolutional networks for semantic segmentation. In Proceedings of the IEEE Conference on Computer Vision and Pattern Recognition, Boston, MA, USA, 7–12 June 2015; pp. 3431–3440.

86. Ye, M.; Xu, S.; Cao, T. Hvnet: Hybrid voxel network for lidar based 3d object detection. In Proceedings of the IEEE/CVF Conference on Computer Vision and Pattern Recognition, Seattle, WA, USA, 13–19 June 2020; pp. 1631–1640.

87. Du, L.; Ye, X.; Tan, X.; Feng, J.; Xu, Z.; Ding, E.; Wen, S. Associate-3Ddet: Perceptual-to-conceptual association for 3D point cloud object detection. In Proceedings of the IEEE/CVF Conference on Computer Vision and Pattern Recognition, Seattle, WA, USA, 13–19 June 2020; pp. 13329–13338.

88. Liu, Z.; Zhao, X.; Huang, T.; Hu, R.; Zhou, Y.; Bai, X. Tanet: Robust 3d object detection from point clouds with triple attention. In Proceedings of the AAAI Conference on Artificial Intelligence, New York, NY, USA, 7–12 February 2020; pp. 11677–11684.
89. Deng, J.; Shi, S.; Li, P.; Zhou, W.; Zhang, Y.; Li, H. Voxel R-CNN: Towards High Performance Voxel-based 3D Object Detection. *arXiv* **2021**, arXiv:cs.CV/2012.15712.
90. Sun, P.; Kretzschmar, H.; Dotiwalla, X.; Chouard, A.; Patnaik, V.; Tsui, P.; Guo, J.; Zhou, Y.; Chai, Y.; Caine, B.; et al. Scalability in perception for autonomous driving: Waymo open dataset. In Proceedings of the IEEE/CVF Conference on Computer Vision and Pattern Recognition, Seattle, WA, USA, 13–19 June 2020; pp. 2446–2454.
91. Li, Z.; Yao, Y.; Quan, Z.; Xie, J.; Yang, W. Spatial information enhancement network for 3D object detection from point cloud. *Pattern Recognit.* **2022**, *128*, 108684. [CrossRef]
92. Liu, S.; Huang, W.; Cao, Y.; Li, D.; Chen, S. SMS-Net: Sparse multi-scale voxel feature aggregation network for LiDAR-based 3D object detection. *Neurocomputing* **2022**, *501*, 555–565. [CrossRef]
93. Sun, J.; Ji, Y.M.; Wu, F.; Zhang, C.; Sun, Y. Semantic-aware 3D-voxel CenterNet for point cloud object detection. *Comput. Electr. Eng.* **2022**, *98*, 107677. [CrossRef]
94. Liu, M.; Ma, J.; Zheng, Q.; Liu, Y.; Shi, G. 3D Object Detection Based on Attention and Multi-Scale Feature Fusion. *Sensors* **2022**, *22*, 3935. [CrossRef]
95. Wang, L.; Song, Z.; Zhang, X.; Wang, C.; Zhang, G.; Zhu, L.; Li, J.; Liu, H. SAT-GCN: Self-attention graph convolutional network-based 3D object detection for autonomous driving. *Knowl.-Based Syst.* **2022**, *259*, 110080. [CrossRef]
96. Fan, L.; Wang, F.; Wang, N.; Zhang, Z. Fully Sparse 3D Object Detection. *arXiv* **2022**, arXiv:2207.10035.
97. Hu, J.S.; Kuai, T.; Waslander, S.L. Point density-aware voxels for lidar 3d object detection. In Proceedings of the IEEE/CVF Conference on Computer Vision and Pattern Recognition, New Orleans, LA, USA, 19–24 June 2022; pp. 8469–8478.
98. Hosang, J.; Benenson, R.; Dollár, P.; Schiele, B. What makes for effective detection proposals? *IEEE Trans. Pattern Anal. Mach. Intell.* **2015**, *38*, 814–830. [CrossRef]
99. Wang, Z.; Jia, K. Frustum convnet: Sliding frustums to aggregate local point-wise features for amodal 3d object detection. In Proceedings of the 2019 IEEE/RSJ International Conference on Intelligent Robots and Systems (IROS), Macau, China, 3–8 November 2019; pp. 1742–1749.
100. Song, S.; Lichtenberg, S.P.; Xiao, J. Sun rgb-d: A rgb-d scene understanding benchmark suite. In Proceedings of the IEEE Conference on Computer Vision and Pattern Recognition, Boston, MA, USA, 7–12 June 2015; pp. 567–576.
101. McCrae, S.; Zakhor, A. 3D object detection for autonomous driving using temporal LiDAR data. In Proceedings of the 2020 IEEE International Conference on Image Processing (ICIP), Virtual Conference, 25–28 October 2020; pp. 2661–2665.
102. Wang, Y.; Fathi, A.; Kundu, A.; Ross, D.A.; Pantofaru, C.; Funkhouser, T.; Solomon, J. Pillar-based object detection for autonomous driving. In Proceedings of the European Conference on Computer Vision, Glasgow, UK, 23–28 August 2020; pp. 18–34.
103. Fan, L.; Pang, Z.; Zhang, T.; Wang, Y.X.; Zhao, H.; Wang, F.; Wang, N.; Zhang, Z. Embracing single stride 3d object detector with sparse transformer. In Proceedings of the IEEE/CVF Conference on Computer Vision and Pattern Recognition, New Orleans, LA, USA, 19–24 June 2022; pp. 8458–8468.
104. Tong, G.; Peng, H.; Shao, Y.; Yin, Q.; Li, Z. ASCNet: 3D object detection from point cloud based on adaptive spatial context features. *Neurocomputing* **2022**, *475*, 89–101. [CrossRef]
105. Zhang, J.; Liu, H.; Lu, J. A semi-supervised 3D object detection method for autonomous driving. *Displays* **2021**, *71*, 102117. [CrossRef]
106. Caine, B.; Roelofs, R.; Vasudevan, V.; Ngiam, J.; Chai, Y.; Chen, Z.; Shlens, J. Pseudo-labeling for Scalable 3D Object Detection. *arXiv* **2021**, arXiv:2103.02093.
107. Ding, Z.; Hu, Y.; Ge, R.; Huang, L.; Chen, S.; Wang, Y.; Liao, J. 1st Place Solution for Waymo Open Dataset Challenge–3D Detection and Domain Adaptation. *arXiv* **2020**, arXiv:2006.15505.
108. Bai, Z.; Wu, G.; Barth, M.J.; Liu, Y.; Sisbot, A.; Oguchi, K. PillarGrid: Deep Learning-based Cooperative Perception for 3D Object Detection from Onboard-Roadside LiDAR. *arXiv* **2022**, arXiv:2203.06319.
109. Dosovitskiy, A.; Ros, G.; Codevilla, F.; Lopez, A.; Koltun, V. CARLA: An open urban driving simulator. In Proceedings of the Conference on Robot Learning, Mountain View, CA, USA, 13–15 November 2017; pp. 1–16.
110. Lin, J.; Yin, H.; Yan, J.; Ge, W.; Zhang, H.; Rigoll, G. Improved 3D Object Detector Under Snowfall Weather Condition Based on LiDAR Point Cloud. *IEEE Sens. J.* **2022**, *22*, 16276–16292. [CrossRef]
111. Pitropov, M.; Garcia, D.E.; Rebello, J.; Smart, M.; Wang, C.; Czarnecki, K.; Waslander, S. Canadian adverse driving conditions dataset. *Int. J. Robot. Res.* **2021**, *40*, 681–690. [CrossRef]
112. Shi, S.; Guo, C.; Jiang, L.; Wang, Z.; Shi, J.; Wang, X.; Li, H. Pv-rcnn: Point-voxel feature set abstraction for 3d object detection. In Proceedings of the IEEE/CVF Conference on Computer Vision and Pattern Recognition, Seattle, WA, USA, 13–19 June 2020; pp. 10529–10538.
113. Shi, S.; Wang, Z.; Shi, J.; Wang, X.; Li, H. From points to parts: 3d object detection from point cloud with part-aware and part-aggregation network. *IEEE Trans. Pattern Anal. Mach. Intell.* **2020**, *43*, 2647–2664. [CrossRef] [PubMed]
114. Graham, B.; van der Maaten, L. Submanifold sparse convolutional networks. *arXiv* **2017**, arXiv:1706.01307.
115. Yang, Z.; Sun, Y.; Liu, S.; Shen, X.; Jia, J. Std: Sparse-to-dense 3d object detector for point cloud. In Proceedings of the IEEE/CVF International Conference on Computer Vision, Seoul, Republic of Korea, 27 October–2 November 2019; pp. 1951–1960.

116. Yu, H.X.; Wu, J.; Yi, L. Rotationally Equivariant 3D Object Detection. In Proceedings of the IEEE/CVF Conference on Computer Vision and Pattern Recognition, New Orleans, LA, USA, 19–24 June 2022; pp. 1456–1464.

117. Qi, C.R.; Litany, O.; He, K.; Guibas, L.J. Deep hough voting for 3d object detection in point clouds. In Proceedings of the IEEE/CVF International Conference on Computer Vision, Seoul, Republic of Korea, 27 October–2 November 2019; pp. 9277–9286.

118. Lu, H.; Chen, X.; Zhang, G.; Zhou, Q.; Ma, Y.; Zhao, Y. SCANet: Spatial-channel attention network for 3D object detection. In Proceedings of the ICASSP 2019-2019 IEEE International Conference on Acoustics, Speech and Signal Processing (ICASSP), Brighton, UK, 12–17 May 2019; pp. 1992–1996.

119. Liu, Z.; Zhang, Z.; Cao, Y.; Hu, H.; Tong, X. Group-free 3d object detection via transformers. In Proceedings of the IEEE/CVF International Conference on Computer Vision, Montreal, QC, Canada, 10–17 October 2021; pp. 2949–2958.

120. Yang, Z.; Sun, Y.; Liu, S.; Jia, J. 3dssd: Point-based 3d single stage object detector. In Proceedings of the IEEE/CVF Conference on Computer Vision and Pattern Recognition, Seattle, WA, USA, 13–19 June 2020; pp. 11040–11048.

121. He, C.; Zeng, H.; Huang, J.; Hua, X.S.; Zhang, L. Structure aware single-stage 3d object detection from point cloud. In Proceedings of the IEEE/CVF Conference on Computer Vision and Pattern Recognition, Seattle, WA, USA, 13–19 June 2020; pp. 11873–11882.

122. Gustafsson, F.K.; Danelljan, M.; Schon, T.B. Accurate 3D Object Detection using Energy-Based Models. In Proceedings of the IEEE/CVF Conference on Computer Vision and Pattern Recognition, Nashville, TN, USA, 20–25 June 2021; pp. 2855–2864.

123. Zheng, W.; Tang, W.; Chen, S.; Jiang, L.; Fu, C.W. CIA-SSD: Confident IoU-Aware Single-Stage Object Detector From Point Cloud. *arXiv* **2020**, arXiv:2012.03015.

124. Shi, W.; Rajkumar, R. Point-gnn: Graph neural network for 3d object detection in a point cloud. In Proceedings of the IEEE/CVF Conference on Computer Vision and Pattern Recognition, Seattle, WA, USA, 13–19 June 2020; pp. 1711–1719.

125. Zhou, D.; Fang, J.; Song, X.; Liu, L.; Yin, J.; Dai, Y.; Li, H.; Yang, R. Joint 3d instance segmentation and object detection for autonomous driving. In Proceedings of the IEEE/CVF Conference on Computer Vision and Pattern Recognition, Seattle, WA, USA, 13–19 June 2020; pp. 1839–1849.

126. Caron, M.; Bojanowski, P.; Joulin, A.; Douze, M. Deep clustering for unsupervised learning of visual features. In Proceedings of the European Conference on Computer Vision (ECCV), Munich, Germany, 8–14 September 2018; pp. 132–149.

127. Mao, J.; Niu, M.; Bai, H.; Liang, X.; Xu, H.; Xu, C. Pyramid r-cnn: Towards better performance and adaptability for 3d object detection. In Proceedings of the IEEE/CVF International Conference on Computer Vision, Montreal, QC, Canada, 10–17 October 2021; pp. 2723–2732.

128. Yang, J.; Shi, S.; Wang, Z.; Li, H.; Qi, X. ST3D: Self-training for Unsupervised Domain Adaptation on 3D Object Detection. In Proceedings of the IEEE/CVF Conference on Computer Vision and Pattern Recognition, Nashville, TN, USA, 20–25 June 2021; pp. 10368–10378.

129. Kesten, R.; Usman, M.; Houston, J.; Pandya, T.; Nadhamuni, K.; Ferreira, A.; Yuan, M.; Low, B.; Jain, A.; Ondruska, P.; et al. Lyft Level 5 Perception Dataset 2020. 2019. Available online: https://level-5.global/level5/data/ (accessed on 27 October 2022).

130. Hegde, D.; Patel, V. Attentive Prototypes for Source-free Unsupervised Domain Adaptive 3D Object Detection. *arXiv* **2021**, arXiv:2111.15656.

131. Zheng, W.; Tang, W.; Jiang, L.; Fu, C.W. SE-SSD: Self-Ensembling Single-Stage Object Detector From Point Cloud. In Proceedings of the IEEE/CVF Conference on Computer Vision and Pattern Recognition, Nashville, TN, USA, 20–25 June 2021; pp. 14494–14503.

132. Wang, J.; Gang, H.; Ancha, S.; Chen, Y.T.; Held, D. Semi-supervised 3D Object Detection via Temporal Graph Neural Networks. In Proceedings of the 2021 International Conference on 3D Vision (3DV), London, UK, 1–3 December 2021; pp. 413–422.

133. Patil, A.; Malla, S.; Gang, H.; Chen, Y.T. The h3d dataset for full-surround 3d multi-object detection and tracking in crowded urban scenes. In Proceedings of the 2019 International Conference on Robotics and Automation (ICRA), Montreal, QC, Canada, 20–24 May 2019; pp. 9552–9557.

134. Zhang, L.; Dong, R.; Tai, H.S.; Ma, K. PointDistiller: Structured Knowledge Distillation Towards Efficient and Compact 3D Detection. *arXiv* **2022**, arXiv:2205.11098.

135. Wang, J.; Lin, X.; Yu, H. POAT-Net: Parallel Offset-attention Assisted Transformer for 3D Object Detection for Autonomous Driving. *IEEE Access* **2021**, *9*, 151110–151117. [CrossRef]

136. Ren, S.; Pan, X.; Zhao, W.; Nie, B.; Han, B. Dynamic graph transformer for 3D object detection. *Knowl.-Based Syst.* **2022**, *259*, 110085. [CrossRef]

137. Theodose, R.; Denis, D.; Chateau, T.; Frémont, V.; Checchin, P. A Deep Learning Approach for LiDAR Resolution-Agnostic Object Detection. *IEEE Trans. Intell. Transp. Syst.* **2021**, *23*, 14582–14593. [CrossRef]

138. Xiao, P.; Shao, Z.; Hao, S.; Zhang, Z.; Chai, X.; Jiao, J.; Li, Z.; Wu, J.; Sun, K.; Jiang, K.; et al. PandaSet: Advanced Sensor Suite Dataset for Autonomous Driving. In Proceedings of the 2021 IEEE International Intelligent Transportation Systems Conference (ITSC), Indianapolis, IN, USA, 19–21 September 2021; pp. 3095–3101.

139. Nagesh, S.; Baig, A.; Srinivasan, S. Structure Aware and Class Balanced 3D Object Detection on nuScenes Dataset. *arXiv* **2022**, arXiv:2205.12519.

140. Zhu, B.; Jiang, Z.; Zhou, X.; Li, Z.; Yu, G. Class-balanced grouping and sampling for point cloud 3d object detection. *arXiv* **2019**, arXiv:1908.09492.

141. Wang, M.; Chen, Q.; Fu, Z. LSNet: Learned Sampling Network for 3D Object Detection from Point Clouds. *Remote Sens.* **2022**, *14*, 1539. [CrossRef]

142. Hahner, M.; Sakaridis, C.; Bijelic, M.; Heide, F.; Yu, F.; Dai, D.; Van Gool, L. Lidar snowfall simulation for robust 3d object detection. In Proceedings of the IEEE/CVF Conference on Computer Vision and Pattern Recognition, New Orleans, LA, USA, 19–24 June 2022; pp. 16364–16374.
143. Chen, Y.; Liu, S.; Shen, X.; Jia, J. Fast point r-cnn. In Proceedings of the IEEE/CVF International Conference on Computer Vision, Seoul, Republic of Korea, 27 October–2 November 2019; pp. 9775–9784.
144. Xu, Q.; Zhou, Y.; Wang, W.; Qi, C.R.; Anguelov, D. Spg: Unsupervised domain adaptation for 3d object detection via semantic point generation. In Proceedings of the IEEE/CVF International Conference on Computer Vision, Montreal, QC, Canada, 10–17 October 2021; pp. 15446–15456.
145. Li, X.; Zhang, T.; Wang, S.; Zhu, G.; Wang, R.; Chang, T.H. Large-Scale Bandwidth and Power Optimization for Multi-Modal Edge Intelligence Autonomous Driving. *arXiv* **2022**, arXiv:2210.09659.

MDPI
St. Alban-Anlage 66
4052 Basel
Switzerland
www.mdpi.com

Sensors Editorial Office
E-mail: sensors@mdpi.com
www.mdpi.com/journal/sensors